STATICS AND STRENGTH OF MATERIALS

PRENTICE-HALL, INTERNATIONAL, INC., *London*
PRENTICE-HALL OF AUSTRALIA, PTY. LTD., *Sydney*
PRENTICE-HALL OF CANADA, LTD., *Toronto*
PRENTICE-HALL OF INDIA PRIVATE LTD., *New Delhi*
PRENTICE-HALL OF JAPAN, INC., *Tokyo*

STATICS AND STRENGTH OF MATERIALS

Irving J. Levinson

Director, Instructional
Systems Administration
Oakland Community College
Union Lake, Michigan

Prentice-Hall, Inc.
Englewood Cliffs, N.J.

13-844506-0
Library of Congress Catalog Card Number: 78-141775
Printed in the United States of America

This book is lovingly dedicated to two people—one person is from the "now" and the other from the "yesterdays."

TO MARILYN T. JANTZ

who, while busy herself, has taken precious time to devotedly help me

TO GILBERT W. BOYD 1905–1969

an outstanding and creative teacher who worked so hard at his trade; he had the right to expect the same from his students.

PREFACE

Citizens who have a few years behind them will, perhaps, remember a remark made by the late Fred Allen during one of his weekly radio broadcasts—long before TV—and during his running feud with his friend Jack Benny; he said:

"an empty cab drove up, and Jack Benny got out."

The meaning in this vitriolic quip is analogous to the role that *statics*, which deals with rigid bodies, plays in the subject of *strength of deformable materials;* statics in itself is a rather empty topic with limited engineering application. However, an understanding of strength of materials is impossible without a working knowledge of statics—something like the alliance of arithmetic and algebra. This book helps provide the background—statics—so that the basic knowledge gained can be then applied to analysis needed to design real life beams and bridges.

Answers and hints, hopefully correct, are supplied to the questions and problems. Although correct answers to textbook problems are important, they serve only to indicate to the student that his approach to and/or attack of "the problem" is one of the right ones.

I. J. LEVINSON

Union Lake, Michigan

vii

CONTENTS

Chapter **1**

INTRODUCTION

Perhaps no science is more basic to the understanding of nature than mechanics. In 350 B.C., Aristotle, an energetic Greek philosopher, attempted to explain the lever. This is when it all began.

The science of mechanics was developed by mathematicians and physicists. These men were mainly interested in a logical explanation of their observations. The lever and the pulley, free fall, and the movement of planets were studied at great length. Each investigator added to the store of knowledge, either with a new theory or with a correction in the theories of his predecessor. Isaac Newton climaxed the study in 1687 with his discovery of universal gravitation and his statement of three laws of motion. His discovery is the mechanics of today.

The importance of the science can be appraised by an accounting of its use. Today, not a building or a bridge or an automobile or an airplane is constructed without some prior analysis based on the principles of mechanics.

1-1 The Divisions of Mechanics: Statics and Dynamics

Statics is the study of bodies at rest—in a state of balance with their surroundings. A force analysis is the eventual goal in statics. Through the application of the principles of statics we answer questions such as: What load will the column have to support? What is the tension in the bridge cable? What is the mechanical advantage of the block and tackle?

Dynamics is, first, the study of the geometry of motion; and second, the study of the forces required to produce motion. The former is called *kinematics* and the latter, *kinetics*. The ultimate goal in dynamics is the determination of the forces required to produce motion and change in motion.

1-2 The Mathematics of Mechanics

Mechanics is an analytical subject; it makes extensive use of mathematics in all of its forms: algebra, geometry, and trigonometry. Although it is not the purpose of this book to dwell on mathematics for mathematics' sake, one phase, trigonometry, is used so frequently that it deserves some special attention.

Right triangles. A right triangle is a closed three-sided figure that has one right angle. The side of the triangle that is opposite the right angle is called the *hypotenuse*. The other two sides are named in relation to either of the two remaining angles. If θ (the Greek letter theta) is selected as the angle in question, then side BC in Fig. 1-1 is referred to as the *opposite side* and

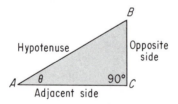

Figure 1-1

side AC, as the *adjacent side*. The six possible ratios of the sides are called the *trigonometric functions* and are given the following names:

$$\text{sine } \theta = \frac{\text{opposite side}}{\text{hypotenuse}} = \frac{O}{H}$$

$$\text{cosine } \theta = \frac{\text{adjacent side}}{\text{hypotenuse}} = \frac{A}{H}$$

$$\text{tangent } \theta = \frac{\text{opposite side}}{\text{adjacent side}} = \frac{O}{A}$$

$$\text{cotangent } \theta = \frac{\text{adjacent side}}{\text{opposite side}} = \frac{A}{O}$$

$$\text{secant } \theta = \frac{\text{hypotenuse}}{\text{adjacent side}} = \frac{H}{A}$$

$$\text{cosecant } \theta = \frac{\text{hypotenuse}}{\text{opposite side}} = \frac{H}{O}$$

These six *trigonometric functions* show that for a given angle θ, the ratios of the lengths of the sides of a right triangle are constant.

The functions, abbreviated *sin*, *cos*, and *tan*, of some of the most frequently used angles are given in Table 1-1. Table 1 of the Appendix lists the functions of all angles and their decimal parts.

Table 1-1

Angle	Sine	Cosine	Tangent
0°	0	1.000	0
30°	0.500	0.866	0.577
45°	0.707	0.707	1.000
60°	0.866	0.500	1.732
90°	1.000	0	infinity

There are five variables in a right triangle: three sides and two angles. If any two of these five quantities are known, the remaining three can easily be determined.

EXAMPLE 1: A 20-ft ladder leans against a wall, as shown in Fig. 1-2. Determine the distance d from the foot of the ladder to the wall.

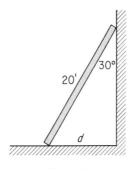

Figure 1-2

Solution: The sine of 30 deg defines the ratio of the desired length d, and the hypotenuse.

$$\sin 30° = \frac{d}{20}$$

therefore

$$d = 20 \sin 30° = 20(0.500) = 10 \text{ ft}$$

The tangent functions are used extensively when accurate angles are to be drawn or measured. For example, to draw a line at 58.5 deg with some reference line, find the tangent of 58.5 deg in the table of trigonometric functions and construct a right triangle, Fig. 1-3, so that the ratio of the opposite side to the adjacent side is equal to this numerical value, in this case

$$\tan 58.5° = 1.632$$

The accuracy of the constructed angle depends upon the length of the sides drawn. Thus the legs of a triangle drawn in the ratio of 16.32 to 10 would give a far more precise angle than one drawn in the ratio of 1.632 to 1.

To measure rather than construct an angle, use a convenient adjacent length and measure the opposite side; then find the angle in the table. This is one way surveyors measure angles.

Figure 1-3

The cosine law. There are many applications in mechanics where two sides and an included angle of a triangle are known and the third side is to be computed. This calculation is carried out most conveniently by the *cosine law* which states: *the square of any side of a triangle is equal to the sum of the squares of the other two sides minus twice the product of those two sides and the cosine of the angle included by them.* For the triangle shown in Fig. 1-4, side c in terms of sides a and b, and the angle θ is

$$c^2 = a^2 + b^2 - 2ab \cos \theta$$

$$c = \sqrt{a^2 + b^2 - 2ab \cos \theta} \tag{1-1}$$

If the included angle θ is 90 deg, then the cosine law reduces to the well-known theorem of Pythagoras: *the square of the hypotenuse of a right triangle is the sum of the squares of the two remaining sides.*

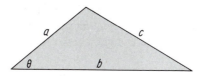

Figure 1-4

$$c^2 = a^2 + b^2$$

and

$$c = \sqrt{a^2 + b^2}$$

The example that follows will demonstrate the use of the cosine law.

EXAMPLE 2: Determine the length of side c, of the triangle shown in Fig. 1-5.

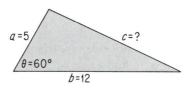

Figure 1-5

Solution: Direct substitution into the cosine law gives

$$c = \sqrt{a^2 + b^2 - 2ab \cos \theta}$$
$$= \sqrt{(5)^2 + (12)^2 - 2(5)12 \cos 60°}$$
$$= \sqrt{25 + 144 - 60} = \sqrt{109} = 10.4$$

EXAMPLE 3: Determine the length of side c of the triangle shown in Fig. 1-6.

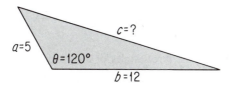

Figure 1-6

Solution: The cosine of any angle lying between 90 deg and 180 deg is negative and equal to

$$\cos \theta = -\cos(180° - \theta)$$

therefore

$$\cos 120° = -\cos 60°$$

substitution into the cosine law gives

$$c = \sqrt{a^2 + b^2 - 2ab \cos \theta}$$
$$= \sqrt{(5)^2 + (12)^2 + 2(5)12 \cos 60°}$$
$$= \sqrt{25 + 144 + 60} = \sqrt{229} = 15.1$$

The sine law. Another useful concept in trigonometry that has extensive use in mechanics states: *in any triangle the sides are proportional to the sines of the opposite angles.* For the triangle shown in Fig. 1-7 the statement of the sine law is

$$\frac{a}{\sin \beta} = \frac{b}{\sin \theta} = \frac{c}{\sin \alpha}$$

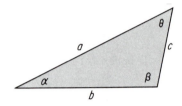

Figure 1-7

The sine law is used to find a third side of a triangle when two sides and an opposite angle are known or to find a side when one side and two angles are known.

EXAMPLE 4: Determine the lengths of sides a and c in the triangle shown in Fig. 1-8.

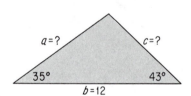

Figure 1-8

Solution: The sum of the three angles of a triangle equals 180 deg; therefore

$$\theta + 35° + 43° = 180°$$
$$\theta = 102°$$

Application of the sine law gives

$$\frac{a}{\sin 43°} = \frac{12}{\sin 102°}$$

The sine of an angle θ, lying between 90 deg and 180 deg is positive and equal to

$$\sin \theta = \sin (180° - \theta)$$

therefore

$$\sin 102° = \sin (180° - 102°) = \sin 78°$$

The sine functions are found in the tables and a is computed.

$$a = \frac{12 \sin 43°}{\sin 78°} = \frac{12(0.682)}{(0.978)} = 8.37$$

The process is repeated to find side c, thus

$$\frac{a}{\sin 43°} = \frac{c}{\sin 35°}$$

$$c = \frac{8.37 \sin 35°}{\sin 43°} = \frac{8.37(0.574)}{0.682} = 7.04$$

1-3 The Conversions of Units

Units are used to define the size of physical quantities. Feet, inches, and miles are units of length; seconds, minutes, and hours are units of time; ounces, pounds, and tons are units of weight, and so on. It is a familiar fact that there are 12 inches in a foot, 3 feet in a yard, 60 seconds in a minute, 2000 pounds in a ton; these and similar equalities are called *conversion factors*. These factors are used to convert from one system of measurement to another. For example, to change a measurement of 6 in. into units of feet requires that the measurement be multiplied by the ratio 1 ft/12 in. Since 1 ft. is equal to 12 in., the ratio changes merely the units of the given length and not the length itself.

$$6 \text{ in.} \times \frac{1 \text{ ft}}{12 \text{ in.}} = 0.5 \text{ ft}$$

The units are canceled, just as numbers would be canceled when a series of fractions is multiplied.

EXAMPLE 5: If a cubic inch of steel weights 0.29 lb, what is the weight of a plate 3.5 ft long, 1.75 ft wide, and 1 in. thick?

Solution: The desired quantity, the weight, is the product of the volume and the weight per unit of volume. The latter is called the *density*.

$$W = 3.5 \text{ ft} \times 1.75 \text{ ft} \times 1 \text{ in.} \times \frac{12 \text{ in.}}{\text{ft}} \times \frac{12 \text{ in.}}{\text{ft}} \times \frac{0.29 \text{ lb}}{\text{in.}^3}$$

$$W = 3.5(1.75)12(12)(0.29) \text{ lb} = 256 \text{ lb}$$

EXAMPLE 6: Express a speed of 30 mph (miles per hour) in the units of fps (feet per second).

Solution:

$$30\frac{\text{mi}}{\text{hr}} = 30\frac{\text{mi}}{\text{hr}} \times \frac{5280 \text{ ft}}{\text{mi}} \times \frac{1 \text{ hr}}{60 \text{ min}} \times \frac{1 \text{ min}}{60 \text{ sec}}$$

$$= \frac{30(5280)}{60(60)} \frac{\text{ft}}{\text{sec}} = 44 \text{ fps}$$

QUESTIONS, PROBLEMS, AND ANSWERS

1-1. The hypotenuse of a right triangle is 20 in. long, and one angle is 60 deg. Find the lengths of the sides that are opposite and adjacent to this angle.

Ans. 17.3 in.; 10 in.

1-2. Two sides of a right triangle are 6 in. and 4 in. Determine the angle between the hypotenuse and the shorter side.

Ans. 56.3°.

1-3. Determine the length of the hypotenuse of a right triangle if the legs have equal lengths of 12 in.

Ans. 17.0 in.

1-4. A 30-ft ladder leans at an angle of 35 deg with the vertical against a wall. Determine the distance from the top of the ladder to the ground.

Ans. 24.6 ft.

1-5. A ship sails 100 mi in a direction N20°E and then 50 mi N60°E as shown. How far is the ship from its starting point?

Ans. 142 mi.

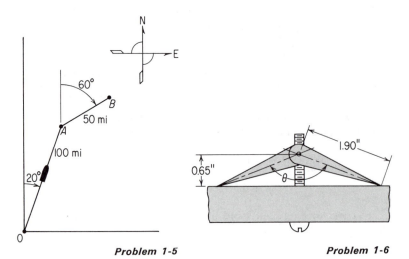

Problem 1-5   *Problem 1-6*

1-6. Find the angle between the wings of the toggle bolt shown.
Ans. 140°.

1-7. A roof truss has the dimensions shown. Determine the angles *DBC* and *BCD*.
Ans. 71.6°; 63.4°.

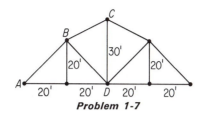

Problem 1-7

1-8. A television antenna that is 1000 ft high is secured by a double series of guy wires as shown. Determine the distance from the anchor A to the antenna and the angle θ between the lower guy wires and the antenna.

Ans. 700 ft; 66.7°.

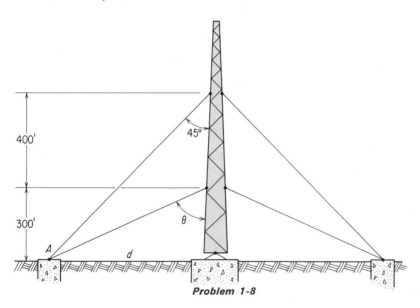

Problem 1-8

1-9. A rectangle is 15 in. high and 35 in. long. Determine the length of the diagonal and the angle between it and the shorter side.

Ans. 38.1 in.; 66.8°.

1-10. Find the lengths of the equal sides of an isosceles triangle and the base angles if the base is 10 in. long and the height is 20 in.

Ans. 20.6 in.; 76.0°.

1-11. A clamping device for a jig is shown in the figure. Find the difference in elevation between points A and B.

Ans. 1.25 in.

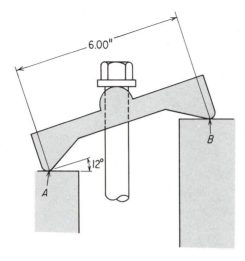

1-12. Determine the span L of the bridge illustrated if all six panels have equal widths.

Ans. 381 ft.

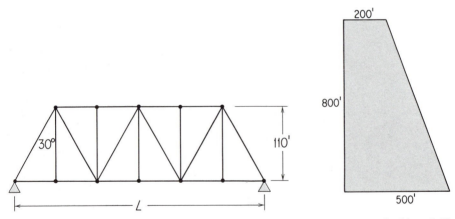

Problem 1-12   **Problem 1-13**

1-13. A parking lot has the shape shown. Determine the number of square yards of paving material required.

Ans. 31,100 sq yd.

1-14. Use the method of tangents to draw a line that forms an angle of 57.2 deg with a horizontal reference line.

1-15. Draw lines, by the method of tangents, at angles of 24.2, 69.7, and 11.7 deg with a horizontal reference line.

1-16. Determine the length of the third side c of the triangle illustrated if a, b, and θ are

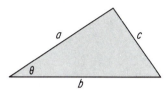

(a) $a = 5$ in.,	$b = 10$ in.,	$\theta = 30°$
(b) $a = 6$ in.,	$b = 6$ in.,	$\theta = 20°$
(c) $a = 14$ ft,	$b = 7$ ft,	$\theta = 70°$
(d) $a = 2$ in.,	$b = 4$ in.,	$\theta = 120°$
(e) $a = 15$ in.,	$b = 25$ in.,	$\theta = 135°$
(f) $a = 10.3$ in.,	$b = 6.4$ in.,	$\theta = 127°$

Problem 1-16

Ans. (a) 6.2 in.; (b) 2.09 in.; (c) 13.3 ft; (d) 5.29 in.; (e) 37.3 in.; (f) 15 in.

1-17–1-19. Determine the lengths of the diagonals BD and AC of the parallelogram.

Ans. **1-17.** 8.5 in.; 18.5 in.
     1-18. 15.9 ft; 9.17 ft.
     1-19. 13.7 in.; 7.82 in.

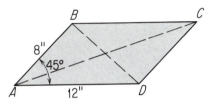

Problem 1-17

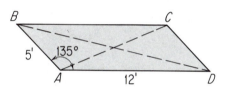

Problem 1-18

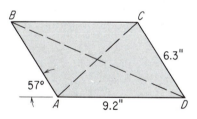

Problem 1-19

1-20. A fence is to extend along the long diagonal of a plot of land that has the shape of a parallelogram. The sides are 150 ft and 300 ft respectively and the acute angle between them is 55 deg. Determine the required length of fence, and the area of the field.

Ans. 405 ft; 36,900 sq ft.

1-21. A man walks the zig-zag path from A to F. How far is he from his starting point?

Ans. 54.9 yd.

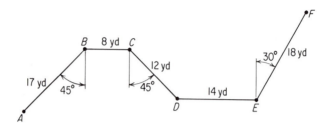

Problem 1-21

1-22. Determine in the triangle illustrated, the missing sides and missing angles if

(a) $a = 10$ in., $\theta = 30°$, $\beta = 40°$
(b) $a = 6$ in., $c = 12$ in., $\theta = 52°$
(c) $b = 5$ in., $c = 15$ in., $\theta = 29°$
(d) $a = 12$ ft, $b = 8$ ft, $\beta = 32°$
(e) $b = 10.3$ in., $c = 12.7$ in., $\alpha = 36.5°$
(f) $a = 5$ mi, $b = 8$ mi, $c = 10$ mi

Ans. (a) $\alpha = 110°$; $b = 6.89$ in.; $c = 5.36$ in.
(b) $\alpha = 23.2°$; $\beta = 104.8°$; $b = 14.7$ in.
(c) $\alpha = 141.7°$; $\beta = 9.3°$; $a = 19.2$ in.
(d) $\alpha = 52.7°$; $\beta = 95.3°$; $c = 15$ ft.
(e) $\theta = 89.3°$; $\beta = 54.2°$; $a = 7.55$ in.
(f) $\alpha = 29.7°$; $\beta = 52.4°$; $\theta = 97.9°$.

Problem 1-22

1-23. A ship must pass between the two buoys A and B. Radar indicates compass directions to the buoys and their respective distances from the point of observation. Find the distance d.

Ans. 82.1 yd.

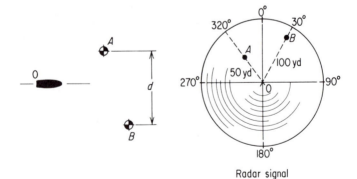

Problem 1-23

1-24. A weight is suspended from cables as shown. Determine the angle θ between cables *AC* and *BC*.

Ans. 95.4°.

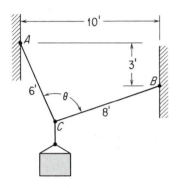

Problem 1-24   **Problem 1-25**

1-25. A rigging boom is supported by means of cable *AB* as shown. Determine the length of the cable and the angle that it makes with the boom.

Ans. 4.11 ft; 46.9°.

1-26. A crankshaft, connecting rod, and piston are shown in the figure. Find the distance from the top of the piston to "top-dead-center," the highest point in the piston's path.

Ans. 0.738 in.

Problem 1-26

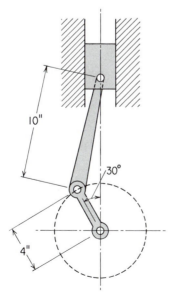

1-27. A rectangular bin 12 ft by 12 ft by 20 ft holds 170 tons of sand. How much does the sand weigh in units of pounds per cubic yard?

Ans. 3188 lb per cu yd.

1-28. A pump discharges water at the rate of 100 gpm (gallons per minute). What rate does this represent in cubic feet per second? (1 gal = 231 cu in.)

Ans. 0.223 cu ft per sec.

1-29. Sound travels at a speed of 1100 fps (feet per second) in air. Express this speed in units of miles per hour and yards per minute.

Ans. 750 mi per hr; 22,000 yd per min.

1-30. A "grain" is equal to $\frac{1}{7000}$ pound. What is the weight in ounces of a 215-grain bullet?

Ans. 0.491 oz.

1-31. An acre is equivalent to 43,560 sq ft. How many gallons of water fall on a 100-acre farm for a rainfall of 0.2 in.?

Ans. 543,000 gal.

1-32. The water level of a reservoir that has 10 sq mi of surface dropped 1.5 in. in a week. If the reservoir supplies a city of a million and a quarter inhabitants, how many gallons per day per person is consumed? (Note: 1 gal = 231 cu in.)

Ans. 99.3 gal.

1-33. Find the area of the figure shown.

Ans. 186.78 sq ft.

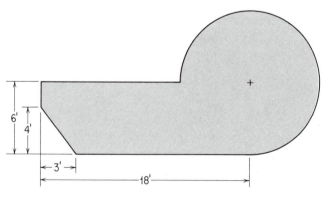

Problem 1-33

1-34. A solid right-circular cylinder has a radius of R centimeters and a length of

L feet. Find the surface area and the volume in square inches and cubic inches, respectively. (Note: 1 in. = 2.54 cm).

Ans. 0.97 R^2 + 59.3 RL sq in.; 5.84 R^2L cu in.

1-35. In the figure, what fraction of the area of the large circle is occupied by the small circles.

Ans. $\frac{7}{9}$.

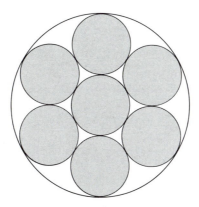

Problem 1-35

1-36. Find the topside surface area of the 4 ft by 4 ft sheet of corrugated siding shown.

Ans. 25.12 sq ft.

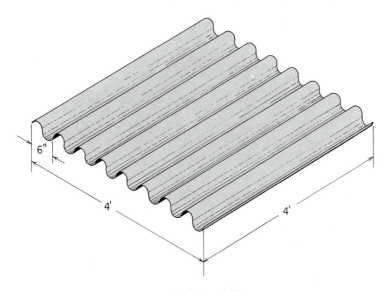

Problem 1-36

1-37. A top view of a roof is shown. All surfaces of the roof have a $\frac{5}{12}$ pitch (5 in. rise in every 12 in. taken horizontally). Find the roof area.

Ans. 13,100 sq. ft.

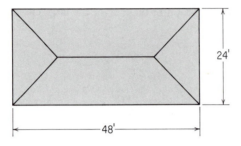

Problem 1-37

1-38. In the equation $y = ax + bx^2 + cx^3$ both x and y are given in feet. What are the units of a, b and c?

Ans. a: none, *b:* 1/ft, *c:* 1/sq ft.

1-39. In the equation $y = \sqrt{ax^3}$ both x and y are given in inches. What are the units of a?

Ans. 1/in.

1-40. The equation for the modulus of elasticity E in terms of stress σ and strain ϵ is $E = \sigma/\epsilon$. If both E and σ have the dimensions of pounds per square inch, what are the dimensions of ϵ?

Ans. None.

1-41. In the figure shown, the center of the 10-in. circle and the center of the 16-in. circle are 25 in. apart. How long is AB?

Ans. Hint: A problem that can be solved graphically and can be solved analytically; try both ways.

Problem 1-41

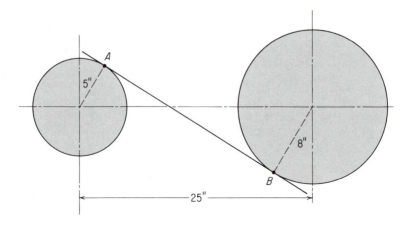

Chapter **2**

FORCE SYSTEMS: COMPONENTS, RESULTANTS, EQUIVALENCE

2-1 The Nature of a Force

A force is described by the effects that it produces. Objects stand still or move, rise or fall, spin or vibrate, because of the effect of forces.

Newton summarized the effect of forces in three laws. The first law states that a force is required to change the motion of a body. Thus, a body at rest will remain at rest and a body in motion will move uniformly in a straight line unless acted upon by a force.

The second law states that if a force acts on a body, the motion will change in speed or direction or both. This change in motion is proportional to the magnitude of the force and is in the direction of the force.

The third law states that if one body exerts a force on a second body, the second body must exert an equal but opposite force on the first. It is the principle of this law which completely discourages the manufacture of "sky hooks."

To summarize: a force is a directed action that tends to change the state of motion of a body; it is always accompanied by an opposing reaction of equal magnitude.

17

Since the "weight" of a body is the force of the earth's gravity, the units of force are defined in units of weight. In American engineering practice the standard *unit of force* is either the pound, the kip (kilopound = 1000 pounds), or the ton (2000 pounds).

2-2 Magnitude, Direction, and Line of Action

Forces can be described graphically, as in Fig. 2-1, by an arrow whose length represents the magnitude or quantity of the force. Point *A*, the tail of the arrow, indicates the point of application of the force, and the head of the arrow indicates the direction of the force's action. The line of action is indicated by the dashed line in the figure.

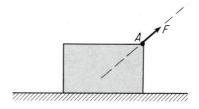

Figure 2-1

An important principle of statics, *the transmissibility of a force*, states that a force can be applied at any point on its line of action without a change in the external effects. Thus, a train of railroad cars would react equally in its motion to being pulled or pushed by a locomotive.

EXAMPLE 1: Locate, in Fig. 2-2, the point of intersection of the action line of the force and the base line of the rectangle, and then find the perpendicular distance from point *O* to the line of action of the force.

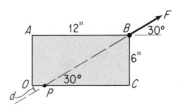

Figure 2-2

Solution: The line of action of the force cuts the base of the rectangle at an angle of 30 deg at *P*. Side *PC* can be found, since

$$\tan 30° = 6/\overline{PC}$$

$$\overline{PC} = \frac{6}{\tan 30°} = \frac{6}{0.577} = 10.4 \text{ in.}$$

The perpendicular distance d is computed in a similar manner.

$$\overline{OP} = 12 - \overline{PC} = 1.6 \text{ in.}$$

$$d = \overline{OP} \sin 30° = 1.6(0.5) = 0.8 \text{ in.}$$

2-3 Vectors

Many physical quantities, such as force, are *directed magnitudes*. These are called *vector quantities* or simply *vectors*. Some quantities, such as length and time, are specified by a single number; these are called *scalar quantities* or *scalars*.

"Two plus two equals three," is a realistic possibility when vectors are added. The reason is that magnitudes and directions are involved in the process of addition. To illustrate: Imagine two cities, A and C, 3 mi apart as shown in Fig. 2-3. One could travel from A to point B and then from B to the destination C. In this instance vector $\overrightarrow{AB}$ has been added to vector $\overrightarrow{BC}$; $\overrightarrow{AC}$ is the *resultant* of the addition. Symbolically, $\overrightarrow{AB}$ means "vector AB."

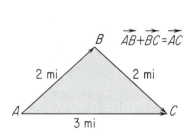

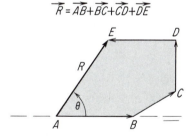

Figure 2-3   Figure 2-4

Figure 2-4 illustrates how four vectors are added to give the resultant vector R. This resultant represents the combined effects of the vector quantities and is completely defined by its magnitude and angular position θ; the latter is measured with respect to some known reference line.

The vector addition just described was performed by the *string polygon method*. The vectors are drawn joining one another, head to tail fashion; the resultant closes the string of vectors.

A *graphical solution* is obtained when the vectors are drawn accurately to some convenient scale, e.g., "Let 50 lb of force equal 1 in." Angles are measured with a protractor or by the *method of tangents* described in Chapter 1. It is important to note that the order in which vectors are added does not in any way alter the direction or the magnitude of the resultant. Thus

$$\vec{R} = \vec{AB} + \vec{BC} = \vec{BC} + \vec{AB}$$

EXAMPLE 2: Find the magnitude and direction of the resultant R of the three forces shown in Fig. 2-5. Use the graphical method.

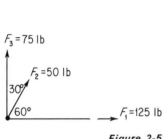

$F_3 = 75$ lb
$F_2 = 50$ lb
30°
60°
$F_1 = 125$ lb

Figure 2-5

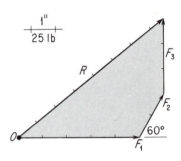

1"
25 lb
F_3
R
F_2
O
60°
F_1

Figure 2-6

Solution: Select a force scale by allowing 1 in. = 25 lb and draw the vectors "head to tail" starting at a convenient origin O.

The resultant R and the angle θ, when measured, are found to be

$$R = 190 \text{ lb}$$

and

$$\theta = 38° \angle$$

A second way of adding vectors is called the *parallelogram method*. In this procedure the vectors F_1 and F_2 are brought to a common origin O, and a parallelogram is constructed as shown in Fig. 2-7. The resultant R is the diagonal of this parallelogram. The cosine law is used to find the resultant.

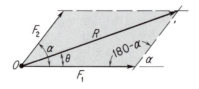

F_2
R
α
θ
$180 - \alpha$
α
O
F_1

Figure 2-7

$$R^2 = F_1^2 + F_2^2 - 2F_1F_2 \cos(180° - \alpha) = F_1^2 + F_2^2 + 2F_1F_2 \cos \alpha$$

$$R = \sqrt{F_1^2 + F_2^2 + 2F_1F_2 \cos \alpha} \qquad (2\text{-}1)$$

The directional angle θ, computed by the sine law, is

$$\sin \theta = \frac{F_2}{R} \sin(180° - \alpha) \qquad (2\text{-}2)$$

When more than two vectors are to be added by this method, a parallelogram is constructed between the third vector and the resultant of any two. The fourth vector is added in a similar manner to the resultant of the first three vectors, and so forth.

EXAMPLE 3: Determine the resultant of the forces F_1 and F_2, shown in Fig. 2-7, by the parallelogram method. Let $F_1 = 20$ lb, $F_2 = 40$ lb, and $\alpha = 30$ deg.

Solution: Find the resultant R and directional angle θ by means of Eqs. (2-1) and (2-2).

$$R = \sqrt{F_1^2 + F_2^2 + 2F_1F_2 \cos \alpha} = \sqrt{(20)^2 + (40)^2 + 2(20)40 \cos 30°}$$

$$R = \sqrt{3386} = 58.2 \text{ lb}$$

$$\sin \theta = \frac{F_2}{R} \sin(180° - \alpha) = \frac{40}{58.2} \sin 30° = 0.344$$

$$\theta = 20.1°$$

2-4 The Addition of Vectors by the Method of Components

The previous section has shown how two vectors, A and B, can be combined to form a resultant R. The converse is also true: the resultant R can be replaced by two equivalent vectors A and B. These two vectors are termed the *components of the resultant* or simply *components*. Since each component is a vector in its own right, it too can be expressed in terms of other equivalent vectors. Thus, it is conceivable that the resultant R could be the sum of an

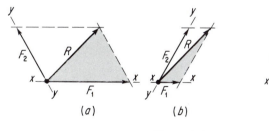

(a)   (b)   (c)

Figure 2-8

infinite number of components. In each of the three cases shown in Fig. 2-8, the vector R has a *pair* or *set* of components, F_1 and F_2.

Figure 2-8(c) is a special case in that the lines of action x-x and y-y are at right angles to each other. As a consequence, F_1 and F_2 are referred to as *rectangular* or *orthogonal components*.

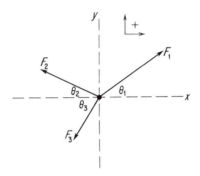

Figure 2-9

To add the forces F_1, F_2, and F_3 by the method of components, a rectangular coordinate system x-y is employed as shown in Fig. 2-9. Each force vector is resolved into its x- and y-components. The x-components are added algebraically and the y-components are added algebraically. The original force system of F_1, F_2, and F_3 is thereby reduced to a pair of forces, R_x along the x-axis, and R_y along the y-axis:

$$R_x = \sum F_x = (F_1)_x + (F_2)_x + (F_3)_x + \cdots$$
$$R_y = \sum F_y = (F_1)_y + (F_2)_y + (F_3)_y + \cdots \qquad (2\text{-}3)$$

The resultant, the vector sum of R_x and R_y, is found by the theorem of Pythagoras, thus

$$R = \sqrt{(R_x)^2 + (R_y)^2} \qquad (2\text{-}4)$$

where the tangent of the angle that the resultant R makes with the x-axis is

$$\tan \theta = \frac{R_y}{R_x} = \frac{\sum F_y}{\sum F_x} \qquad (2\text{-}5)$$

In Fig. 2-9, the positive directions are taken as "up" for y-components and to the "right" for x-components.

EXAMPLE 4: Find the resultant of the force system shown in Fig. 2-10.

Solution: Find the components R_x and R_y of the resultant by an algebraic summation of the orthogonal components of the forces.

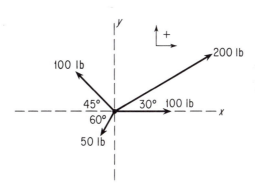

Figure 2-10

$$R_x = \sum F_x = 100 + 200 \cos 30° - 100 \cos 45° - 50 \cos 60°$$

$$R_x = 100 + 173.2 - 70.7 - 25 = +177.5 \text{ lb}$$

$$R_y = \sum F_y = 200 \sin 30° + 100 \sin 45° - 50 \sin 60°$$

$$R_y = 100 + 70.7 - 43.3 = +127.4 \text{ lb}$$

$$R = \sqrt{(R_x)^2 + (R_y)^2} = \sqrt{(177.5)^2 + (127.4)^2} = \sqrt{47{,}700}$$

$$R = 218 \text{ lb}$$

$$\theta = \arctan \frac{R_y}{R_x} = \arctan \frac{127.4}{177.5} = 35.7°$$

Expressed symbolically, the answer is

2-5 Subtraction of Vectors

By definition, the negative of a given vector is represented by an arrow whose direction has been changed by 180°. Thus, to subtract vector $\vec{F}_2$ from vector $\vec{F}_1$, reverse $\vec{F}_2$ and add it to $\vec{F}_1$ by any of the three methods described.

$$\vec{R} = \vec{F}_1 - \vec{F}_2 = \vec{F}_1 + (-\vec{F}_2) \tag{2-6}$$

2-6 Space Forces

The preceding discussion was concerned with components and resultants of groups of force vectors that occupied a single plane. In each instance, two rectangular components, R_x and R_y, completely identified the force. In contrast, a *three-dimensional*, or *space force* is a vector that occupies

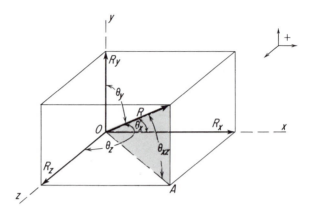

Figure 2-11

space, instead of lying in a plane, and it has three rectangular components.[1]

To aid in visualizing a three-dimensional force, a rectangular box is drawn with the force R as the diagonal of this box. This is illustrated in Fig. 2-11. The far edges of the box lie on the three rectangular axes x, y, and z, and R_x, R_y, and R_z are the rectangular components of R. A study of the figure indicates that there are three right triangles, each having R as the hypotenuse and each lying in a different plane. The triangles are distorted by the perspective of the drawing. The angle θ_x, θ_y, and θ_z define the inclination of R with respect to the rectangular axis of each right triangle. The diagonal $\overline{OA}$ represents the componet of R in the x-z plane.

$$(\overline{OA})^2 = R_x^2 + R_z^2$$

Since R is the resultant of OA and R_y, it follows that

$$R^2 = (\overline{OA})^2 + R_y^2 = R_x^2 + R_y^2 + R_z^2 \tag{2-7}$$

$$R = \sqrt{R_x^2 + R_y^2 + R_z^2} \tag{2-8}$$

The angle between the resultant and each rectangular component can be expressed as a cosine function; these are called *direction cosines*.

$$\cos \theta_x = \frac{R_x}{R}, \qquad R_x = R \cos \theta_x$$

$$\cos \theta_y = \frac{R_y}{R}, \qquad R_y = R \cos \theta_y \tag{2-9}$$

$$\cos \theta_z = \frac{R_z}{R}, \qquad R_z = R \cos \theta_z$$

[1]As an example of a space vector, imagine that to arrive at a given point, you have walked two blocks north, three blocks east, and then up ten flights of stairs. The straight line drawn from where you were to where you are is a space vector.

When the components in terms of their direction cosines are substituted into Eq. (2-7), the sum of the squares of the direction cosines is found to equal unity.

$$R^2 = R_x^2 + R_y^2 + R_z^2 = R^2 \cos^2 \theta_x + R^2 \cos^2 \theta_y + R^2 \cos^2 \theta_z$$

$$\cos^2 \theta_x + \cos^2 \theta_y + \cos^2 \theta_z = 1$$

In some instances, the direction cosines are conveniently obtained by employing the coordinates of the force. For example, imagine that the force in Fig. 2-11 passes through the origin O and the coordinates (3, 4, 12); this means that if a rectangular box is drawn to a convenient scale, the x-component is 3 units long, the y-component 4 units long, and the z-component 12 units long. The resultant would have a scale length of

$$d = \sqrt{3^2 + 4^2 + 12^2} = 13$$

The direction cosines of the force are

$$\cos \theta_x = \tfrac{3}{13}$$

$$\cos \theta_y = \tfrac{4}{13}$$

$$\cos \theta_z = \tfrac{12}{13}$$

EXAMPLE 5: Determine the resultant of two space forces, $F_1 = 100$ lb and $F_2 = 200$ lb, having coordinates (2, 3, 4) and $(-3, 5, -2)$ respectively. Both forces pass through the origin of the coordinate system.

Solution: Set up a table and determine the x-, y-, and z-component of each force systematically, using the direction cosines.

Table 2. 1

$F_1 = 100$ lb	$F_2 = 200$ lb
$d_1 = \sqrt{2^2 + 3^2 + 4^2} = 5.39$	$d_2 = \sqrt{(-3)^2 + (5)^2 + (-2)^2} = 6.16$
Direction Cosines	*Direction Cosines*
$\cos \theta_x = 2/5.39 = 0.371$	$\cos \theta_x = -3/6.16 = -0.487$
$\cos \theta_y = 3/5.39 = 0.557$	$\cos \theta_y = 5/6.16 = +0.812$
$\cos \theta_z = 4/5.39 = 0.742$	$\cos \theta_z = -2/6.16 = -0.325$
Force Components	*Force Components*
$(F_1)_x = 100(0.371) = 37.1$ lb	$(F_2)_x = 200(-0.487) = -97.4$
$(F_1)_y = 100(0.557) = 55.7$ lb	$(F_2)_y = 200(0.812) = 162.4$
$(F_1)_z = 100(0.742) = 74.2$ lb	$(F_2)_z = 200(-0.325) = -65.0$

Then add the components:

$$R_x = \sum F_x = 37.1 - 97.4 = -60.3 \text{ lb}$$

$$R_y = \sum F_y = 55.7 + 162.4 = 218.1 \text{ lb}$$

$$R_z = \sum F_z = 74.2 - 65.0 = 9.2 \text{ lb}$$

and find R from Eq. (2-8):

$$R = \sqrt{R_x^2 + R_y^2 + R_z^2} = \sqrt{(-60.3)^2 + (218.1)^2 + (9.2)^2}$$
$$= 226 \text{ lb}$$

The inclination of the resultant is specified by its direction cosines.

$$\cos \theta_x = \frac{-60.3}{226} = -0.267$$

$$\cos \theta_y = \frac{218.1}{226} = 0.965$$

$$\cos \theta_z = \frac{9.2}{226} = 0.041$$

2-7 The Resultant of Two Parallel Forces

Two forces having the same line of action are said to be *collinear*. The resultant of collinear forces that are equal in magnitude but opposite in direction is zero. Thus, by adding the collinear forces F to the parallel forces F_1

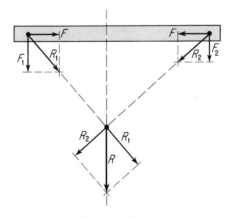

Figure 2-12

and F_2, in Fig. 2-12, the original force system is unchanged. The resultant R_1 of F_1 and F is added to the resultant R_2 of F_2 and F, giving the final resultant R. A problem of this type is best solved graphically.

2-8 Moment of a Force

A force has a moment about a given axis when it tends to produce rotation about that axis. The magnitude of the moment is defined as the product of the force and the perpendicular distance from the line of action of the force to the axis of rotation. In Fig. 2-13, the moment M is the product $F \cdot d$. Since a moment tends to produce rotation, the sign $(+)$ or $(-)$ is one of choice; clockwise or counterclockwise moments take opposite signs.

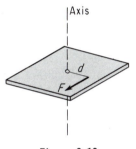

Figure 2-13

An important theorem of statics states that *the moment of a force is equal to the sum of the moments of the components.* The example that follows will demonstrate this theorem.

EXAMPLE 6: Determine the moment about point O of the 100-lb force that acts at point A on the bracket shown in Fig. 2-14 by: (a) finding the product of the force and its perpendicular distance from O; (b) taking moments of the rectangular com-

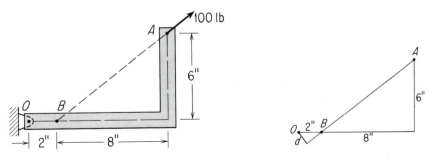

Figure 2-14

ponents of the force at A; (c) taking moments of the rectangular components of the force at B.

Solution:

Part (a). First determine the distance $\overline{AB}$, the hypotenuse of the right triangle.

$$\overline{AB} = \sqrt{6^2 + 8^2} = 10 \text{ in.}$$

Next, find the distance d by using the ratio of the corresponding sides of the two similar triangles.

$$\frac{d}{2} = \frac{6}{10}$$

$$d = 1.2 \text{ in.}$$

The moment about point O of the 100-lb force is

$$M = F \cdot d = 100(1.2) = 120 \text{ lb in.}$$

Part (b). Since the slope of the force is given, the x- and y-components are

$$F_x = F \cos \theta = 100 \times \tfrac{4}{5} = 80 \text{ lb}$$

and

$$F_y = F \sin \theta = 100 \times \tfrac{3}{5} = 60 \text{ lb}$$

If a counterclockwise moment is positive, the algebraic sum of the moments about O is

$$M = 60(10) - 80(6) = 120 \text{ lb in.}$$

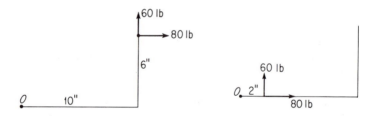

Part (c). The x- and y-components also act at point B, which is on the action line of the force. The 80-lb force has a moment equal to zero, since it passes through O. The moment of the 60-lb force is

$$M = 60(2) = 120 \text{ lb in.}$$

2-9 Resultants of Force Systems: The Use of the Principle of Moments

The resultant force that acts on a body has the same external effect as the forces it replaces. R_x represents the sum of the x-components of force; R_y, the sum of the y-components; R_z, the sum of the z-components. Symbolically, this is stated as

$$R_x = \sum F_x \qquad R_y = \sum F_y \qquad R_z = \sum F_z \qquad (2\text{-}10)$$

The moment of the resultant about all three rectangular axes must, in a similar fashion, be equal to the sum of the moments of the individual forces about each rectangular axis. In Fig. 2-15 a group of parallel forces, F_1, F_2, and F_3, act on the beam. The resultant R has the magnitude of $\sum F_y$, and its moment about any arbitrary point O must be equal to the sum of the moments of the individual forces about point O.

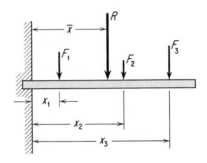

Figure 2-15

$$R = \sum F_y = F_1 + F_2 + F_3$$

and
$$M_0 = R\bar{x} = F_1 x_1 + F_2 x_2 + F_3 x_3 \qquad (2\text{-}11)$$

Two moment equations are required to locate the resultant R in the force system of Fig. 2-16. These are

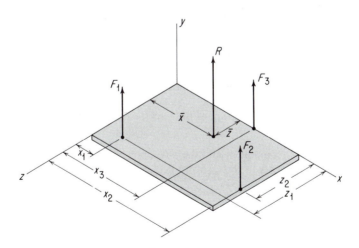

Figure 2-16

$$M_{0z} = R\bar{x} = F_1 x_1 + F_2 x_2 + F_3 x_3$$

and $\qquad M_{0x} = R\bar{z} = F_1 z_1 + F_2 z_2 + F_3 z_3$

The forces in this figure are parallel to the y-axis and therefore have a zero moment about this axis.

EXAMPLE 7: Determine the magnitude and location of the resultant of the parallel forces acting on the truss shown in Fig. 2-17.

Solution: The magnitude of the resultant is the algebraic sum of the parallel forces. Assume "down" to be the positive y-direction.

$$R_y = \Sigma F_y = 2 + 5 + 10 + 3 = 20 \text{ kips}$$

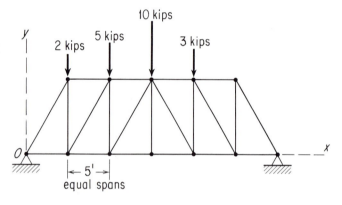

Figure 2-17

The moment of the resultant is equal to the sum of the moments of the individual forces. If clockwise moments are assumed to be positive, then

$$M_0 = R\bar{x} = \Sigma F \cdot x$$

$$20x = 2(5) + 5(10) + 10(15) + 3(20)$$

$$x = \frac{270}{20} = 13.5 \text{ ft to the right of } O$$

EXAMPLE 8: Find the resultant of the four forces that act on the bracket shown in Fig. 2-18.

Solution: The x-component and the y-component of the resultant are the sum of the force components in these two respective directions.

$$R_x = \Sigma F_x = 55 - 15 = 40 \text{ lb} \rightarrow$$

$$R_y = \Sigma F_y = 70 - 40 = 30 \text{ lb} \uparrow$$

The resultant R is the vector sum of R_x and R_y; thus

$$R = \sqrt{R_x^2 + R_y^2} = \sqrt{(40)^2 + (30)^2} = 50 \text{ lb}$$

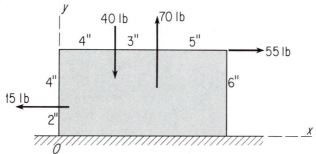

Figure 2-18

The inclination of the resultant is

$$\theta = \text{arc tan} \frac{\sum F_y}{\sum F_x} = \text{arc tan} \frac{30}{40}$$

$$\theta = 36.9° \angle$$

The moment of the resultant must be equal to the sum of the moments of the forces acting on the bracket. Assume clockwise to be the positive direction.

$$R \cdot d = \sum \text{Moments}_0$$

$$50 \cdot d = 55(6) - 70(7) + 40(4) - 15(2) = -30 \text{ lb in.}$$

$$d = -\frac{30}{50} = -0.6 \text{ in.}$$

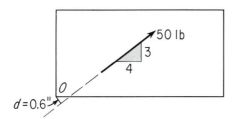

The negative sign indicates that the moment of the 50-lb resultant about O is counterclockwise or opposite to the assumed positive direction.

2-10 Couples

The force system shown in Fig. 2-19 consists of two equal but oppositely directed parallel forces separated by a distance d. This system, called a *couple*, has a resultant force equal to zero but a resultant moment equal to $F \cdot d$. Since the resultant of a couple is a *pure moment*, rotation of the couple in its plane does not alter its effect on the body. By computing the moment of the two forces about point O in Fig. 2-19 (assuming clockwise to be positive), it is also found that the moment of a couple does not depend upon its position in the plane of the body; thus

$$\sum M_0 = -Fx + F(x + d) = F \cdot d$$

The sum of two or more couples is simply the algebraic sum of their moments.

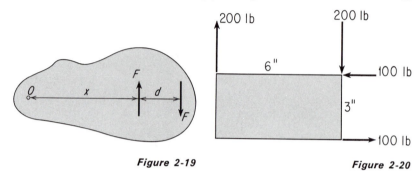

Figure 2-19   Figure 2-20

EXAMPLE 9: Compute the moment of the two couples that act on the body shown in Fig. 2-20.

Solution: Select clockwise as the positive direction. The resultant moment is the algebraic sum of the individual moments:

$$\sum M = 200(6) - 100(3) = 900 \text{ lb in.}$$

QUESTIONS, PROBLEMS, AND ANSWERS

2-1. Draw each of the force vectors illustrated to scale. Allow 1 in. of length to equal 100 lb of force.

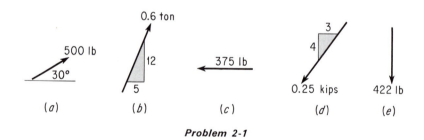

Problem 2-1

2-2. Two forces, F_1 and F_2, act on the beam as shown. Determine the vertical distance above F_1 of the point of intersection of the action lines of these two forces.

Ans. 6.67 ft.

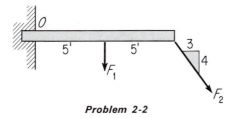

Problem 2-2

2-3. Determine the point of intersection of the action line of F_2 and the vertical wall in Prob. 2-2.

Ans. 13.3 ft.

2-4. Find the perpendicular distance from the action line of F_2 to point O in Prob. 2-2.

Ans. 8 ft.

2-5. A force acts as shown on the corner of the beam section. Find the perpendicular distance d from the action line of the force to the geometric center of the beam.

Ans. 3.00 in.

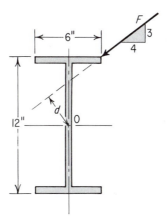

Problem 2-5

Problem 2-6

2-6. The lines of action of the two forces illustrated pass through point *P*. Determine the inclinations θ_1 and θ_2 of each force, measured from the horizontal reference line.

Ans. 73.3°; 55°.

2-7 to 2-12. Determine the magnitudes and directions of the resultants in the figures by the string polygon method.

Ans.   **2-7.** 510 lb; 101.3° $\angle$.
   2-8. 101 lb; 46.4° $\angle$.
   2-9. 5.85 k; 57.8° $\angle$.
   2-10. 548 lb; 98.6° $\angle$.
   2-11. 1310 lb; 107° $\angle$.
   2-12. 316 lb; 74.9° $\angle$.

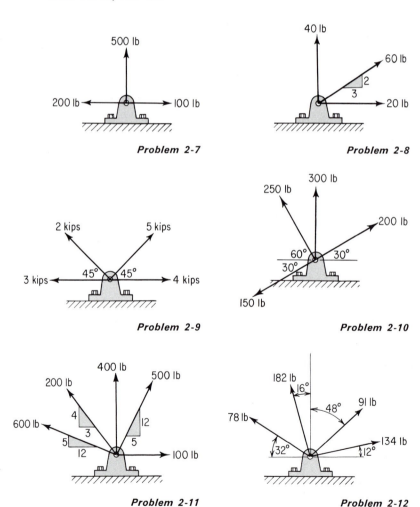

Problem 2-7

Problem 2-8

Problem 2-9

Problem 2-10

Problem 2-11

Problem 2-12

2-13. Three forces act on the tee crank as shown. Find the magnitude and direction of the resultant by the string polygon method. Use a force scale of 100 lb = 1 in.

Ans. 757 lb; 152.1° ∠.

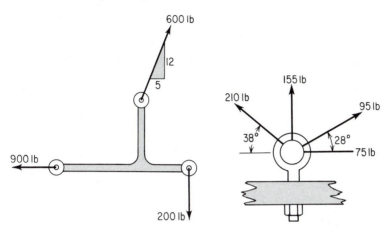

Problem 2-13   **Problem 2-14**

2-14. Four forces act on an "eye" hook. Select a convenient force scale and determine the magnitude and direction of the resultant by the string polygon method.

Ans. 329 lb; 90° ∠.

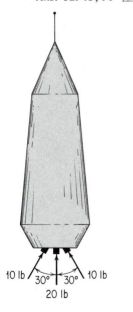

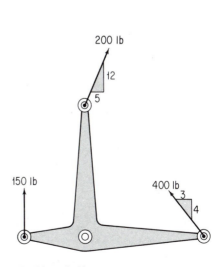

Problem 2-15   **Problem 2-16**

2-15. A system of maneuvering jets exerts forces on the rocket as shown. Determine the magnitude and direction of the resultant by the string polygon method.

Ans. 74.7 lb; 105.5° ∠.

2-16. Three forces act on the tee crank. Find the resultant of these forces by the parallelogram method by first combining any two forces and then combining this resultant with the third force.

Ans. 675 lb; 76° ∆.

2-17. A rope that has been looped around a bracket supports the two loads indicated. Find the resultant of these two forces.

Ans. 375 lb; 49.6° ∇.

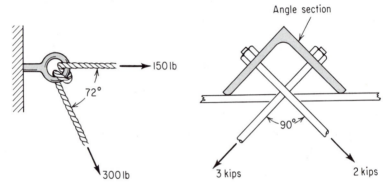

Angle section

150 lb

72°

300 lb

90°

3 kips

2 kips

Problem 2-17   *Problem 2-18*

2-18. A section of a tie rod brace is illustrated. Find the resultant of the two tie rod forces and angle between the resultant and the vertical.

Ans. 3.61 k; 11.3° cw from vertical.

2-19. A section of a bucket elevator is shown. Find the resultant of the two chain forces.

Ans. 1200 lb.

2-20. The resultant of the cable force *F* and the 5000-lb weight acts along the boom of the derrick. Find the force *F*.

Ans. 12,700 lb.

2-21. A uniform weight of 20 tons is lifted by the plate clamps and chains as shown. Find the force in each chain.

Ans. 11.5 tons.

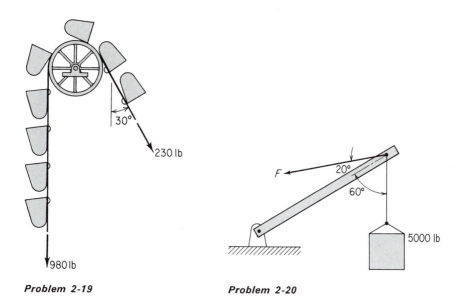

30°

230 lb

F ◄— 20°

60°

5000 lb

980 lb

Problem 2-19

Problem 2-20

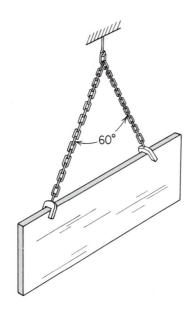

60°

Problem 2-21

2-22. The "chain grab" illustrated has a lifting capacity of 3 tons. Find the force in the chain. Hint: the resultant of the two equal chain forces at point A is 3 tons.

Ans. 1.73 tons.

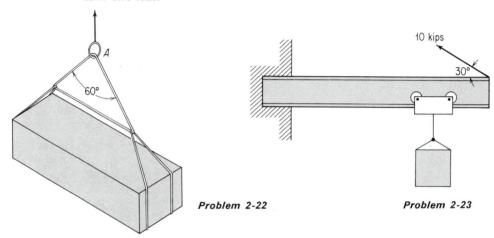

Problem 2-22   **Problem 2-23**

2-23. Determine the horizontal and vertical components of the 10-kip force acting on the jib crane.

Ans. 8.66 k; 5.00 k.

2-24. A reel is pulled up a ramp by a winch as illustrated. The tensile force in the horizontal cable is 5 tons. Find the components of the force tangent to and normal to the ramp.

Ans. 4.70 tons; 1.71 tons.

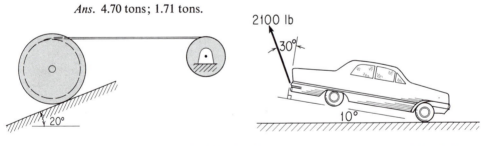

Problem 2-24   **Problem 2-25**

2-25. An automobile is being towed by a truck as indicated. Find the horizontal and vertical components of the 2100-lb force in the cable.

Ans. 718 lb; 1970 lb.

2-26. Determine the components, of a 220-lb force, that are perpendicular and parallel to a line making an angle of 28 deg with the force.

Ans. 103 lb; 194 lb.

2-27. A uniform belt force of 325 lb is exerted throughout the entire tandem drive shown. Find the direction and magnitude of the resultant force on each of the four pulleys A, B, C, and D.

Ans. $A_x = 625 \rightarrow$, $A_y = 125 \downarrow$.
$\quad\;\; B_x = 105 \leftarrow$, $B_y = 385 \uparrow$.
$\quad\;\; C_x = 195 \leftarrow$, $C_y = 585 \downarrow$.
$\quad\;\; D_x = 325 \leftarrow$, $D_y = 325 \uparrow$ all in pounds.

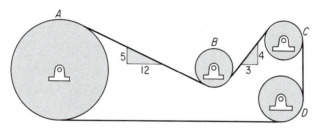

Problem 2-27

2-28. Show that the vector sum of the four resultant pulley forces in Prob. 2-27 is zero.

2-29. A vertical force of 20 lb can be resolved into two components, one of 40 lb and the other of 50 lb. Determine the angle each makes with the horizontal.

Ans. $18.2°$; $40.5°$.

2-30–2-35. Use the method of components to determine the magnitude and direction of the resultant of the forces which act at a common point on a body as shown in the figures.

Ans. **2-30.** *Hint:* $R_x = 112$ lb $\rightarrow$; $R_y = 412$ lb $\uparrow$.
$\qquad$ **2-31.** *Hint:* $R_x = 8.6$ lb $\leftarrow$; $R_y = 20.8$ lb $\uparrow$.
$\qquad$ **2-32.** *Hint:* $R_x = 8$ k $\leftarrow$; $R_y = 2$ k $\downarrow$.
$\qquad$ **2-33.** *Hint:* $R_x = 90$ lb $\rightarrow$; $R_y = 125$ lb $\uparrow$.
$\qquad$ **2-34.** *Hint:* $R_x = 2.57\ F \leftarrow$; $R_y = 1.317\ F \uparrow$.
$\qquad$ **2-35.** *Hint:* $R_x = 408$ lb $\leftarrow$.

2-36. Plastics can be laminated as illustrated by rolling several layers together. Small tensile forces keep the materials taut. Find the resultant of the three tensile forces.

Ans. 230 lb $\leftarrow$.

2-37. The clamshell bucket, used in earth excavations, is opened and closed by means of a piston. When it is in the position shown, the forces in AB and BC

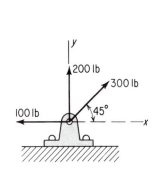

Problem 2-30

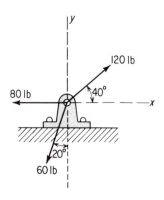

Problem 2-31

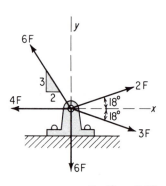

Problem 2-32

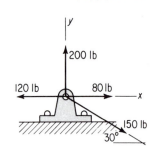

Problem 2-33

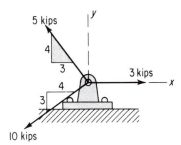

Problem 2-34

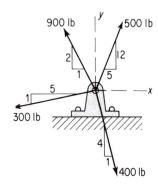

Problem 2-35

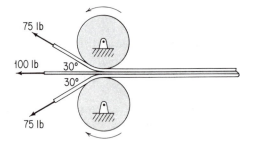

Problem 2-36

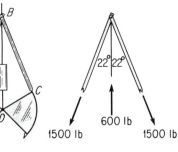

Problem 2-37

are 1500 lb each, and in the piston *BD*, 600 lb. Determine the resultant of these three forces.

Ans. 2182 lb ↓ .

2-38. A 200-lb force pointing down and to the left makes an angle of 45 deg with the horizontal. A second 200-lb force points to the right at an angle of zero degrees with the horizontal. Determine the magnitude and direction of the resultant of these two forces.

Ans. 153 lb; 67.5° ↘.

2-39. Determine the resultant of the two forces that act on the truss illustrated. Where does the resultant intersect member *AB*?

Ans. 14 k, 75.5° ↘; 1.3 ft to right of *A*.

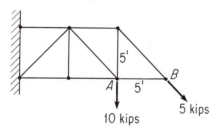

Problem 2-39

2-40. Two forces act as shown; find the force vector representing (a) $\vec{F_1} + \vec{F_2}$; (b) $\vec{F_1} - \vec{F_2}$; (c) $\vec{F_2} - \vec{F_1}$.

Ans. (a) 90.1 lb, 33.7° ∠.
(b) 90.1 lb, 33.7° ↘.
(c) 90.1 lb, 33.7° ↘.

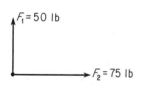

Problem 2-40

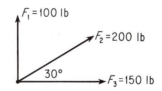

Problem 2-41

2-41. Three forces act at a point as shown. Determine the force vector representing (a) $\vec{F_1} + \vec{F_2} + \vec{F_3}$; (b) $\vec{F_1} - \vec{F_2} + \vec{F_3}$; (c) $\vec{F_1} - \vec{F_2} - \vec{F_3}$; (d) $\vec{F_3} - \vec{F_2} - \vec{F_1}$.

Ans. (a) 380 lb, 318° ∠.
(b) 23.2 lb, ←.
(c) 323 lb, ←.
(d) 201 lb, 83.4° ↗.

2-42. With the aid of a three-dimensional coordinate system, sketch a space force that passes through the origin and has coordinates of (2,2,2).

2-43. Make a perspective drawing of a space force having components $F_x = 10$ lb, $F_y = 30$ lb, $F_z = 50$ lb. Determine the magnitude of the resultant and its direction cosines.

2-44. What are the direction cosines of a space force that passes through the origin and has coordinates of (2, 5, −3)?

Ans. $\cos \theta_x = 0.324$, $\cos \theta_y = 0.811$, $\cos \theta_z = -0.487$.

2-45. Find the components of the resultant of two space forces: $F_1 = 200$ lb, and $F_2 = 100$ lb. Both forces pass through the origin. The coordinates of F_1 and F_2 are (1, 3, −2) and (2, 5, 1) respectively.

Ans. $R_x = 90$ lb; $R_y = 252$ lb; $R_z = 88.7$ lb.

2-46. Find the magnitude and direction cosines of the resultant of the two space forces.

Ans. 293 lb; $\cos \theta_x = \frac{107}{293}$, $\cos \theta_y = \frac{103}{293}$, $\cos \theta_z = -\frac{253}{293}$.

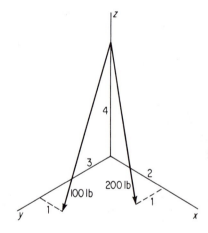

Problem 2-46

2-47. Use a graphical method to find the magnitude and location of the resultant of the two parallel forces which act on the beam illustrated.

Ans. 10 k ↓, 11.2 ft to right of *A*.

2-48. Parallel forces act on the column shown. Find the magnitude and location of the resultant by graphical means.

Ans. 400 lb to the right, 23.5 ft above *A*.

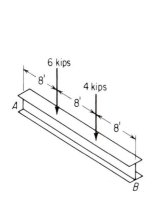

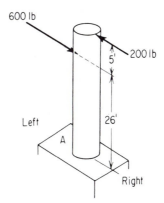

Problem 2-47   **Problem 2-48**

2-49. Find the magnitude and location of the resultant in Prob. 2-48 if both forces act to the right.

Ans. 800 lb to the right, 27.3 ft above *A*.

2-50. The overhead crane shown has a rated capacity of 500 tons. Find the moment of this load about end *A* if (a) $d = 3$ ft; (b) $d = 5$ ft; (c) $d = 12$ ft; (d) $d = 20$ ft.

Ans. (c) 6000 ton ft.

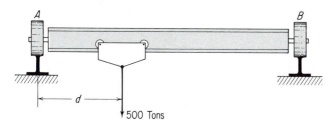

Problem 2-50

2-51. The hairpin conveyor truck illustrated is used to move sheet steel coils to various places in a stamping plant. If the coil weighs 2 tons find the wheel reactions at *A* and *B* and the bending moment in the arm at *C*.

Ans. $R_A = 1.1$ ton; $R_B = 0.9$ ton; $M_C = 5$ ton ft.

2-52. Determine the moment of the 3-kip load about point *O*.

2-53–2-58. Find the magnitude, direction, and location of the resultant force which acts on the beam shown in the respective figures.

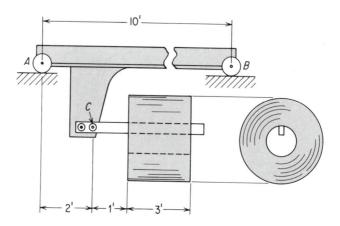

Problem 2-51

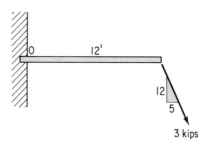

Problem 2-52

Ans. **2-52.** 33.2 kip ft.
  2-53. 5 k ↓ ; 6 ft to right of *A*.
  2-54. 3.5 ton ↓ ; 2 ft to right of *A*.
  2-55. Resultant force is zero, resultant moment is 600 lb ft.
  2-56. 6870 lb, 77.1° ↗ ; 8.41 ft to right of *A*.
  2-57. 10.5 k, 77.3° ∡ ; 9.38 ft to right of *A*.
  2-58. 1580 lb, 85.8° ↘ ; 8.11 ft to right of *A*.

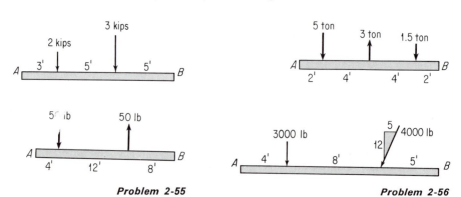

Problem 2-55   **Problem 2-56**

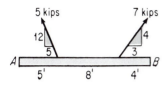

Problem 2-57

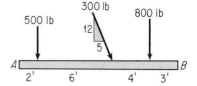

Problem 2-58

2-59–2-61. Find the magnitude and direction of the resultant force which acts on the plate shown in the respective illustrations. Locate the resultant with respect to a point of intersection on a line collinear with AD.

Ans. **2-59.** $R_x = 350$ lb $\rightarrow$, $R_y = 100$ lb $\downarrow$; 21 in. to right of A.

2-60. $R_x = 180$ lb $\leftarrow$, $R_y = 25$ lb $\uparrow$; 73.9 in. to right of A.

2-61. $R_x = 5$ k $\leftarrow$, $R_y = 9.5$ k $\downarrow$; 12.6 in. to right of A.

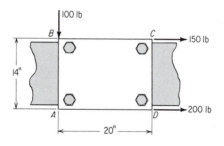

Problem 2-59

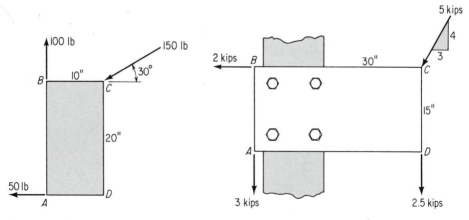

Problem 2-60   Problem 2-61

2-62–2-67. Determine the magnitude and the direction of the resultant couple which acts on the system shown in the respective illustrations.

Ans. **2-62.** Zero.
2-63. 990 lb in. ↻.
2-64. 40 lb ft ccw looking downward
2-65. 250 lb ft ↻.
2-66. 1050 lb in. ↻.
2-67. 4800 lb in. cw looking toward the right.

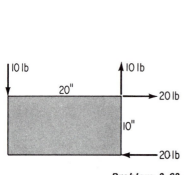

Problem 2-62

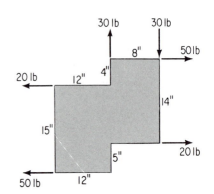

Problem 2-63

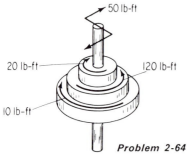

Problem 2-64

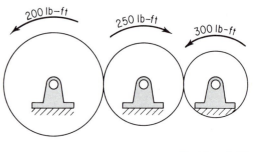

Problem 2-65

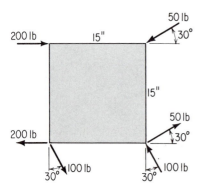

Problem 2-66

Levinson P2-66

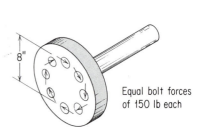

Equal bolt forces
of 150 lb each

Problem 2-67

2-68. Sketch the addition of two vectors $A = 2$ units and $B = 2$ units such that:

(a) $A + B = 4$
(b) $A + B = 3$
(c) $A + B = 2$
(d) $A + B = 1$
(e) $A + B = 0$

2-69. List the three specifications necessary to completely define a vector.

Ans. Magnitude, direction, and line of action.

2-70. The resultant of collinear forces that are equal in magnitude but opposite in direction is _____.

Ans. Zero.

2-71. Couples are vector quantities. True or false?

Ans. True, they have magnitude, direction, and a line of action.

2-72. The *right-hand rule* is usually employed to graphically describe the line of action of a couple; let the fingers of the right hand form a fist with the thumb extended and let the circular position of the fingers represent the rotation of the couple—the thumb will graphically represent the couple vector. Sketch on the cube shown each of the couple represented by the three respective vectors.

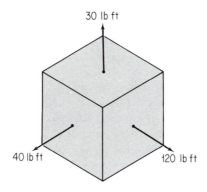

Problem 2-72

Chapter 3

CENTER OF GRAVITY

An infinite variety of man-made forces can act on a body; the magnitudes, directions, and locations of these forces can be changed at will by placing a cable here or by tightening a bolt there. One natural force, however, which is equally important, is *weight*—the earth's force of attraction for the body. In this instance, nature prescribes how much this is to be and through which point on the body it may be assumed to act. This point of action is called the *center of gravity*.

3-1 Experimental Determination of the Center of Gravity

The weight of a body is a resultant force that is equal to the sum of the weights of all the particles that comprise the body. The location of this resultant *W* may be determined experimentally by suspending the weight first from one point and then from another as shown in Fig. 3-1.

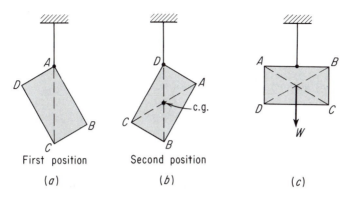

First position
(a)

Second position
(b)

(c)

Figure 3-1

Since the weight force acts downward, its line of action must be somewhere on line AC when suspended from point A. When the body is suspended from point D, the weight force acts somewhere on line DB. The intersection of line AC with line DB is the center of gravity of the body. The weight force always acts somewhere on a vertical line drawn downward from the point of suspension.

In many instances it is not convenient to suspend the body and then draw intersecting lines on it. Another method must be employed. Suppose, for example, that it is important to know the location of the center of gravity of a particular automobile. It can be determined by first weighing the front axle with the car tipped upwards through some convenient angle θ as shown in Fig. 3-2(a). The moment of the weight force W about the rear axle must equal the moment of the scale force $(W_f)_1$, about that same point.

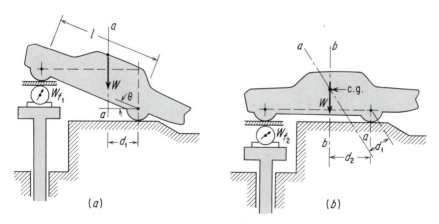

(a)

(b)

Figure 3-2

$$Wd_1 = (W_f)_1 l \cos \theta$$

and d_1 can be computed

$$d_1 = \frac{(W_f)_1 l \cos \theta}{W}$$

In a similar manner, with the car horizontal

$$d_2 = \frac{(W_f)_2 l}{W}$$

Lines a–a and b–b can be drawn on the car, or trigonometric calculations can be made and the center of gravity thus located.

EXAMPLE 1: Assume the car in Fig. 3-2 weighs 3000 lb and has a wheel base l of 10 ft. When the car is tipped at an angle of 30 deg, the recorded weight $(W_f)_1$ is 1200 lb; when it is horizontal, the recorded weight $(W_f)_2$ is 1500 lb. Locate the center of gravity of the car with respect to the rear axle.

Solution: The distances d_1 and d_2 are first computed:

$$d_1 = \frac{(W_f)_1 l \cos \theta}{W} = \frac{1200(10)(0.866)}{3000}$$

$$d_1 = 3.46 \text{ ft}$$

and
$$d_2 = \frac{(W_f)_2 l}{W} = \frac{1500(10)}{3000}$$

$$d_2 = 5 \text{ ft}$$

The location of the center of gravity above the rear axle is next computed. Trigonometric relationships are applied to the triangles that are drawn as shown in Fig. 3-3. The distance $\overline{BC}$ is given by

$$\overline{BC} = \frac{d_1}{\cos \theta} = \frac{3.46}{0.866}$$

where
$$\overline{BC} = 4 \text{ ft}$$

and
$$\overline{AB} = 5 - 4 = 1 \text{ ft}$$

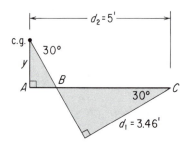

Figure 3-3

The quantity y is then computed

$$y = \frac{\overline{AB}}{\tan 30^\circ} = \frac{1}{0.577} = 1.73 \text{ ft}$$

The center of gravity, therefore, is located 5 ft ahead of the rear axle and 1.7 ft above it.

3-2 Center of Gravity of Grouped Particles

A group of small particles with weights w_1, w_2, and w_3 is arranged in a straight line as shown in Fig. 3-4. The single resultant force W, which represents this system of particles, must: first, be equal to the sum of the individ-

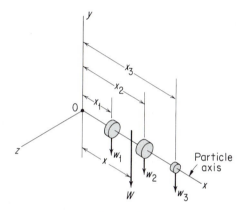

Figure 3-4

ual weights $W = w_1 + w_2 + w_3$, and second, have a moment about any arbitrary point equal to the sum of the moments of the individual weights. The lever arm $\bar{x}$ which locates the resultant, is the single *coordinate* of the center of gravity of this group.

$$W\bar{x} = w_1 x_1 + w_2 x_2 + w_3 x_3 \tag{3-1}$$

$$\bar{x} = \frac{w_1 x_1 + w_2 x_2 + w_3 x_3}{W}$$

When the particles are grouped to occupy a plane rather than a line, two moment equations must be written, since two coordinates are required to locate the center of gravity. If the group occupies space rather than a plane, three equations must be written in order to find the three coordinates of the center of gravity.

EXAMPLE 2: Locate the center of gravity of the grouped particles shown in Fig. 3-5.

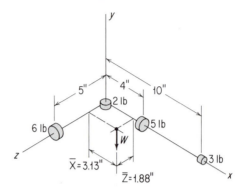

Figure 3-5

Solution: The total weight of the group is $W = 6 + 2 + 5 + 3 = 16$ lb. Since the system occupies a plane, two coordinates of the center of gravity must be found, one with respect to the x-axis and the other with respect to the z-axis. To find $\bar{x}$, moments are taken about the z-axis.

$$W\bar{x} = w_1 x_1 + w_2 x_2 + w_3 x_3 + w_4 x_4$$

$$16\bar{x} = 6(0) + 2(0) + 5(4) + 3(10)$$

$$\bar{x} = \frac{20 + 30}{16} = 3.13 \text{ in.}$$

Next, moments are summed with respect to the x-axis and $\bar{z}$ is computed

$$W\bar{z} = w_1 z_1 + w_2 z_2 + w_3 z_3 + w_4 z_4$$

$$16\bar{z} = 6(5) + 2(0) + 5(0) + 3(0)$$

$$\bar{z} = \frac{30}{16} = 1.88 \text{ in.}$$

EXAMPLE 3: Determine the center of gravity of two weights located in space as shown in Fig. 3-6.

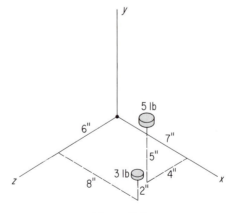

Figure 3-6

Solution: Since the system occupies space, three coordinates of the center of gravity must be found. The total weight of the system is

$$W = w_1 + w_2 = 3 + 5 = 8 \text{ lb}$$

Moments are taken with respect to the z-axis and $\bar{x}$ is found.

$$W\bar{x} = w_1 x_1 + w_2 x_2$$
$$8\bar{x} = 3(8) + 5(7)$$
$$\bar{x} = \frac{24 + 35}{8} = 7.38 \text{ in.}$$

In a similar manner, moments are taken with respect to the x-axis and $\bar{z}$ is computed.

$$W\bar{z} = w_1 z_1 + w_2 z_2$$
$$8\bar{z} = 3(6) + 5(4)$$
$$\bar{z} = \frac{18 + 20}{8} = 4.75 \text{ in.}$$

To find $\bar{y}$, it must be imagined that the entire system is turned about with the x-y axis horizontal. Moments are then taken with respect to the x-axis

$$W\bar{y} = w_1 y_1 + w_2 y_2$$
$$8\bar{y} = 3(2) + 5(5)$$
$$\bar{y} = \frac{6 + 25}{8} = 3.88 \text{ in.}$$

The center of gravity is located at $\bar{x} = 7.38$ in., $\bar{y} = 3.88$ in., $\bar{z} = 4.75$ in.

3-3 Centroids and Centers of Gravity of Two-Dimensional Figures

The *centroid* of a geometrical figure corresponds to the *center of gravity* of a homogeneous body of the same form. The terms centroid and center of gravity are often used interchangeably; they differ in location, however, when the body is not homogeneous. For example, if a rod were made by joining a length of aluminum to an equal length of steel, the centroid would be located at the geometric center of the bar, while the center of gravity would be past the geometric center close to the heavy end.

Most irregularly shaped figures can be subdivided into shapes whose centroids are known. The plate shown in Fig. 3-7 consists of two rectangular areas A_1 and A_2. Since both of these areas are symmetrical, the centroid of each is located at the intersection of the diagonals as shown. By definition, the moment of the total area about any line must be equal to the sum of the moments of the portions of area about the same line. Thus, $\bar{x}$ is found by taking moments about the y-axis.

$$(A_1 + A_2)\bar{x} = A_1 x_1 + A_2 x_2$$

and

$$\bar{x} = \frac{A_1 x_1 + A_2 x_2}{(A_1 + A_2)} \tag{3-2}$$

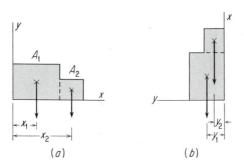

Figure 3-7

The second coordinate of the centroid is found by taking moments about the x-axis, thus

$$(A_1 + A_2)\bar{y} = A_1 y_1 + A_2 y_2$$

$$\bar{y} = \frac{A_1 y_1 + A_2 y_2}{A_1 + A_2} \tag{3-3}$$

Partial volumes are substituted for partial areas in Eqs. (3-2) and (3-3) when determining the centroids of solid geometric figures. In this case, three coordinates are needed to specify completely the position of the center of gravity.

$$\bar{x} = \frac{v_1 x_1 + v_2 x_2 + \cdots}{v_1 + v_2 + \cdots} \tag{3-4}$$

$$\bar{y} = \frac{v_1 y_1 + v_2 y_2 + \cdots}{v_1 + v_2 + \cdots} \tag{3-5}$$

$$\bar{z} = \frac{v_1 z_1 + v_2 z_2 + \cdots}{v_1 + v_2 + \cdots} \tag{3-6}$$

To aid in the computations that follow, centroids of several frequently encountered geometric figures are given in Table 3-1. The examples that follow will illustrate both the use of the table and the method of finding centroids.

EXAMPLE 4: Determine the x- and y-coordinates of the centroid of the figure shown in Fig. 3-8.

Solution: Three simple areas are represented in the composite figure: a rectangle of area A_1, a triangle of area A_2, and a circle of area A_3. Moments of these areas about the x-axis and the y-axis will locate the centroid. Since the circular area A_3 represents material that is "missing" from the figure, its moment is negative. Moments taken about the y-axis give

$$(A_1 + A_2 - A_3)\bar{x} = A_1 x_1 + A_2 x_2 - A_3 x_3$$

Table 3-1. Centroids

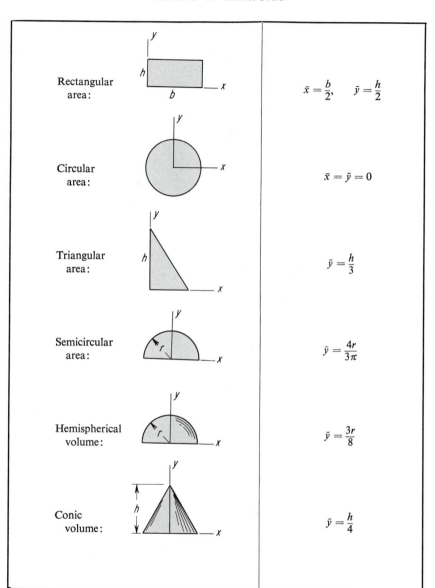

Rectangular area:	$\bar{x} = \dfrac{b}{2}, \quad \bar{y} = \dfrac{h}{2}$
Circular area:	$\bar{x} = \bar{y} = 0$
Triangular area:	$\bar{y} = \dfrac{h}{3}$
Semicircular area:	$\bar{y} = \dfrac{4r}{3\pi}$
Hemispherical volume:	$\bar{y} = \dfrac{3r}{8}$
Conic volume:	$\bar{y} = \dfrac{h}{4}$

Structural Shapes

	Size (in.)	Weight per ft (lb)	Area (in.^{2})	w (in.)	d (in.)	$\bar{x}$ (in.)	$\bar{y}$ (in.)
Wide Flange beam:	12×12	120	35.31	12.32	13.12	0	0
	10×10	100	29.43	10.35	11.12	0	0
Channel:	12×3	30	8.79	3.17	12.00	0.68	0
	6×2	13	3.81	2.16	6.00	0.52	0
Equal Leg angle:	$6 \times 6 \times \frac{1}{2}$	19.6	5.75	6	6	1.68	1.68
	$3 \times 3 \times \frac{3}{8}$	7.2	2.11	3	3	0.89	0.89

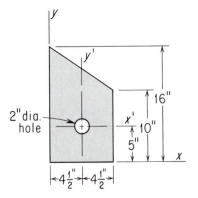

Figure 3-8

where
$$\bar{x} = \frac{[9(10)]\frac{9}{2} + [\frac{1}{2}(9)6]\frac{9}{3} - [\pi(1)^2]\frac{9}{2}}{9(10) + \frac{1}{2}(9)6 - \pi(1)^2}$$

$$= \frac{405 + 81 - 14.13}{90 + 27 - 3.14} = \frac{471.87}{113.86}$$

$$\bar{x} = 4.14 \text{ in. to the right of } y\text{-axis}$$

Moments taken about the x-axis give

$$(A_1 + A_2 - A_3)\bar{y} = A_1 y_1 + A_2 y_2 - A_3 y_3$$

where
$$\bar{y} = \frac{[9(10)]5 + [\frac{1}{2}(9)6][10 + \frac{6}{3}] - [\pi(1)^2]5}{113.86}$$

$$\bar{y} = \frac{450 + 324 - 15.7}{113.86} = \frac{758.3}{113.86}$$

$$\bar{y} = 6.66 \text{ in. above } x\text{-axis}$$

Alternate Solution: By selecting a set of rectangular axes x'-y' whose origin is at the centroid of the rectangle, the computations can be appreciably reduced, since the moment of both the rectangle and the circle would be zero:

$$\bar{x}' = \frac{A_1 x_1 + A_2 x_2 - A_3 x_3}{A_1 + A_2 - A_3}$$

$$= \frac{0 + \frac{1}{2}(9)6(1.5) - 0}{113.86} = 0.36 \text{ in. to the left of } y'\text{-axis}$$

Therefore $\bar{x} = 4.5 - 0.36 = 4.14$ in. to the right of y-axis

In a similar manner

$$\bar{y}' = \frac{\frac{1}{2}(9)(6)[5 + 2]}{113.86} = 1.66 \text{ in. above } x'\text{-axis}$$

Therefore $\bar{y} = 5 + 1.66 = 6.66$ in. above x-axis

EXAMPLE 5: Determine the center of gravity of the steel oven door shown in Fig. 3-9. A heat-resistant glass insert 1 ft by 1 ft is located in the offset hole as illustrated. Assume that both the glass and the steel have a uniform thickness of $\frac{1}{4}$ in. Glass of this thickness weighs 3.5 lb per sq ft and steel this thick weighs 10 lb per sq ft.

Solution: The body is symmetrical about a line 1 ft above the x-axis; therefore, by inspection, $\bar{y}$ is equal to 1 ft. Next take moments about the y-axis to locate the x-coordinate of the center of gravity.

The total weight of the door is

$$W = \text{weight of steel} + \text{weight of glass}$$
$$= (\text{volume} \times \text{density})_{\text{steel}} + (\text{volume} \times \text{density})_{\text{glass}}$$
$$= [2(3) - 1(1)]10 + [1(1)]3.5$$
$$= 50 + 3.5 = 53.5 \text{ lb}$$

Moments of the weights, summed about the y-axis, are

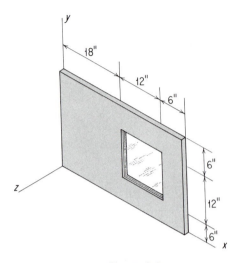

Figure 3-9

$$53.3\bar{x} = 2(3)10(1.5) - 1(1)10(2) + 1(1)3.5(2)$$

and

$$\bar{x} = \frac{90 - 20 + 7}{53.5} = \frac{77}{53.5}$$

$$\bar{x} = 1.44 \text{ ft to the right of } y\text{-axis}$$

Note: In the computation, subtract the moment of the missing steel from the moment of the entire steel plate.

QUESTIONS, PROBLEMS, AND ANSWERS

3-1. A frame is suspended first from corner A and then from corner B as shown. Make a sketch of the part and graphically locate the center of gravity with respect to point B.

Ans. 1.81 in. right and 2.06 in. above point B when the 4.4 in. side is horizontal.

3-2. A scale indicates a weight of 2000 lb when it is placed under the two front wheels of an automobile. Placed under the rear wheels, the indicated weight is 1600 lb. Locate, with respect to the front wheels, the vertical action line that passes through the center of gravity. Assume the car to have a wheel base of 120 in. and to be horizontal during both weighings.

Ans. 53.3 in.

3-3. Locate the center of gravity with respect to corner A of the 5000-lb machine

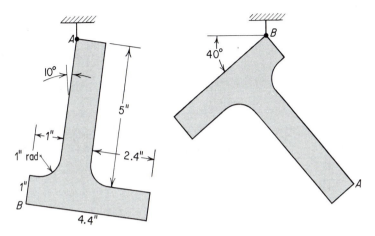

Problem 3-1

illustrated if a scale reading of 3500 lb is indicated when $\theta = 0°$ and 3000 lb when $\theta = 20°$.

Ans. 2.4 ft left to A and 2.20 ft above base.

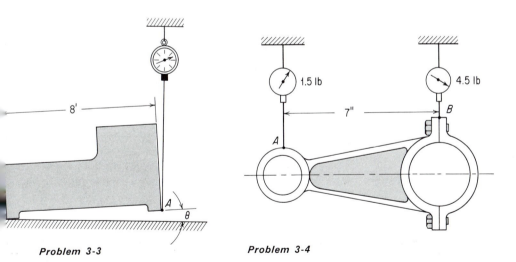

Problem 3-3   **Problem 3-4**

3-4. Two spring scales are used to weigh the connecting rod shown. Determine the center of gravity of the rod if the left scale reads 1.5 lb and the right scale 4.5 lb.

Ans. 5.25 in. to right of A.

3-5 to 3-9 Locate the coordinates of the center of gravity of the grouped weights indicated.

Ans. **3-5.** $\bar{x} = 20$ in.

3-6. $\bar{x} = \frac{40}{7}$ in.; $\bar{z} = \frac{50}{7}$ in.

3-7. $\bar{x} = 4.95$ in.; $\bar{z} = 4.73$ in.

3-8. $\bar{x} = \frac{70}{9}$ in.; $\bar{y} = \frac{10}{3}$ in.; $\bar{z} = \frac{4}{3}$ in.

3-9. $\bar{x} = \frac{20}{7}$ in.; $\bar{y} = -\frac{10}{7}$ in.; $\bar{z} = \frac{100}{7}$ in.

3-10 to 3-13 Determine the x- and y-coordinates of the centroid of the plane geometric forms shown in the respective illustrations.

Ans. **3-10.** $\bar{x} = \frac{35}{3}$ in.; $\bar{y} = \frac{25}{6}$ in.

3-11. $\bar{x} = 6.33$ in.; $\bar{y} = 10.67$ in.

3-12. $\bar{x} = 3.87$ in.; $\bar{y} = 3.05$ in.

3-13. $\bar{x} = 11.8$ in.; $\bar{y} = 7.12$ in.

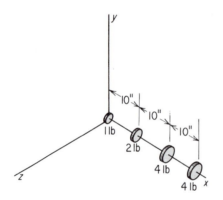

Problem 3-5

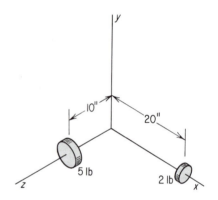

Problem 3-6

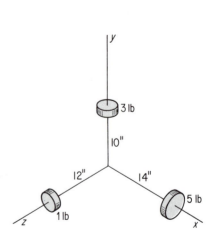

Problem 3-7

Problem 3-8

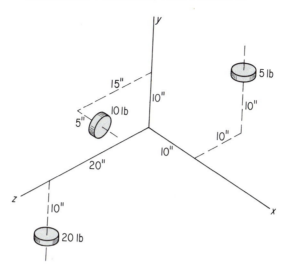

Problem 3-9

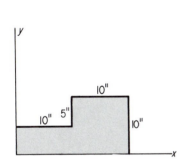

Problem 3-10

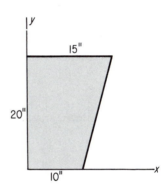

Problem 3-11

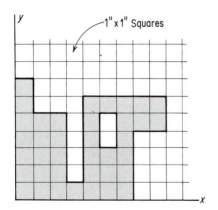

Problem 3-12

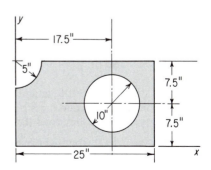

Problem 3-13

3-14. Determine the centroid of a beam fabricated by welding two 6-in. by 6-in. by $\frac{1}{2}$-in. angles to a flat plate as shown.

Ans. $\bar{y} = 15.3$ in.

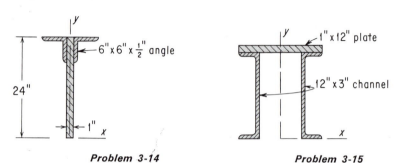

Problem 3-14   *Problem 3-15*

3-15. Two 12-in. by 3-in. by 30-lb-per-ft channels are welded to a plate as indicated. Determine the centroid of the composite section.

Ans. $y = 8.64$ in.

3-16. A 12-in. by 3-in. structural channel is welded to a 10-in. by 10-in. wide-flanged beam as shown. Locate the centroid of the composite area.

Ans. $\bar{y} = 6.99$ in.

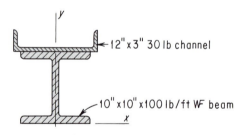

Problem 3-16

3-17. A steel sphere with a diameter of 4 in. is fastened to the end of a square aluminum rod 1 in. by 1 in. by 18 in. long. Locate the center of gravity of the composite section. The densities of steel and aluminum are 0.28 lb per cu in. and 0.095 lb per cu in. respectively.

Ans. 18.3 in. from free end.

3-18. A steel shaft 3 ft long and weighing 25 lb has a 15-lb brass gear fastened to one end. A second brass gear weighing 30 lb is located at the other end. Find the center of gravity of the assembly.

Ans. 1.82 ft from light gear.

3-19. A steel block has a 5-in.-diameter hole drilled through it as illustrated. Locate the center of gravity of the block.

Ans. $\bar{x} = 9.57$ in.; $\bar{y} = 6.90$ in.; $\bar{z} = 5$ in.

3-20. A welded machine frame has the shape indicated. Find the center of gravity of the frame if the steel has a uniform thickness throughout.

Ans. 13.6 in. above base at point of symmetry.

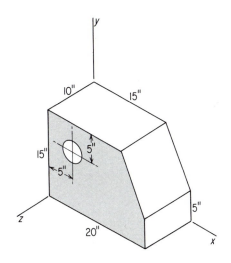

Problem 3-19

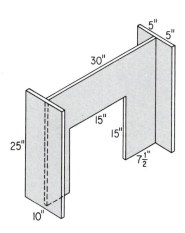

Problem 3-20

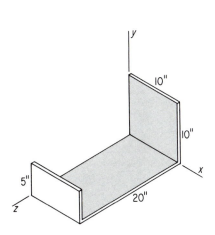

Problem 3-21

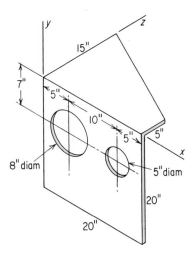

Problem 3-22

3-21. A thin rectangular plate is bent to form the bracket shown. Locate the center of gravity of the bracket.

Ans. $\bar{x} = 5$ in.; $\bar{y} = 1.79$ in.; $\bar{z} = 8.57$ in.

3-22. A piece of thin sheet metal is stamped to form the bracket illustrated. Locate the center of gravity of the stamping.

Ans. $\bar{x} = 9.66$ in.; $\bar{y} = -6.62$ in.; $\bar{z} = 3.62$ in.

Chapter 4

EQUILIBRIUM

A body at rest is a body in equilibrium. The forces and moments that act are completely balanced by counteracting forces and counteracting moments. The mathematical conditions for equilibrium are very simply stated:

$$\sum F = 0$$
$$\sum M = 0 \tag{4-1}$$

This chapter will show how the problem of equilibrium is approached and how the forces and moments required to maintain equilibrium are determined for two-dimensional systems and for space systems.

4-1 The Free-Body Diagram

It would be a hopeless task to assemble a complicated piece of machinery without a blueprint showing what part goes where and in what order. A similar situation exists in mechanics; an analysis of forces and moments cannot be made unless there is a pictorial representation of where and how these forces and moments act on the body. This pictorial representation is called the *free-body diagram*. Its construction is perhaps the most important single step in the solution of statics as well as dynamics problems.

In sketching the free-body diagram, the primary thing to remember is that forces are exerted on bodies either because they are in *contact* with other bodies or because the bodies are acted upon by a *remote* force of attraction like that of gravity or magnetism. To illustrate: a weight W is suspended from a group of cables as shown in Fig. 4-1(a). What are the forces in the cables? The answer lies hidden in the original figure, but it is apparent in Fig. 4-1(b) and Fig. 4-1(c). Each of the cables is cut and the junction B removed and isolated. The three forces T_{AB}, T_{BC}, T_{BD} are in equilibrium; their resultant is zero. In a similar manner isolation of the weight shows that the force W and the cable force T_{BD} are in equilibrium. These are the two free-body diagrams that would be considered in the solution of the problem.

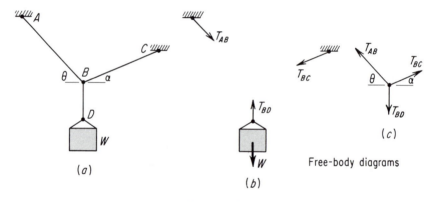

Figure 4-1

A cylinder weighing W lb, shown in Fig. 4-2, is supported by cable AB and a smooth inclined surface C. The free-body diagram shows that the reaction R_C of the surface on the cylinder is normal to both the cylinder and the plane. This is characteristic of contact forces between smooth surfaces.

Still another type of free-body reaction is that found in the *pin-connected support*. This is a mechanical device capable of resisting motion normal to the pin; it cannot, however, resist a moment that acts about the pin axis. To illustrate: consider the ladder in Fig. 4-3 to be pin-connected at B. The top of the ladder rests against a smooth vertical surface and a weight W_C hangs from point C. The weight of the ladder W_L is assumed to act at its *center of gravity*. The ladder isolated as a free body is shown in Fig. 4-3(b).

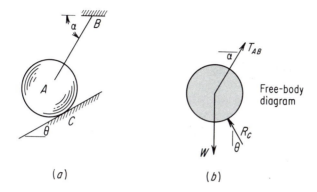

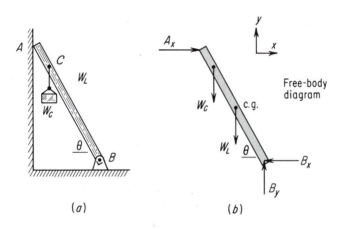

Figure 4-2

Figure 4-3

The weight forces W_C and W_L act downward; A_x is the wall's reaction to the "push" of the ladder; and B_x and B_y are the reactions at the pin.

In some instances, directions of pin reactions—up or down, to the right or to the left—will not be readily apparent. *This does not mean they are to be omitted.* The rule is simply to assume a direction. If the assumption is wrong, the answer for that particular reaction will carry a minus sign, indicating that the reaction should be reversed.

EXAMPLE 1: Draw the free-body diagram of the rod AC, the pulley D, and the weight W shown in Fig. 4-4(a). The rod is pin-connected at A and the pulley pin-connected at D. Assume the surface at C to be perfectly smooth and neglect both the weight of the rod and the weight of the pulley, as they are too small to consider.

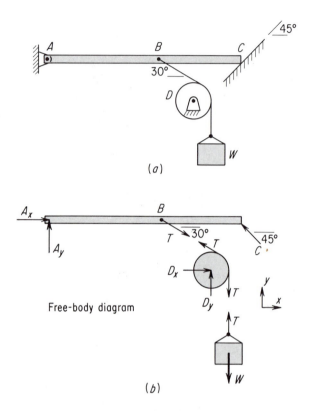

(a)

(b)

Figure 4-4

Solution:

The weight: A gravity force W and a tensile force T are the two forces that act on the weight.

The pulley: The components of the pin reaction are D_x and D_y. Since the cable is "cut" above and below the pulley, the cable force T appears twice.

The bar: The directions of the pin forces at A are not immediately apparent, so it is merely assumed that they act as shown. The force C, the reaction of the smooth surface on the bar, is normal to the surface.

EXAMPLE 2: Isolate and draw the free-body diagram of the bar AB, the pulley C, and the weight W of the system shown in Fig. 4-5.

Solution:

The weight: A cable force T and a gravity force W act on the weight.

The pulley: The cable tension T acts in both the horizontal and vertical direc-

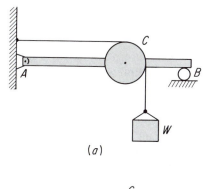

(a)

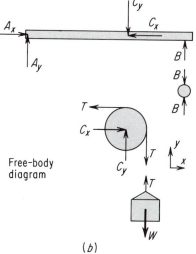

Free-body diagram

(b)

Figure 4-5

tions as shown, while the pin forces C_x and C_y act at the center of the pulley. It is readily apparent that these pin forces are drawn in the proper direction to maintain equilibrium; C_x opposes the horizontal force T and C_y opposes the vertical force T.

The bar: Since the smooth roll support at B is free to move horizontally, it can exert only a vertical force on the bar. The reaction of the pulley on the bar is opposite to that of the bar on the pulley. In other words, C_x acts to the right on the pulley and to the left on the bar; in a similar manner, C_y acts vertically upward on the pulley and vertically downward on the bar. A_x and A_y, the pin reactions, are the two remaining forces that act on the bar.

4-2 Equilibrium of Two-Dimensional Force Systems

The two-dimensional body shown in Fig. 4-6 is acted upon by a group of forces, each of which can be expressed in terms of two similarly directed components. The body and the forces that act upon it are referred to as a *two-dimensional force system*. To maintain equilibrium, *the resultant force must be zero and the resultant moment of the forces about any arbitrary point O must be zero.* Since the resultant is the vector sum of force components, the first necessary condition for equilibrium is mathematically stated as

$$R = \sqrt{(\sum F_x)^2 + (\sum F_y)^2} = 0 \qquad (4\text{-}2)$$

where

$$\sum F_x = 0 \qquad (4\text{-}3)$$

$$\sum F_y = 0 \qquad (4\text{-}4)$$

The second necessary condition for equilibrium, mathematically stated, is

$$\sum M_0 = 0 \qquad (4\text{-}5)$$

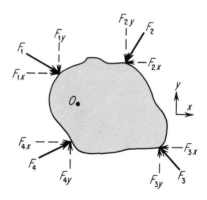

Figure 4-6

4-3 Simplification of the Equations of Equilibrium for Three Special Cases

Case I: Collinear force system. When no more than two forces act upon a body, these forces must: (1) be parallel and be on the same line of action, (2) be oppositely directed, and (3) have equal magnitudes. If the forces are parallel but have different action lines, the condition $\sum M_o = 0$ cannot be satisfied; if the action lines are not parallel, then $\sum F_x = \sum F_y = 0$ cannot be satisfied.

Case II: Concurrent force systems. When no more than three nonparallel forces act on a body (Fig. 4-7), their lines of action must be concurrent— they must pass through a common point. If this were not true, the second condition of equilibrium, that of $\sum M_0 = 0$, could not be satisfied. Another way of looking at this is to imagine two of the three forces to be added vectorally, thereby reducing the system to one composed of only two forces; as shown in Case I, two forces acting on a body must be collinear.

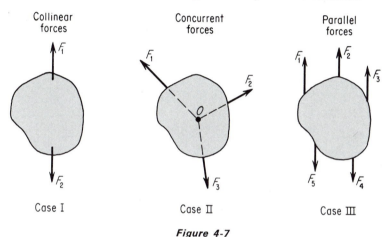

Collinear forces

Concurrent forces

Parallel forces

Case I   Case II   Case III

Figure 4-7

Case III: Parallel force systems. This is perhaps the simplest of force systems, since one orthogonal direction, either x or y, is eliminated from the computations. The equations of equilibrium reduce to $\sum F = 0$ and $\sum M_0 = 0$. The examples that follow will illustrate the method of approach and analysis of two-dimensional force systems.

EXAMPLE 3: A 100-lb weight is supported by two cables as shown in Fig. 4-8. Determine the tensile force in each of the cables.

Solution: Cut the cables and draw the free-body diagram of joint B as shown in Fig. 4-8(b). Then write the equations of equilibrium, $\sum F_x = 0$ and $\sum F_y = 0$, for the three forces.

$$\sum F_x = 0$$

$$T_{BC} \cos 45° - T_{AB} \cos 30° = 0$$

$$0.707 T_{BC} = 0.866 T_{AB} \tag{a}$$

and

$$\sum F_y = 0$$

$$T_{AB} \sin 30° + T_{BC} \sin 45° - 100 = 0$$

$$0.5 T_{AB} + 0.707 T_{BC} = 100 \tag{b}$$

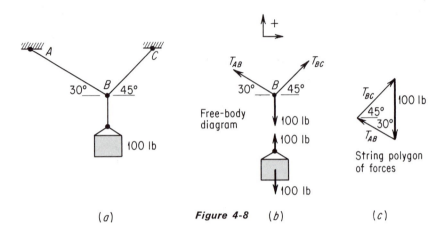

(a)   **Figure 4-8**   (b)   (c)

Solving Eqs. (a) and (b) simultaneously gives

$$T_{AB} = \frac{0.707}{0.866}T_{BC} = 0.816\,T_{BC}$$

$$[(0.5)0.816 + 0.707]T_{BC} = 100$$

$$1.115T_{BC} = 100$$

$$T_{BC} = 89.7\text{ lb}$$

and $$T_{AB}\frac{0.707}{0.866}(89.7) = 73.2\text{ lb}$$

Alternate Solution: The string polygon of vector forces must close, since the resultant is zero. First draw the 100-lb force to scale; then draw lines parallel to T_{BC} and T_{AB} to form the closed triangle of forces shown in Fig. 4-8(c). The lengths of the legs of a triangle, when measured to scale, are equal to the unknown forces.

$$T_{AB} = 73\text{ lb}\quad\text{and}\quad T_{BC} = 90\text{ lb}$$

EXAMPLE 4: Determine the reactions at the carriage wheels A and B of the double trolley crane shown in Fig. 4-9. Assume the weight of the main beam, 1000 lb, to be concentrated at its center of gravity.

Solution: Draw the free-body diagram as shown in Fig. 4-9(b). Since this is a system of parallel forces, two equations of equilibrium, $\sum M_0 = 0$ and $\sum F_y = 0$, are all that are required to calculate the desired quantities. As a check, write a third equation in which moments are taken about some arbitrary point other than O; if the solution is correct, this equation will equal zero.

$$\sum F_y = 0$$

$$R_A + R_B - 7000 - 1000 - 3000 = 0$$

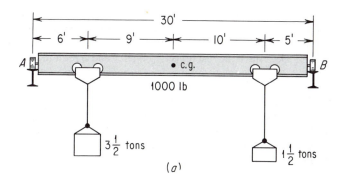

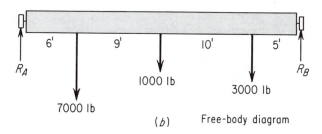

(b)   Free-body diagram

Figure 4-9

$$R_A + R_B = 11,000 \text{ lb}$$

$$\sum M_A = 0$$

$$7000(6) \times 1000(15) + 3000(25) - 30R_B = 0$$

$$R_B = \frac{42,000 + 15,000 + 75,000}{30} = \frac{132,000}{30}$$

$$R_B = 4400 \text{ lb}$$

Substitution of R_B into Eq. (a) gives

$$R_A = 11,000 - R_B = 11,000 - 4400$$

$$R_A = 6600 \text{ lb}$$

Check: Moments are taken about point B

$$\sum M_B = 0$$

$$R_A 30 - 7000(24) - 10000(5) - 3000(5) = 0$$

$$6600(30) - 7000(24) - 1000(15) - 3000(5) = 0$$

$$198,000 - 168,000 - 15,000 - 15,000 = 0$$

EXAMPLE 5: The portable jib crane shown in Fig. 4-10 weighs 500 lb and is supported on a column by the slotted bracket A and the smooth fulcrum B. The weight of the crane may be assumed to act somewhere on a vertical line passing through point C. Determine the reactions at points A and B required to support the crane.

Solution: The pin A resists both horizontal and vertical downward motion; two force components, A_x and A_y, therefore, are shown in the free-body diagram. The fulcrum B, assumed to be smooth, exerts a single force, B_x, on the crane. Since there are three unknown quantities to be found, three equations of equilibrium are required. These are $\sum F_x = 0$, $\sum F_y = 0$, and $\sum M_0 = 0$.

$$\sum F_x = 0$$
$$B_x - A_x = 0$$

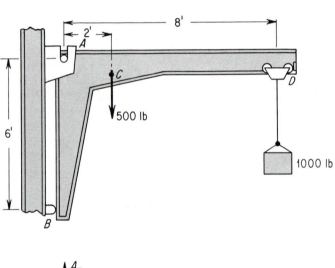

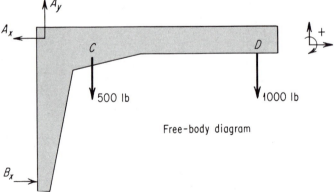

Free-body diagram

Figure 4-10

Therefore $\qquad\qquad\qquad A_x = B_x \qquad\qquad\qquad$ (a)

and $\qquad\qquad\qquad \sum F_y = 0$

$$A_y - 500 - 1000 = 0$$

$$A_y = 1500 \text{ lb}$$

A moment summation is next taken about point A.

$$\sum M_A = 0$$

$$500(2) + 1000(8) - 6B_x = 0$$

$$B_x = \frac{9000}{6} = 1500 \text{ lb} \qquad\qquad\qquad \text{(b)}$$

From Eq. (a) $\qquad\qquad A_x = B_x = 1500$

The reaction at A is the vector sum of A_x and A_y, where

$$A = \sqrt{(A_x)^2 + (A_y)^2} = \sqrt{(1500)^2 + (1500)^2}$$

$$A = 2121 \text{ lb}$$

4-4 Equilibrium of Forces in Space

The resultant of a group of space forces was shown in Chapter 2 to have, at most, three rectangular components: R_x, R_y, and R_z. It was also shown that the resultant could conceivably have a moment about all three axes: M_x, M_y, and M_z. These six quantities, when set equal to zero, form the basis of three-dimensional equilibirum.

$$\begin{array}{ccc} R_x = 0, & & M_x = 0 \\ R_y = 0, & \text{and} & M_y = 0 \\ R_z = 0, & & M_z = 0 \end{array} \qquad (4\text{-}6)$$

One could expect, in the most complicated problem in three-dimensional equilibrium, to compute as many as six unknown quantities. Fortunately, however, in a good many situations these six equations can be reduced to as few as three by inspection alone.

The examples that follow will illustrate the types of problems that are encountered and the approach to their solution.

EXAMPLE 6: A weight of 300 lb is raised by means of a force F applied to the crank as shown in Fig. 4-11. Neglect the weights of all the members and find the reactions at bearings A and B and the force F.

Solution: The free-body diagram, which has been superimposed on an x-, y-, z-coordinate system, indicates the absence of x-components of force. This reduces the six equations of equilibrium to the following five:

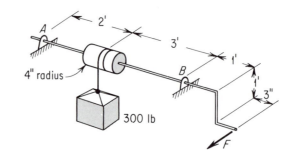

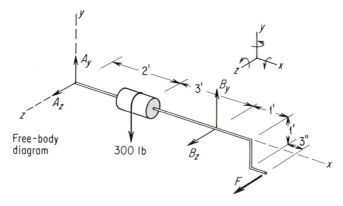

Figure 4-11

$$\sum F_y = 0, \quad \sum M_x = 0$$
$$\sum F_z = 0, \quad \sum M_y = 0, \quad \sum M_x = 0$$

Determine the crank force F by equating moments about the x-axis.

$$\sum M_x = 0$$
$$-F(12) + 300(4) = 0$$
$$F = 100 \text{ lb}$$

Next, write a moment equation to include the free-body forces that tend to produce a rotation about the z-axis. These include the weight force of 300 lb and the bearing reaction B_y.

$$\sum M_z = 0$$
$$5B_y - 300(2) = 0$$

Therefore $B_y = 120 \text{ lb}$

Then write a moment equation for the forces that tend to produce rotation in the system about the y-axis. These forces incluce B_z and F; the latter, found in the first step, is 100 lb.

$$\sum M_y = 0$$

$$F(6.25) + 5B_z = 0$$

where $\quad\quad\quad B_z = \dfrac{-100(6.25)}{5} = -125 \text{ lb}$

The minus sign indicates that B_z is improperly directed and should be reversed after the problem is completed. Two unknowns, A_y and A_z, remain to be determined, and two of the original five equations have yet to be used—a force summation in the y-direction and a force summation in the z-direction.

$$\sum F_y = 0$$

$$A_y + B_y - 300 = 0$$

Substitute the value of B_y from the previous step and compute A_y.

$$A_y + 120 - 300 = 0$$

$$A_y = 180 \text{ lb}$$

Finally, for the z-components of force

$$\sum F_z = 0$$

$$A_z + B_z + F = 0$$

Substitution in this equation of the known values of B_z and F gives

$$A_z - 125 + 100 = 0$$

$$A_z = 25 \text{ lb}$$

The tabulated answers are

$$F = 100 \text{ lb}$$

$$A_x = 0, \quad\quad\quad B_x = 0$$

$$A_y = 180 \text{ lb}, \quad\quad B_y = 120 \text{ lb}$$

$$A_z = 25 \text{ lb}, \quad\quad B_z = -125 \text{ lb}$$

Indicate positive and negative directions for the force components on the free-body diagram.

EXAMPLE 7: Determine the tension in the three cables, AD, BD and CD, that support the 500-lb weight shown in Fig. 4-12.

Solution: Three equations of equilibrium are required: $\sum F_x = 0$, $\sum F_y = 0$, and $\sum F_z = 0$. The moment equations are automatically satisfied since the three forces meet at a common point D.

The method of *direction-cosines* described in Chapter 2 will be used to determine the components of space force T_{BD}.

$$(T_{BD})_x = T_{BD} \cos \theta_x = T_{BD} \dfrac{-3}{\sqrt{(-3)^2 + (-4)^2 + (12)^2}}$$

$$(T_{BD})_x = -\dfrac{3}{13} T_{BD}$$

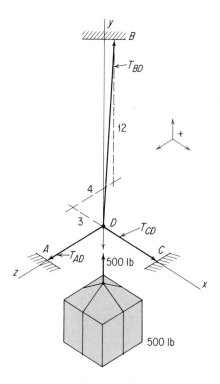

Figure 4-12

In a similar manner $\qquad (T_{BD})_y = T_{BD} \cos \theta_y = \dfrac{12}{13} T_{BD}$

and $\qquad\qquad\qquad (T_{BD})_z = T_{BD} \cos \theta_z = -\dfrac{4}{13} T_{BD}$

The force components in each of the rectangular directions are set equal to zero.

$$\sum F_x = 0$$

$$-\frac{3}{13} T_{BD} + T_{CD} = 0 \qquad\qquad\qquad\text{(a)}$$

$$\sum F_y = 0$$

$$\frac{12}{13} T_{BD} - 500 = 0 \qquad\qquad\qquad\text{(b)}$$

$$\sum F_z = 0$$

$$-\frac{4}{13} T_{BD} + T_{AD} = 0 \qquad\qquad\qquad\text{(c)}$$

From Eq. (b) $T_{BD} = 500(13)/12 = 542$ lb. The value T_{BD} when substituted into Eqs. (a) and (c) gives T_{CD} and T_{AD}. From Eq. (a)

$$T_{CD} = \frac{3}{13}T_{BD} = \frac{3}{13}(542) = 125\text{ lb}$$

From Eq. (c)

$$T_{AD} = \frac{4}{13}T_{BD} = \frac{4}{13}(542) = 167\text{ lb}$$

QUESTIONS, PROBLEMS, AND ANSWERS

Note: In Problems 4-1–4-8 indicate the unknown forces by appropriate symbols; it is not necessary to determine these forces.

4-1. Draw the free-body diagram of the 250-lb cylinder illustrated. Assume the contact surfaces to be perfectly smooth.

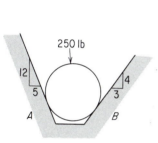

Problem 4-1   **Problem 4-2**

4-2. A ladder of negligible weight, as shown, is supported by a smooth vertical wall, a smooth horizontal surface, and a stake driven into the ground at *B*. A 450-lb weight acts at point *C*. Draw the free-body diagram of the ladder.

4-3. As illustrated, a weight of 500 lb is supported by a cable attached to end *B* of the horizontal bar; the bar in turn is supported by a pin at *A* and a smooth roller at *C*. Draw the free-body diagram of the bar *AB*.

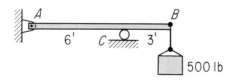

Problem 4-3

4-4. Isolate beams *AB* and *ED* in the figure and draw the free-body diagram of each. Smooth rollers act at points *B* and *C*.

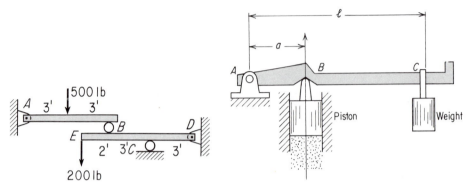

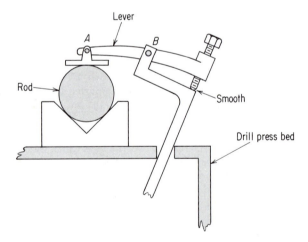

Problem 4-4   *Problem 4-5*

4-5. The lever shown is pin-supported at one end. The weight can be moved along the axis to balance the pressure force on the piston. Draw the free-body diagram of the lever and indicate the forces in terms of the variables.

4-6. The illustration indicates the essential features of a drilling machine clamp. The pins at *A* and *B* are smooth. Draw the free-body diagrams of the rod and the lever.

Problem 4-6

4-7. A device used to draw a variable curve is illustrated. If all hinges are to be pin-connected, draw the free-body diagram of members *A*, *B*, and *C*.

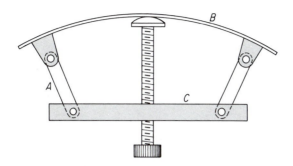

Problem 4-7

4-8. The arrangement of blocks and tackles shown is called a "bell purchase." Draw a free-body diagram of each pulley and show that the force F is one seventh of the weight W.

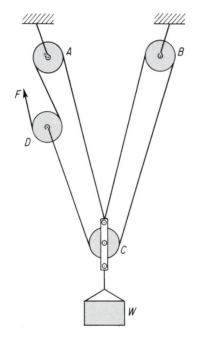

Problem 4-8

4-9. Determine the tensile force in cord AB and in cord BC that support the 100-lb weight illustrated in the figure.

Ans. 167 lb; 133 lb.

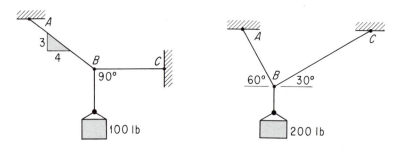

Problem 4-9   Problem 4-10

4-10. Determine the tensile force in each of the cords that support the 200-lb weight in the figure.

Ans. 173 lb; 100 lb.

4-11. Find the horizontal force F that must be applied to the 50-lb weight indicated in order that the cord AB makes a 20-deg angle with the vertical.

Ans. 18.2 lb.

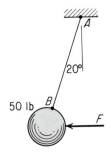

Problem 4-11

4-12. The bolt of the holding jig shown is tightened to produce a tensile force of 200 lb. Find the forces at A, B, and C that act on the cylinder.

Ans. $A = 50$ lb; $B = C = 28.9$ lb.

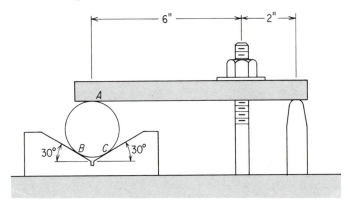

Problem 4-12

4-13. Determine the weight W that a man is capable of supporting if he applies a force $F = 50$ lb directed as shown on the handles of a truck.

Ans. 91.7 lb.

4-14. Find the reaction on the axle A of the truck described in Prob. 4-13.

Ans. 142 lb.

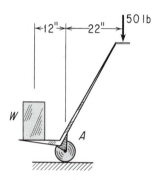

Problem 4-13

4-15. The hairpin conveyor indicated is used to transport tires in an assembly plant. If each tire weighs 30 lb find the bolt reaction at A and at B. Assume that friction between the components is negligible.

Ans. $A = 222$ lb ↓ ; $B = 342$ lb ↑ .

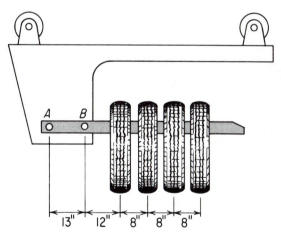

Problem 4-15

4-16. Find the tension in each of the chains of the notched-beam conveyor transport shown.

Ans. Right: 2510 lb; left: 1070 lb.

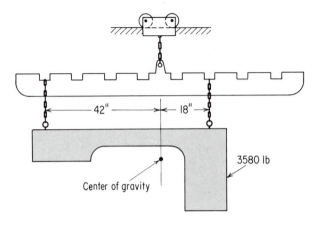

Problem 4-16

4-17. A double-lever plug cock is shown. If a steady force of 20 lb is maintained as constant tension and the resisting moment within the cock is 10 lb ft, find the required tension in the cable at B to close the cock

Ans. 25.2 lb.

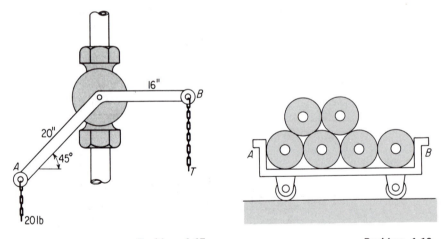

Problem 4-17   **Problem 4-18**

4-18. Six identical 500-lb coils of steel are stacked on a truck as shown. Determine the reactions at A and B on the truck.

Ans. $A = B = 144$ lb.

4-19. Determine the weight W required to prevent the pulleys indicated from rotating, and the bearing reactions at A and B. The pulleys A and B weigh 60 lb and 30 lb, respectively.

Ans. $W = 50$ lb; $R_A = 149$ lb; $R_B = 164$ lb.

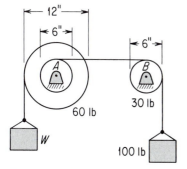

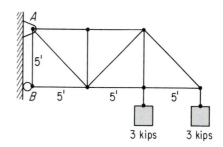

Problem 4-19   Problem 4-20

4-20. The wall truss shown is pin-supported at A and roller-supported at B. Determine the reactions at the pin and at the roller if the truss supports two 3-kip loads as shown. Neglect the weight of the truss in the calculations.

Ans. $A_y = 6$ kips ↑ ; $A_x = 15$ kips ← ; $B = 15$ kips →.

4-21. Find the force F required to pull the 1000-lb spool of wire illustrated over the 2-in. obstruction. Assume the weight of the spool to act through its geometric center.

Ans. 402 lb.

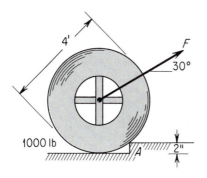

Problem 4-21

4-22. Determine the lifting force at A that can be exerted by the crowbar shown.

Ans. 831 lb.

Problem 4-22

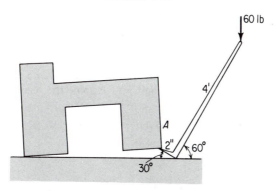

4-23. Determine the reactions at pin *A* and at roller *B* for the system shown. The weight of the beam, 500 lb, may be assumed to act at the center of gravity *C*.

Ans. $A = 450$ lb ↑ ; $B = 1050$ lb ↑ .

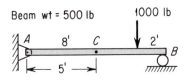

Problem 4-23

4-24. Find the reactions at pin *A* and roller *B* of the system shown. The weight of the beam, 600 lb, acts through point *C*, the center of gravity.

Ans. $A_x = 1000$ lb ← ; $A_y = 1540$ lb ↑ ; $B = 1790$ lb ↑ .

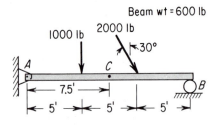

Problem 4-24

4-25. Determine the reactions at *A*, *B*, *C*, and *D* in the system shown. The weight of the beams may be neglected in the computations.

Ans. $A = 300$ lb ↓ ; $B = 1300$ lb ↑ ; $C = 1625$ lb ↑ ; $D = 325$ lb ↓ .

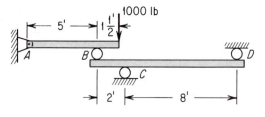

Problem 4-25

4-26. The beam shown is supported by a pin at *A* and a smooth surface at *B*. If the weight of the beam is negligible, determine the reactions at these two points.

Ans. $A_x = 173$ lb → ; $A_y = 100$ lb ↑ ; $B = 200$ lb; $30°$ ↘ .

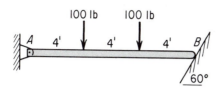

Problem 4-26

4-27. The tractor illustrated weighs 10,000 lb. The weight is equally distributed, half on the front wheels and half on the rear wheels. When the tractor is coupled to the trailer, the front wheels scale out to 6000 lb and the rear wheels to 18,000 lb. Determine the distance d that locates the kingpin P.

Ans. 10.3 in.

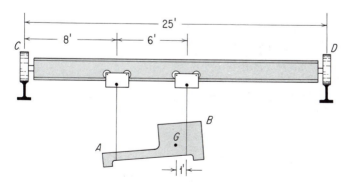

Problem 4-27

4-28. The double-trolley crane indicated picks up a machine weighing 5 tons. If the weight of the machine is concentrated at point G, determine the tension in cables A and B, and the reactions at the crane wheels C and D.

Ans. $A = 0.83$ ton; $B = 4.17$ tons; $C = 2.40$ tons; $D = 2.60$ tons.

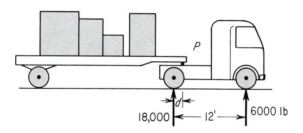

Problem 4-28

4-29. Determine the magnitudes of the equal counterweights W that would just balance the 1000-lb elevator shown.

Ans. 250 lb.

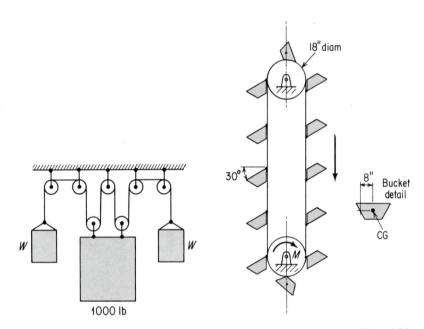

18" diam

30°

8" Bucket
    detail

CG

W W

1000 lb

Problem 4-29   **Problem 4-30**

4-30. Each bucket of the elevator shown weighs 20 lb and when full and rising carries 100 lb of payload. The buckets discharge at the top and then descend empty. Determine the moment required to operate the conveyor at constant speed in the position shown.

Ans. 664 lb ft.

4-31. Find the force P required to lift the 300-lb weight in the pulley system shown.

Ans. 100 lb.

4-32. What weight W can be lifted by the 75-lb force with the pulley arrangement indicated?

Ans. 300 lb.

4-33. The double-strut deck truss shown supports the loads indicated. Determine the pin and roller reactions at A and B.

Ans. $A_x = 3$ kips $\rightarrow$; $A_y = 2.17$ kips $\uparrow$; $B = 3.83$ kips.

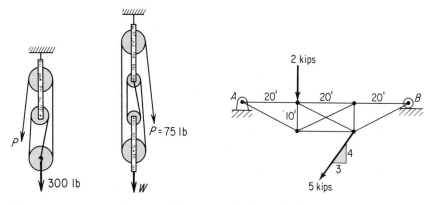

Problem 4-31   **Problem 4-32**   **Problem 4-33**

4-34. A force of 500 lb is required at the punch head in the position shown. Assume all joints to be pin-connected and determine for the position shown the necessary driving couple M and the pin reaction at C.

Ans. 794 lb ft; 6000 lb ↓ .

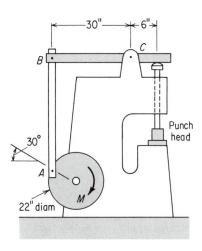

Problem 4-34

4-35. The recorded weights at the axles A, B, and C of the haul-away tractor and trailer illustrated are 6000 lb, 16,000 lb, and 18,000 lb, respectively. Determine the magnitude and line of action of the weight of the trailer and its cargo if the uncoupled tractor weighs 10,000 lb, equally distributed over the front and rear wheels.

Ans. 11.2 ft ahead of C; 30,000 lb.

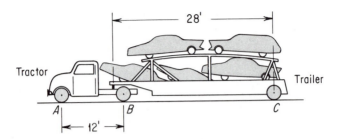

Problem 4-35

4-36. Trucks are often transported in the piggy-back style shown. Determine the reaction at points *A*, *B*, and *C* if each truck weighs 4000 lb at the front axle and 4000 lb at the rear axle when in the horizontal position.

Ans. $A = 4320$ lb; $B = 6900$ lb; $C = 4780$ lb.

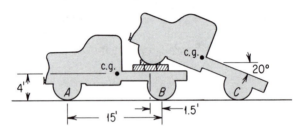

Problem 4-36

4-37. Determine the reactions at *A* and *B* on the bracket shown. Consider the weight of the bracket to be negligible.

Ans. $A_x = 300$ lb $\rightarrow$; $A_y = 200$ lb $\uparrow$; $B = 300$ lb $\leftarrow$.

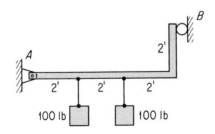

Problem 4-37

4-38. The tensile force exerted by the spring shown is proportional to the displacement of the end *D*. The constant of proportionality is called "the spring modulus." If the spring tension is 100 lb for the position shown and the modulus is 200 lb per in., find the spring tension and pin reaction at *C* for a cam rotation of 180 deg.

Ans. 489 lb; 330 lb (approximate answers).

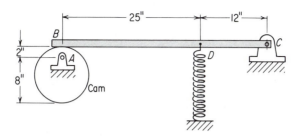

Problem 4-38

4-39. Find the weight W that can be supported by the boom illustrated if the maximum force that the cable AB can exert is 2000 lb. What is the pin reaction at C on the boom when it is supporting the maximum load W?

Ans. $W = 1400$ lb; $C_x = 1730$ lb $\rightarrow$; $C_y = 2400$ lb $\uparrow$.

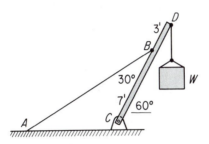

Problem 4-39

4-40. The pulley E and the boom AD are both pin-supported as shown. Determine the tension T in the cable and the pin reaction at A on the boom.

Ans. $T = 1300$ lb; $A_x = 1130$ lb $\rightarrow$; $A_y = 956$ lb $\downarrow$.

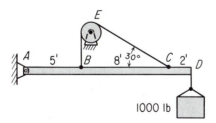

Problem 4-40

4-41. The hydraulic boom and the winch are employed to raise the 5000-lb load as shown. Determine the tension in the cable, the pin reaction on the boom at A, and the force exerted by the hydraulic ram. Neglect the weight of the boom.

Ans. $T = 2500$ lb; $A_x = 4530$ lb $\rightarrow$; $A_y = 1990$ lb $\uparrow$; $F = 5510$ lb.

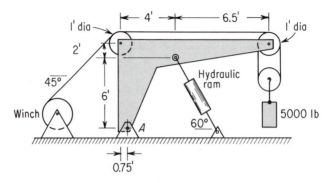

Problem 4-41

4-42. What load W is the portable hoist illustrated capable of lifting if a horizontal force of 30 lb is exerted on the crank at C? What are the reactions on the rollers A and B? Assume that horizontal motion of the hoist is prevented by locked wheels.

Ans. $W = 90$ lb; $A = B = 270$ lb.

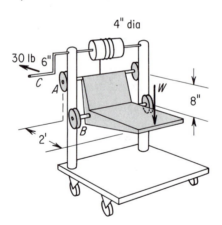

Problem 4-42

4-43. Determine the tension in each of the three cables that support the 100-lb weight in the figure.

Ans. $AD = 123$ lb; $CD = 60.1$ lb; $BD = 40$ lb.

4-44. Find the tension in each of the three cables that support the 3-kip weight shown.

Ans. $AD = 1.88$ kips; $BD = 0.314$ kip; $CD = 1.48$ kips.

4-45. The triangular plate indicated is supported by three cables. Find the tension in each cable if the uniform plate weighs 500 lb.

Ans. $T_A = T_B = T_C = 167$ lb.

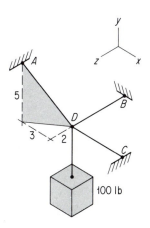

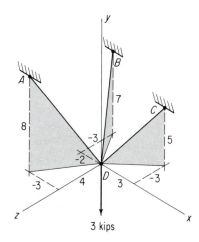

Problem 4-43

Problem 4-44

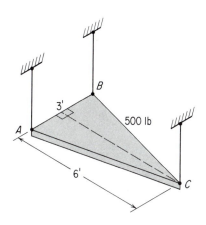

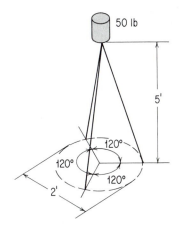

Problem 4-45

Problem 4-46

4-46. Determine the force in each leg of the tripod shown.

Ans. 17 lb.

4-47. Boom DC supports a 500-lb load as shown. Determine the tension in each of the cables and the pin reaction at D.

Ans. $T = 417$ lb; $D_x = 667$ lb; $D_y = 500$ lb.

4-48. A steel I beam weighing 75 lb per ft is supported as shown. Find the force in each of the cables.

Ans. $T_A = T_C = 469$ lb; $T_D = 750$ lb.

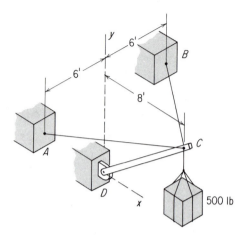

Problem 4-47

4-49. Beams weighing 35 lb per ft are welded to form an ell as shown. Determine the force in each of the cables that support this frame.

Ans. $T_A = 140$ lb; $T_B = 245$ lb; $T_C = 105$ lb.

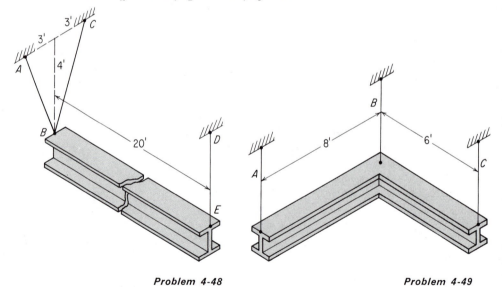

Problem 4-48   Problem 4-49

4-50. Find the bearing reactions at A and D on the shaft shown.

Ans. 566 lb; 267 lb.

4-51. The "jack-shaft" supports three pulleys as shown. Determine the tension T required to maintain equilibrium and the bearing reactions at A and E on the shaft.

Ans. $T = 373$ lb; $A_y = -75$ lb; $A_z = -113$ lb; $E_y = 137$ lb; $E_z = -37.5$ lb.

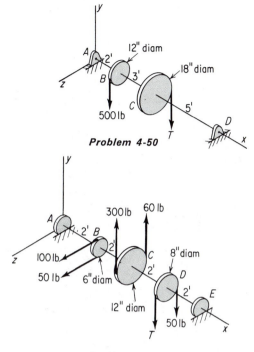

Problem 4-50

Problem 4-51

4-52. A schematic diagram of a single-cylinder air compressor is shown. Determine the belt tension T on the pulley and the bearing reactions at A and B on the crankshaft if the compressed gas in the cylinder exerts a force of $P = 800$ lb on the piston.

Ans. $T = 650$ lb; $A_x = B_x = 0$; $A_y = B_y = 400$ lb; $A_x = 200$ lb; $B_x = 1200$ lb.

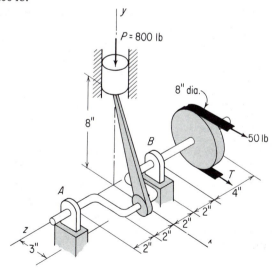

Problem 4-52

Chapter **5**

FORCE
ANALYSIS OF
STRUCTURES

A structure is an assembled group of beams, cables, and columns which is capable of supporting loads while it remains stable in geometric form. The conditions and equations of equilibrium, when they are applied to the structure as a unit, to sections of it, or to the individual members within it, constitute a force analysis.

5-1 Simple Trusses

A *simple truss* is a structure, lying in a single plane, that consists of straight members connected together to form a series of joining triangles. In practice the members are welded, bolted, or riveted together and heavy cover plates, called *gussets*, are used to reinforce each joint.

The first step in the analysis is to make certain simplifying assumptions so that the basic equations of equilibrium can be used to determine the forces that act in the various members.

First, the weights of the individual members are disregarded since they are small compared to the loads they carry. Second, the members are joined: not by riveting, welding, or bolting, but by smooth pins; each joint behaves as a perfect hinge. Third, the loads are applied on the structure only at the pinned joint; thus, each component of the truss is a *two force member*.

5-2 Determination of Tension or Compression

To know whether the members are pulling or pushing is just as important as the force determination. A tension member—one that is pulling on the joint—can be a light bar or cable, while a compression member—one that pushes on the joint—must be heavy and bulky so that it will not buckle under the tremendous loads that it usually carries. The designer must know how the load is applied as well as its magnitude before he can suggest a proper structural member.

Member *AB* of the wall truss shown in Fig. 5-1 is a tension member; isolated as a free-body, it tends to stretch because of the action of the loads that are applied to it. Member *EF* on the truss is a compression member;

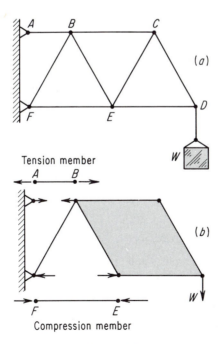

Tension member

Compression member

Figure 5-1

it pushes against joints E and F, which in turn push on it and tend to shrink it. A simple test often answers the question of tension or compression: if a flexible cable can support the load, the member is in tension; if it buckles, the member is in compression. A cable would collapse in position EF.

5-3 Analysis by the Method of Joints

The action lines of the forces that act on a simple truss lie on the axes of the individual members; the magnitude and sense of each force are the unknown quantities. One method of determining these forces is to analyze trigonometrically each joint in the truss. Select some initial point—usually either of the ends, where no more than two unknown forces are involved —and proceed from joint to joint, attacking no more than two unknown quantities at a time. The example that follows will illustrate this procedure.

EXAMPLE 1: Determine the force in each member of the truss shown in Fig. 5-2. The structure, supported by a pin at A and a roller at D, is acted upon by a force of 1000 lb at C.

Solution: Isolate the truss from its supports and compute the reactions at A and D.

The reaction A_x is zero, since it is the only x-component of force acting on the free-body:

$$\sum F_x = 0$$

$$A_x = 0$$

A moment summation about pin D gives the reaction A_y,

$$\sum M_D = 0$$

$$20A_y = 5(1000)$$

$$A_y = 250 \text{ lb}$$

Equilibrium of forces in the y-direction gives the reaction D_y:

$$\sum F_y = 0$$

$$A_y + D_y - 1000 = 0$$

$$D_y = 1000 - 250 = 750 \text{ lb}$$

Two members are joined at pin A and two at pin D; three members come together at B and C, and four at E. Since a joint with more than unknown forces cannot be solved, the obvious place to start is either A or D. If joint A is selected, members AB and AE are cut and the free-body diagram is drawn as shown in Fig. 5-2(c). The conditions of equilibrium, $\sum F_x = 0$ and $\sum F_y = 0$, are then applied and the unknown forces are computed. If the sense of a particular force is not obvious,

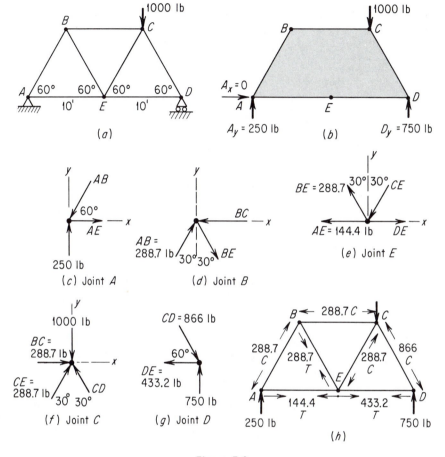

Figure 5-2

it must be assumed. An incorrect guess is indicated by an answer whose magnitude is correct but whose sign is negative.

$$\Sigma F_y = 0$$

$$250 - AB \sin 60° = 0$$

$$AB = \frac{250}{0.866} = 288.7 \text{ lb (compression)}$$

$$\Sigma F_x = 0$$

$$AE - AB \cos 60° = 0$$

$$AE = 288.7(0.5) = 144.4 \text{ lb (tension)}$$

The free-body diagram is next drawn for joint B and the force AB is used to compute BC and BE. BE must pull on the pin to balance the compressive push of AB. BC acts to the left, thereby balancing the horizontal components of force. The equations of equilibrium written for this joint are

$$\Sigma F_y = 0$$
$$288.7 \cos 30° - BE \cos 30° = 0$$
$$BE = 288.7 \text{ lb (tension)}$$

and
$$\Sigma F_x = 0$$
$$288.7 \sin 30° + BE \sin 30° - BC = 0$$
$$BC = 288.7 \text{ lb (compression)}$$

Next, the equations of equilibrium are applied to the free-body diagram of joint E.

$$\Sigma F_y = 0$$
$$288.7 \cos 30° - CE \cos 30° = 0$$
$$CE = 288.7 \text{ lb (compression)}$$

and
$$\Sigma F_x = 0$$
$$DE - 144.4 - 288.7 \cos 60° - CE \cos 60° = 0$$
$$DE = 144.4 + 288.7(0.5) + 288.7(0.5)$$
$$DE = 433.2 \text{ lb (tension)}$$

One unknown force, that in member CD, remains to be determined and joint C will be employed for this computation.

$$\Sigma F_y = 0$$
$$288.7 \cos 30° + CD \cos 30° - 100 = 0$$
$$CD = \frac{1000 - 288.7(0.886)}{0.866}$$

$$CD = \frac{750}{0.866} = 866 \text{ lb (compression)}$$

Since so many calculations, each dependent upon the one preceding it, are involved, some sort of check should be made. The easiest method is to see if the forces balance at joint D.

$$\Sigma F_x = 0$$
$$433.2 - 866 \cos 60° = 0$$
$$433.2 - 433.2 = 0$$
$$\Sigma F_y = 0$$
$$750 - 866 \sin 60° = 0$$
$$750 - 750 = 0$$

A final sketch showing the magnitudes and directions of the forces within the truss completes the problem. This is shown in Fig. 5-2(h).

5-4 Members That Carry No Load

Often a truss that appears at the outset to be very complicated in reality is quite simple since many of its members are not load supporting. From the practical point of view these members do have a purpose; they are used either to stiffen the truss or to keep certain critical compression members from buckling. As far as the equations of equilibrium are concerned, however, the force within them is zero. To illustrate, consider the truss shown in Fig. 5-3. At first glance it appears that there are 29 members to be determined; this would require the writing of 29 simultaneous equations for a complete analysis. Close inspection, fortunately, shows the entire inner maze to be composed of non-load-carrying members. In the final analysis, only three

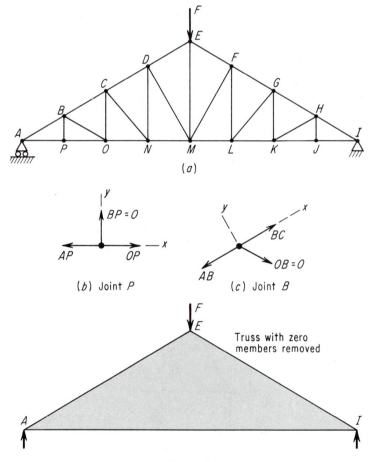

Figure 5-3

members, *AE*, *EI*, and *AI*, actually support loads. The free-body diagram of joint *P* shows *BP* to be the only member that can support a force in the *y*-direction; there is nothing to balance this force, so *BP* must be zero. A similar situation exists at joint *B*, where the component of *OB* along the *y*-axis stands alone: *OB* is zero. If this *analysis by inspection* is continued and the zero members are removed, all that will remain of the truss will be the outer frame *AEI*. Obviously, one of the first steps in the force analysis should be to locate and remove the non-load-carrying members.

5-5 Analysis by The Method of Sections

By isolating as a free-body a section of a truss, it is possible to determine the forces that act in certain members without resorting to the tedious step-by-step method described in the previous sections. Consider again, for example, the truss that was analyzed by the method of joints in Example 1. After the reactions at *A* and *D* have been found, the truss is sectioned along line *a-a* by cutting members *BC*, *BE*, and *AE* as shown in Fig. 5-4. The left hand portion is isolated as a free-body and the equations of equilibrium are applied to it. The force in member *AE* is found by summing moments about point *B*.

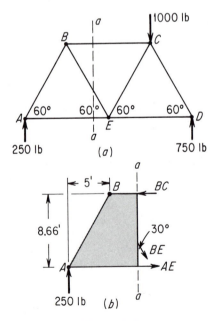

Figure 5-4

$$\sum M_B = 0$$

$$8.66AE - 5(250) = 0$$

$$AE = \frac{1250}{8.66} = 144.3 \text{ lb (tension)}$$

BE is the only one of the three members that has a component capable of balancing the 250 lb upward reaction at A. Therefore, it is a tension member and the load it carries can be found by a force summation in the y-direction.

$$\sum F_y = 0$$

$$250 - BE \cos 30° = 0$$

$$BE = \frac{250}{0.866} = 288.7 \text{ lb (tension)}$$

Finally, the components of force in the horizontal direction must balance.

$$\sum F_x = 0$$

$$AE + BE \sin 30° - BC = 0$$

$$BC = 144.3 + 288.7(0.5) = 288.7 \text{ lb (compression)}$$

The method of sections is particularly useful when a force analysis of only a few members is required. Occasionally both methods, joints and sections, are used together, since some members are more easily determined by one method than the other. It should be noted, however, that the method of sections does not render a solution when more than three members with unknown loads are cut by the section.

5-6 Analysis of Frames

A *frame*, like a truss, is a structure; they differ only in the way the loads are applied. A frame is capable of supporting loads at points other than the extremities of its members. To illustrate, the frame shown in Fig. 5-5 is composed of three pin-connected members. A force P acts on the horizontal member BD. The analysis proceeds, in this case, by first determining the reactions at the supports A and E. Free-body diagrams are then drawn for each member; when necessary, a direction is assumed for the components of force at the various pins. A wrong assumption, just as in truss analysis, will give an answer of correct magnitude, but it will carry a negative sign. The equations of equilibrium are written for each member and the unknown forces are computed.

EXAMPLE 2: Determine the pin reactions at A, B, and C on the members of the frame shown in Fig. 5-6.

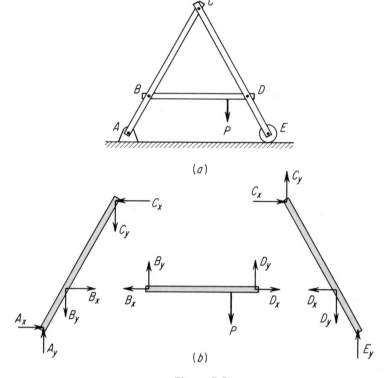

Figure 5-5

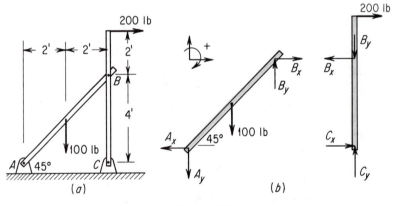

Figure 5-6

Solution: Draw a free-body diagram for each member; assume the force directions at the pins. Attempt to find as many of the reactions as possible by single, rather than simultaneous, equations.

C_x can be found by summing moments with respect to point B on the vertical member.

$$\Sigma \, M_B = 0$$
$$200(2) - 4C_x = 0$$
$$C_x = 100 \text{ lb}$$

A force summation in the x-direction on this same member gives the force B_x.

$$\Sigma \, F_x = 0$$
$$200 + C_x - B_x = 0$$
$$B_x = 200 + 100 = 300 \text{ lb}$$

A force summation in the y-direction shows B_y and C_y to be equal.

$$\Sigma \, F_y = 0$$
$$C_y - B_y = 0$$
$$C_y = B_y$$

Now apply the equations of equilibrium to member AB. Moments taken with respect to point A give the force B_y.

$$\Sigma \, M_A = 0$$
$$4B_x + 2(100) - 4B_y = 0$$
$$B_y = \frac{4(300) + 2(100)}{4} = 350 \text{ lb}$$

Summing forces on member AB in the x- and y-directions completes the analysis.

$$\Sigma \, F_x = 0$$
$$B_x - A_x = 0$$
$$A_x = B_x = 300 \text{ lb}$$
$$\Sigma \, F_y = 0$$
$$B_y - A_y - 100 = 0$$
$$A_y = 350 - 100 = 250 \text{ lb}$$

Summarized, the answers are

$$A_x = 300 \text{ lb}, \qquad B_x = 300 \text{ lb}, \qquad C_x = 100 \text{ lb}$$
$$A_y = 250 \text{ lb}, \qquad B_y = 350 \text{ lb}, \qquad C_y = 350 \text{ lb}$$

As a check, both the horizontal forces and the vertical forces must balance on the frame considered a free-body in itself.

Vertical forces:

$$C_y - A_y - 100 = 0$$
$$350 - 250 - 100 = 0$$
$$0 = 0$$

Horizontal forces:

$$200 + C_x - A_x = 0$$
$$200 + 100 - 300 = 0$$
$$0 = 0$$

QUESTIONS, PROBLEMS, AND ANSWERS

5-1. Determine the force in members AB and BC of the cantilever truss shown. The truss is pin-supported at A and C.

Ans. $AB = 5200$ lb T; $BC = 4800$ lb C.

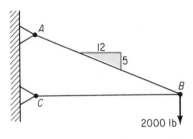

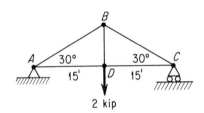

Problem 5-1   **Problem 5-2**

5-2. A simple bridge truss shown is pin-supported at A and roller-supported at C. Determine the force in each member if a 2 kip load is applied at point D.

Ans. $AB = 2$ kips C; $AD = 1.73$ kips T; $BD = 2$ kips T.

5-3. Determine the force in each member of the roof truss shown. The truss is pin-supported at A and roller-supported at F.

Ans. Answers in kips: $A = {}^8/_3$; $F = {}^{10}/_3$; $AB = {}^{10}/_3$ C; $BC = 0$; $AC = CE = 2T$; $BE = {}^5/_6$ T; $BD = {}^5/_2$ C; $DE = {}^2/_3$ C; $EF = {}^5/_2$ T; $DF = {}^{25}/_6$ C.

5-4. The cantilever truss supports a 5000-lb load directed as indicated. Determine the force in each member.

Ans. Hint: $DE = 5$ kips C; $EG = 8.66$ kips C.

5-5. Determine the force in each member of the truss illustrated. A vertical load of 2000 lb acts downward as shown.

Ans. Hint: $BG = DE = EF = 0$.

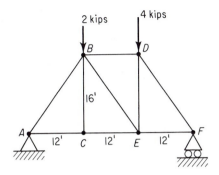

Problem 5-3

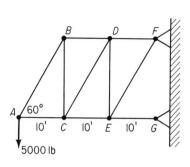

Problem 5-4

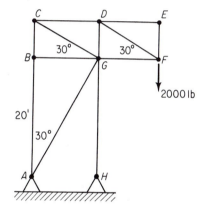

Problem 5-5

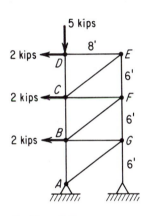

Problem 5-6

5-6. A section of a vertical tower supports the loads shown. Determine the force in each member of the tower.

Ans. Answers in kips: $AB = 9.5$ C; $BC = 6.5$ C; $CD = 5$ C; $DE = 2$ T; $EF = 1.5$ T; $FG = 4.5$ T; $GH = 9$ T; $CE = 2.5$ C; $CF = 4$ T; $BF = 5$ C; $BG = 6$ T; $AG = 7.5$ C

5-7. Determine the force in each member of the sign truss illustrated.

Ans. Hint: All internal members carry zero load.

5-8 to 5-13 Locate by inspection the members of the truss that carry no load in the respective figures.

Ans. **5-8.** *BD.*
5-9. *FH, EH, EI, DI, DJ.*
5-10. *EG, EH, DH, DI, CI, CJ, BJ,* and *BK.*
5-11. *BL, BK, CK, EI, FI,* and *FH.*

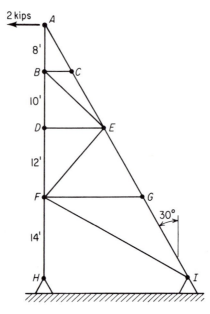

Problem 5-7

5-12. *BM*, *CM*, *CN*, *DN*, *FO*, and *OG*.
5-13. *EH*, *EI*, *DI*, *DJ*, *CJ*, *CK*, *BK*, *BL*, and all like members on right-hand side.

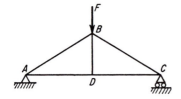

Problem 5-8

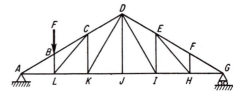

Problem 5-9

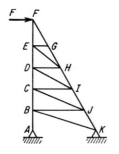

Problem 5-10

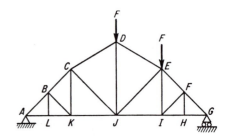

Problem 5-11

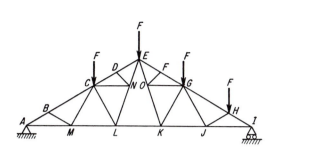

Problem 5-12

Problem 5-13

5-14. Use the method of sections to determine the force in members CD, DH, and GH in the figure.

Ans. $CD = 2.89$ kips C; $DH = 1.44$ kips C; $GH = 3.61$ kips T.

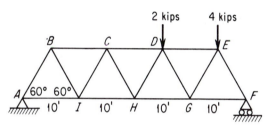

Problem 5-14

5-15. Determine the force in members BC and BF of the cantilever truss illustrated. Use the method of sections.

Ans. $BC = 1000$ lb T; $BF = 1000$ lb C.

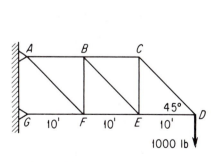

Problem 5-15

Problem 5-16

5-16. Use the method of sections to find the force in members *BC*, *BE*, and *EF* in the cantilever truss shown.

Ans. BC = 1429 lb T; *BE* = 1745 lb T; *EF* = 3485 lb C.

5-17. Use the method of sections to determine the force in members *BC*, *CG*, and *FG* of the overhanging truss illustrated.

Ans. BC = 0.845 kips C; *CG* = 5.36 kips C; *GF* = 1.53 kips T.

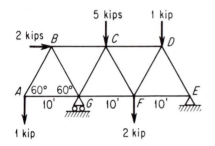

Problem 5-17

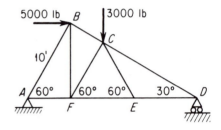

Problem 5-18

5-18. Determine the force in member *CE* in the triangular truss shown.

Ans. 0

5-19. Determine the force in members *BC* and *BK* in the bridge truss illustrated.

Ans. BC = 33 kips C; *BK* = 9.90 kips T.

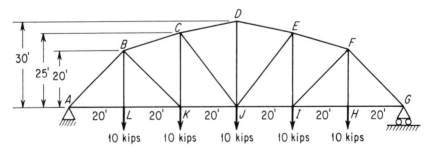

Problem 5-19

5-20. Determine the pin reactions at *A*, *B*, and *C* for the frame loaded as shown.

Ans. Hint: $A_x = 400$ lb →; $A_y = 950$ lb ↓ on vertical member.

5-21. Determine the pin reactions at *A*, *B*, and *C* that act on the wall bracket illustrated.

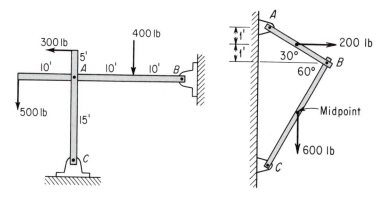

Problem 5-20   **Problem 5-21**

Ans. $A_x = 305$ lb; $A_y = 118$ lb; $B_x = 105$ lb; $B_y = 118$ lb; $C_x = 105$ lb; $C_y = 482$ lb.

5-22. Find the pin reactions at A, B, and C that act on the frame shown.

Hint: $A_x = A_y = B_x = B_y = 125$ lb.

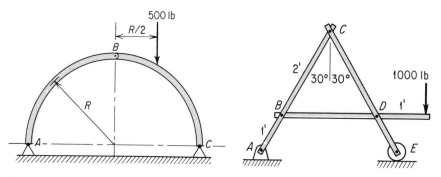

Problem 5-22   **Problem 5-23**

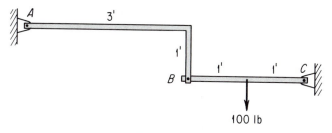

Problem 5-24

5-23. The "A" frame supports a load of 1000 lb as shown. Determine the reactions on the horizontal member at pins B and D. The frame itself is supported by a pin at A and a roller at E.

Ans. $B_x = 145$ lb $\rightarrow$; $B_y = 500$ lb $\downarrow$; $D_x = 145$ lb $\leftarrow$; $D_y = 1500$ lb $\uparrow$.

5-24. Determine the reactions in the figure that act on the double link at the wall supports A and C.

Ans. $A_x = 150$ lb $\leftarrow$; $A_y = 50$ lb $\uparrow$; $C_x = 150$ lb $\rightarrow$; $C_y = 50$ lb $\uparrow$.

Chapter 6

FRICTION

Friction and gravity are the two forces that are most frequently encountered in mechanics, yet they are the least understood. Some 300 years of the combined efforts of scientists and engineers have failed to answer questions concerning their origin and nature.

Friction is both an aid and a hindrance; we are dependent upon it in belt drives, clutches, and brakes, and we seek ways to increase it and thereby increase mechanical efficiency. On the other hand, friction causes bearings, gear teeth, and even space vehicles to wear away; we seek ways to decrease friction to preserve these devices.

6-1 The Laws of Friction

From the practical point of view, three laws describe the retarding force of friction. These laws state that when one solid body slides over another, the frictional force is:

1. *Proportional to the pressure force between the bodies,*
2. *Independent of the area of contact, and*
3. *Independent of the sliding velocity.*

A simple experiment will illustrate these three laws. Body A of weight W rests on body B as shown in Fig. 6-1. A spring scale attached to a rigid support is used to indicate the dragging force of body B on body A.

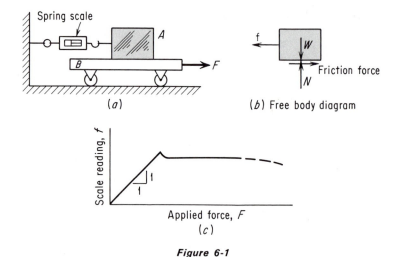

(a)

(b) Free body diagram

(c)

Figure 6-1

The force F is applied to body B and readings of the scale force f are taken. The scale force increases as the applied force increases in a one-to-one ratio until slipping occurs. Once the blocks slide on one another, the scale force drops slightly, and thereafter remains constant and completely independent of the force F that acts on body B. The relationship beween the friction force and the applied force is shown in Fig. 6-1(c). The ratio of the maximum friction force that acts between the surfaces and the normal pressure force N is called the *coefficient of static friction* and is symbolized by the Greek letter *mu* with a subscript s.

$$\mu_s = \frac{f_{\max}}{N} \tag{6-1}$$

The coefficient of kinetic friction, μ_k, is the ratio of the frictional force to the normal force when sliding occurs.

$$\mu_k = \frac{f_{\text{sliding}}}{N} \tag{6-2}$$

Two types of problems involving friction are encountered in mechanics. In the first type, the static coefficient of friction is used to determine whether or not a body will move under the action of a given system of forces or to determine the force necessary to just overcome a state of rest. The second type of problem involves bodies that are moving relative to one another; the coefficient of kinetic friction applies to the force analysis. The three examples that follow will illustrate the general class of problems involving static friction.

EXAMPLE 1: A 100-lb block rests on an inclined plane as shown in Fig. 6-2. Determine the maximum inclination angle θ that can exist without allowing the block to move. The coefficient of static friction between the contact surfaces is $\mu_s = 0.65$.

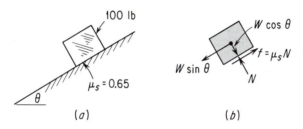

Figure 6-2

Solution: To maintain equilibrium, the component of the weight force must balance the maximum available frictional force

$$W \sin \theta = \mu_s N$$

where N, the normal force, is equal to the component of weight, $W \cos \theta$. Therefore,

$$W \sin \theta = \mu_s W \cos \theta$$

Dividing both sides of this equation by $W \cos \theta$ gives

$$\tan \theta = \mu_s$$

or
$$\theta = \text{arc} \tan 0.65 = 33°$$

EXAMPLE 2: Determine the range of values for the force F so that the 100-lb block shown in Fig. 6-3(a) will slide neither up nor down the inclined surface. The coefficient of static friction between the block and the plane is $\mu_s = 0.40$.

Solution: The force of friction always opposes the direction of motion or the direction of *impending* motion. Two free-body diagrams are therefore required:

1. to describe the force system when impending motion is upward, and
2. to describe the force system when impending motion is downward.

These two diagrams are shown in Fig. 6-3(b) and in Fig. 6-3(c).

To start the block upward, the applied force F must overcome the gravity force $W \sin \theta$ and the friction force $f = \mu_s N$.

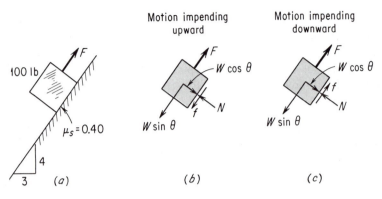

Figure 6-3

$$F = W \sin \theta + \mu_s N = W \sin \theta + \mu_s W \cos \theta$$
$$= 100(\tfrac{4}{5}) + 0.40(100)\tfrac{3}{5} = 104 \text{ lb}$$

For impending downward motion the direction of the friction force reverses; in this case, friction helps support the weight on the plane.

$$F = W \sin \theta - \mu_s W \cos \theta$$
$$= (100)\tfrac{4}{5} - 0.40(100)\tfrac{3}{5} = 56 \text{ lb}$$

The block will not move if F is not less than 56 lb or does not exceed 104 lb.

EXAMPLE 3: The bracket shown in Fig. 6-4(a) is a *self-locking* device; if the weight is far enough outward, the bracket will not slide downward. Determine the minimum distance d at which the weight W can be placed without causing the bracket to slip downward. The coefficient of static friction between the sliding surfaces is $\mu_s = 0.25$.

Solution: The bracket bears on the vertical rod at edges A and B; the reactions at these two points are N_A and N_B. Frictional forces at these two points are upward, since they oppose the impending downward motion. Equilibirum in the x- and y-directions gives

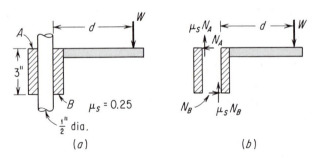

Figure 6-4

$$\sum F_x = 0$$

$$N_A = N_B \tag{a}$$

and $$\sum F_y = 0$$

$$\mu_s N_A + \mu_s N_B - W = 0$$

$$0.25(N_A + N_B) = W \tag{b}$$

Combining Eqs. (a) and (b) gives

$$N_A = N_B = 2W \tag{c}$$

Taking moments about corner B completes the solution:

$$\sum M_B = 0$$

$$Wd + 0.5\mu_s N_A - 3N_A = 0$$

Simplifying gives

$$Wd = [3 - (0.5)(0.25)]N_A = 2.875\ N_A \tag{d}$$

The distance d is computed by combining Eqs. (c) and (d):

$$Wd = 2.88(2)W$$

where $$d = 5.75 \text{ in.}$$

This is the minimum distance to prevent slipping.

6-2 Angle of Friction

In solving many problems involving friction, it is sometimes more convenient to use the resultant of the friction force f and the normal force N rather than the two separately. The angle, shown in Fig. 6-5, between the resultant force R and the normal force N is called the *angle of friction*, and is defined by

$$\tan \alpha = \frac{f}{N} \tag{6-3}$$

The angle of friction has a maximum value when $f = \mu_s N$, thus

$$\tan \alpha_{max} = \frac{\mu_s N}{N} = \mu_s \tag{6-4}$$

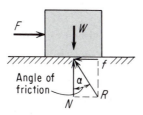

Figure 6-5

Eq. (6-4) states that *the tangent of the angle of friction is equal to the coefficient of friction.*

6-3 Wedges

The angle of friction is most advantageously employed in problems dealing with *wedges*. This is illustrated in the typical wedge problem shown in Fig. 6-6: a force, applied to wedge A, lifts the weight W. The free-body diagram shows A to be acted upon by three forces: the applied force F, the reaction R_1 at the horizontal surface, and the reaction R_2 between wedge A and block B. Both R_1 and R_2 represent the combined effect of the friction force and the normal force between the sliding surfaces. In a similar manner three forces R_2, R_3, and W act on body B.

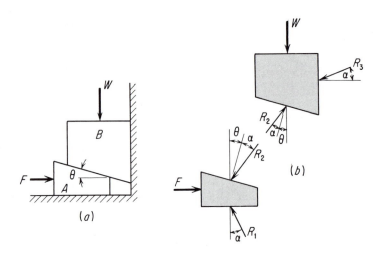

Figure 6-6

EXAMPLE 4: A 5-deg wedge is used to raise a 1000-lb weight as shown in Fig. 6-7(a). Determine the force F required to start the wedge if the coefficient of static friction between all sliding surfaces is $\mu_s = 0.30$.

Solution: The angle of friction is given by Eq. (6-4).

$$\tan \alpha = 0.30$$

$$\alpha = 16.7°$$

The free-body diagram of block A, shown in Fig. 6-7(b), indicates the three forces that are in equilibrium: 1000 lb acting downward, the rope tension T to the right, and the resultant R_1 at an angle of $(\alpha + 5°)$ with the vertical.

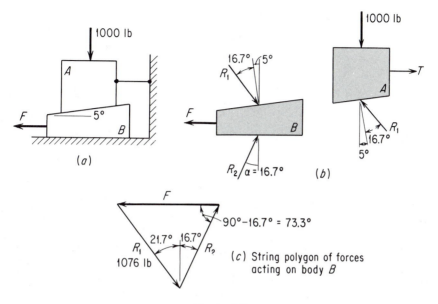

(c) String polygon of forces acting on body B

Figure 6-7

A force summation in the y-direction on body A gives

$$\sum F_y = 0$$

$$R_1 \cos (5° + \alpha) - 1000 = 0$$

$$R_1 = \frac{1000}{\cos 21.7°} = \frac{1000}{0.9291}$$

$$R_1 = 1076 \text{ lb}$$

The required force F is found by applying the sine law to the string polygon of forces acting on wedge B.

$$\frac{F}{\sin (21.7° + 16.7°)} = \frac{1076}{\sin 73.3°}$$

Solving for F gives

$$F = 1076 \frac{\sin 38.4°}{\sin 73.3°} = 1076 \left(\frac{0.6212}{0.9578} \right)$$

where $F = 698 \text{ lb}$

6-4 Belt Friction

The friction developed between a flat belt or a rope and a cylindrical surface is used to *transmit* power, while band brakes, capstans and similar devices make use of belt friction to *retard* motion. A typical free-body

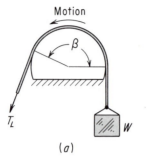

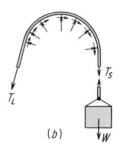

(a)   *(b)*

Figure 6-8

diagram of a rope that has impending motion against a rough cylindrical surface is shown in Fig. 6-8. A force T_L is applied to the rope causing the weight to move upward. The normal reactions of the cylinder on the rope vary from zero to a maximum value and then to zero again. T_L and T_S are the tensions on either side of the rope. The subscripts L and S distinguish the larger tension from the smaller: *the larger tension always opposes impending motion.*

The equation that governs the relationship between T_L and T_S states that

$$T_L = T_S e^{\mu\beta} \qquad (6\text{-}5)$$

where: e = base of the natural logarithm = 2.718
μ = coefficient of friction
β = angle of contact in radians

The term $e^{\mu\beta}$ represents an exponential power of the number 2.718. To determine the magnitude of the exponential term for values of $\mu\beta$ refer to Table II of the Appendix or use a slide rule to determine these values.

EXAMPLE 5: A 100-lb weight is suspended from a rope that passes over a rough cylindrical surface as shown in Fig. 6-9. Determine the force F that must be applied

Figure 6-9

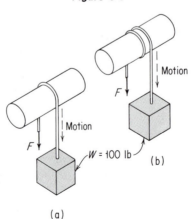

(a)

to the free end of the rope to just prevent the weight from slipping downward if: (a) the rope is in contact with one-half of the cylindrical surface; (b) the rope is wrapped one and one-half times around the cylinder. The coefficient of static friction between the rope and the cylinder is $\mu_s = 0.40$.

Solution: Since the larger tension always opposes the direction of impending motion, $T_L = 100$ lb and T_S is the desired force. For a one-half turn, the contact angle β equals π radians, and for one and one-half turns β equals 3π radians.

The data are substituted into Eq. (6-5):

Part (a):

$$T_L = T_S e^{\mu\beta}$$

$$100 = T_S e^{(0.40)\pi}$$

solving for T_S gives

$$T_S = \frac{100}{e^{1.26}}$$

Table II of the Appendix gives the value of the exponential term as $e^{1.26} = 3.525$;

therefore $$T_S = \frac{100}{3.525} = 28.4 \text{ lb}$$

Part (b): The exponential term assumes a much larger value as β increases

$$e^{\mu\beta} = e^{(0.40)3\pi} = e^{3.77}$$

$$e^{\mu\beta} = 43.31$$

The force required to support the weight in this case is considerably less than that required in Part (a):

$$T_S = \frac{100}{43.31} = 2.31 \text{ lb}$$

It is interesting to note that the size of the bar has no effect on the required force.

EXAMPLE 6: The band brake shown in Fig. 6-10(a) is used to arrest the motion of the flywheel. Determine the tension in either side of the band if a force of $F = 20$ lb is applied at point D as shown. The coefficient of kinetic friction between the wheel and the band is $\mu_k = 0.60$.

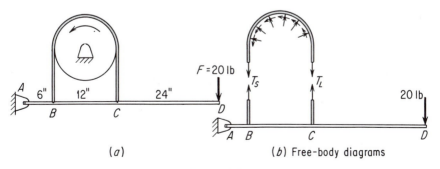

(a)   (b) Free-body diagrams

Figure 6-10

Solution: Two equations are required, since two unknowns, T_L and T_S, are involved in the computation. The belt friction equation governs the ratio of T_L to T_S.

$$\frac{T_L}{T_S} = e^{\mu\beta} = e^{(0.60)\pi}$$

$$T_L = 6.554\,T_S \tag{a}$$

A moment summation about point A on the lever gives a second equation involving the two unknowns.

$$\sum M_A = 0$$

$$6T_S + 18T_L - 42(20) = 0$$

$$T_S + 3T_L = 140 \tag{b}$$

The simultaneous solution of Eqs. (a) and (b) gives the desired answers:

$$T_S + 3(6.554)T_S = 140$$

$$T_S = \frac{140}{20.66} = 6.78 \text{ lb}$$

$$T_L = 6.554(6.78) = 44.4 \text{ lb}$$

6-5 Rolling Resistance

Rolling friction, which is usually referred to as *rolling resistance*, results when the surface that supports a rolling load deforms. Figure 6-11 is an exaggerated illustration of this effect. W is the vertical load acting on the wheel and F the force required to pull the wheel out of the depression or rut. Three forces, F, W, and R are in equilibrium. A moment summation about point A gives

$$\sum M_A = 0$$

$$W \cdot a - F \cdot d = 0$$

Since the depression is usually very small, the distance d may be replaced by the radius r.

$$F = \frac{Wa}{r} \tag{6-6}$$

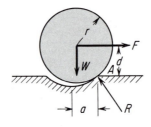

Figure 6-11

The distance a is called the *coefficient of rolling resistance;* it may vary from 0.003 in. for steel on steel to 4 in. or more for steel on soft ground.

EXAMPLE 7: The coefficient of rolling resistance between a 24-in.-in-diameter steel drum and a level pavement is 0.05 in. Determine the force necessary to overcome rolling resistance if the drum weighs 2000 lb.

Solution: The data are substituted into Eq. (6-6).

$$F = \frac{Wa}{r} = \frac{2000(0.05)}{12}$$
$$= 8.33 \text{ lb}$$

6-6 Disk Friction

Another type of frictional force which has great importance in the field of machine design is that which occurs between circular areas that are thrust together as shown in Fig. 6-12. Pivots, disk brakes, clutches, and thrust bearings are examples of mechanisms in which this type of friction is developed.

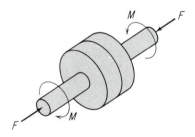

Figure 6-12

Of interest is the magnitude of the moment M that is required to cause motion to impend. It is assumed in the analysis that the contact pressure between the mated surfaces is uniformly distributed. Each component of pressure is equivalent to a normal force; frictional forces, then, act over the entire contact area. The products of these friction forces and their respective lever arms, when summed, is equivalent to the external moment M necessary to cause motion to impend. The value of this moment is given by

$$M = \tfrac{2}{3}\mu FR \qquad (6\text{-}7)$$

where $F =$ thrust force
       $R =$ radius of contact area
       $\mu =$ coefficient of friction

If the shaft were hollow, the contact area would be a ring rather than a complete circle, and the moment M would be given by

$$M = \frac{2}{3}\mu F \frac{(R_0^3 - R_i^3)}{(R_0^2 - R_i^2)} \qquad (6\text{-}8)$$

R_i and R_0 are the inside and outside radii respectively.

EXAMPLE 8: A thrust force of 800 lb acts on the pilot bearing shown in Fig. 6-13. Determine the moment M that must be applied to cause the shaft to rotate. The coefficient of kinetic friction at the contact surface is $\mu_k = 0.10$.

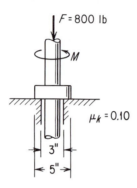

$F = 800$ lb

M

$\mu_k = 0.10$

3"

5"

Figure 6-13

Solution: Since a ring of area is involved, Eq. (6-8) is used to find the moment M:

$$M = \frac{2}{3}\mu F \frac{(R_0^3 - R_i^3)}{(R_0^2 - R_i^2)}$$

$$= \frac{2}{3}(0.10)800\frac{(2.5)^3 - (1.5)^3}{(2.5)^2 - (1.5)^2}$$

$$= 163 \text{ lb in.}$$

QUESTIONS, PROBLEMS, AND ANSWERS

6-1. The 100-lb body shown in three positions indicated is at rest; in each case the coefficient of static friction between the surfaces is 0.25. Determine the frictional forces developed between the body and the surface.

Ans. 20 lb in each position.

6-2. The force F in Prob. 6-1 is increased to 30 lb; find the frictional force developed between each body and the horizontal surface. Are the bodies in equilibrium?

Ans. 25 lb in each case; no.

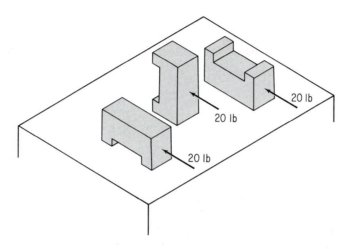

Problem 6-1

6-3. Plot a curve which will show the magnitude of the frictional force developed between the block and the plane shown as the angle of elevation is increased from zero to 60 deg. The coefficient of friction is 0.30.

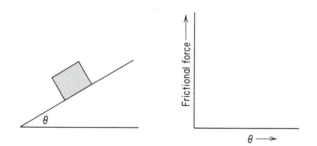

Problem 6-3

6-4. The wooden crate is on the verge of moving downward when the plane is tilted 20 deg as shown. Determine the coefficient of static friction between the contacting surfaces.

Ans. 0.364.

Problem 6-4

6-5. If $\mu_k = 0.9\mu_s$ in Prob. 6-4, by how much can the angle θ be reduced and still have motion downward after the crate overcomes static friction?

Ans. 1.9°.

6-6. Determine whether the block shown is moving or is at rest. The coefficients of friction between the contact surfaces are $\mu_s = 0.22$ and $\mu_k = 0.18$.

Ans. Block is moving down.

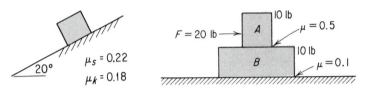

Problem 6-6   **Problem 6-7**

6-7. The two blocks shown are resting upon one another and in turn the two are resting on the rough plane. Will block A move relative to B? Will block B move relative to the plane? Prove your answers by drawing free-body diagrams.

Ans. Yes, in both instances.

6-8. Find the couple C necessary to cause the 200-lb cylinder indicated to begin rotating clockwise. The coefficient of static friction at both A and B is $\mu_s = 0.30$.

Ans. 71.6 lb ft.

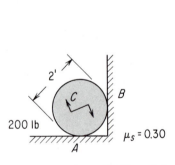

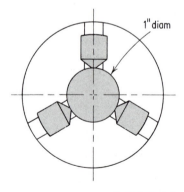

Problem 6-8   **Problem 6-9**

6-9. Determine the twisting moment required to cause the 1-in. cylindrical bar to slip in the three-jaw-face chuck illustrated. Assume a coefficient of static friction of 0.5 and a normal jaw force of 200 lb.

Ans. 150 lb in.

6-10. Solve Prob. 6-9 assuming that the chuck has four equispaced jaws instead of the three as shown.

Ans. 200 lb in.

6-11. A ladder weighing 25 lb supports a man weighing 180 lb at point C as shown. The vertical surface is smooth; the floor is rough. Determine the minimum angle θ at which the ladder will stand without slipping if the coefficient of static friction at A is $\mu_s = 0.30$.

Ans. 67.4°.

Problem 6-11

6-12. Determine the frictional force developed at point A in Prob. 6-11 if θ is 75 deg.

Ans. 39.5 lb.

6-13. Find the maximum value of the distance h in order that the 50-lb block shown will slide without tipping. The coefficient of static friction between the block and the plane is $\mu_s = 0.35$.

Ans. 5.71 in.

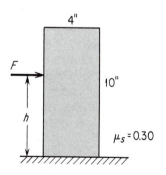

Problem 6-13

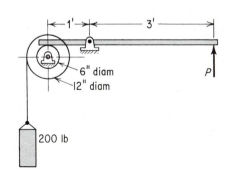

Problem 6-14

6-14. Find the force P required to prevent rotation of the 100-lb wheel shown. The coefficient of static friction between the wheel and the brake shoe is $\mu_s = 0.40$.

Ans. 333 lb.

6-15. A 26-ft ladder of negligible weight rests against a wall as illustrated. Determine how high a 200-lb man can climb before the ladder slips. The coefficient of static friction at both A and B is $\mu_s = 0.25$.

Ans. 15 ft.

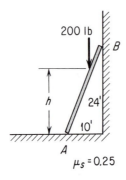

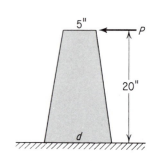

Problem 6-15   **Problem 6-16**

6-16. Determine the smallest allowable width d if the 20-lb block shown is to slide rather than tip as P is gradually increased. The coefficient of friction between the block and the plane is $\mu_s = 0.20$.

Ans. 8 in.

6-17. The 100-lb cylinder is supported by a roller free to turn as shown. Determine the magnitude of the couple required to turn the cylinder if the coefficient of friction between the cylinder and the vertical wall is 0.5.

Ans. 400 lb in.

6-18. Determine the torque C required to spin the cylinder of weight W against the wall as shown. The coefficient of friction is μ at both surfaces.

Ans. $C = RW\mu(1 + \mu)/(1 + \mu^2)$.

6-19. Wedge A as indicated is used to raise the 2000-lb weight that rests on block B. Determine the force P required to cause motion to impend. The coefficient of static friction between all surfaces is $\mu_s = 0.30$.

Ans. 1900 lb (graphical solution).

6-20. Determine the force P required to start the wedge illustrated downward. The coefficient of static friction between all contact surfaces is $\mu_s = 0.15$.

Ans. 41.9 lb.

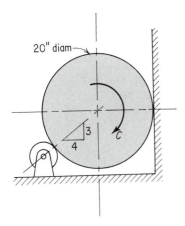

Problem 6-17

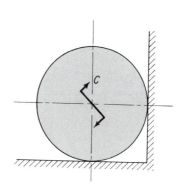

Problem 6-18

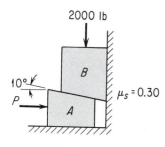

Problem 6-19

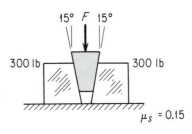

Problem 6-20

6-21. A uniformly distributed load of 100 lb per ft acts on the 12-ft pivoted beam as shown. Determine the force P that must be applied to the 5-deg wedge to cause it just to move. The coefficient of static friction between all surfaces at the wedge is $\mu_s = 0.20$.

Ans. 295 lb.

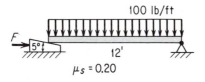

Problem 6-21

6-22. A 10-deg wedge is used to raise the 1000-lb weight as shown. Find the force P if the coefficient of friction for all sliding surfaces is 0.2.

Ans. 108 lb.

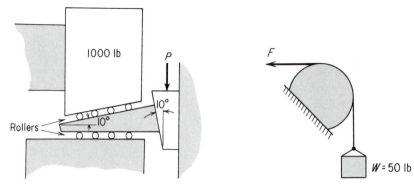

Problem 6-22   Problem 6-23

6-23. Find the force necessary to just hold the 50-lb weight illustrated. The coefficient of kinetic friction between the rope and the cylindrical surface is $\mu_k = 0.20$.

Ans. 36.5 lb.

6-24. Two turns of rope are wrapped around a horizontal, cylindrical bar. A pull of 20 lb on one end of the rope just prevents a 150-lb weight attached to the other end from slipping downward. Determine the coefficient of friction between the rope and the bar.

Ans. 0.16.

6-25. The coefficient of kinetic friction, in the illustration, between the cylindrical surfaces and the rope that supports the 300 lb weight is $\mu_k = 0.25$. Find the force F and the cable tension at A if the weight is on the verge of moving downward.

Ans. 137 lb; 203 lb.

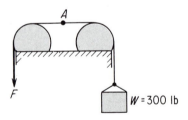

Problem 6-25

6-26. A weight of 500 lb is supported by a force of 50 lb applied to a rope that is passed around a horizontal, cylindrical bar. How many turns are required if the coefficient of friction between the rope and the post is $\mu_s = 1/\pi$?

Ans. 1.15 turns.

6-27. The band brake in the illustration is used to stop the motion of the flywheel. What force *P* must be applied to the lever if the flywheel is rotating counterclockwise and the tension in the vertical portion of the belt is 200 lb? The coefficient of kinetic friction between the belt and the wheel is $\mu_k = 0.20$.

Ans. 19.5 lb.

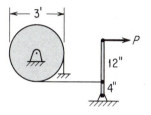

Problem 6-27

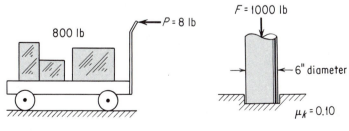

Problem 6-28

6-28. A strap pipe wrench is shown. Find the least value of *d* which is necessary so that the wrench will not slip. Assume the coefficient of friction between the belt and pipe is 0.20 and neglect any frictional force between the handle and the pipe.

Ans. 0.39 *D*.

6-29. Determine the horizontal pull required to move a 1500-ton train along a level track. The coefficient of rolling resistance is 0.01 in. and the wheels are 3 ft in diameter.

Ans. 1670 lb.

6-30. Determine the coefficient of rolling resistance if a force of 8 lb is required to move the hand truck shown at a constant velocity. The wheels are 12 in. in diameter.

Ans. 0.06 in.

6-31. A solid circular shaft 6 in. in diameter pivots in a flat socket as shown. Find

Problem 6-30

Problem 6-31

the frictional moment developed as the shaft rotates. The shaft carries a thrust load of 1000 lb, and the coefficient of kinetic friction μ_k is 0.10.

Ans. 200 lb in.

6-32. If a torque of 35 lb ft is required to move the shaft shown in Prob. 6-31, determine the coefficient of kinetic friction at the contact surfaces.

Ans. 0.21.

6-33. Determine the moment required to rotate the pressure plate shown. The coefficient of friction at the contact surface is 0.15.

Ans. 175 lb in.

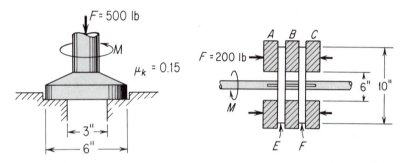

Problem 6-33   **Problem 6-34**

6-34. A double-disk clutch is shown; disks *A*, *B*, and *C* can move axially but cannot rotate, while disks *E* and *F* are keyed to the shaft and rotate with it. Find the moment *M* that must be applied to slip the clutch. The coefficient of kinetic friction between all contact surfaces is $\mu_k = 0.15$.

Ans. 490 lb in.

Chapter 7

MOMENT
OF INERTIA

The *moment of inertia*, or *second-moment*, as it is sometimes called, is a particular mathematical expression that appears in mechanics: it is used in the study of the strength of beams, columns, and torsion bars; in the study of fluids; and in the study of angular motion and vibrations. Because this important quantity is frequently encountered, it is essential to know how it is determined.

7-1 Moment of Inertia of an Area
by Approximation

Moment of inertia I is always computed relative to a reference point, line, or plane. In the case of the small element of area ΔA shown in Fig. 7-1, I is equal to the product of the area and the square of its distance from a particular axis. Thus, the moment of inertia with reference to the x-axis is

$$I_x = y^2(\Delta A) \qquad (7\text{-}1)$$

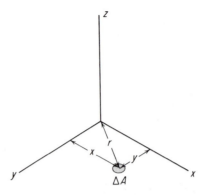

Figure 7-1

With reference to the y-axis the moment of inertia of ΔA is given by

$$I_y = x^2(\Delta A) \tag{7-2}$$

The product of ΔA and the square of its distance from the z-axis, which is perpendicular to the plane of the area, is called the *polar moment of inertia* and is symbolized by the letter J.

$$J = r^2(\Delta A) \tag{7-3}$$

Since r can be expressed in terms of x and y, the polar moment of inertia is related to I_x and I_y; thus

$$J = r^2(\Delta A) = (x^2 + y^2)\Delta A = x^2(\Delta A) + y^2(A\Delta)$$

and $\qquad\qquad\qquad J = I_x + I_y \tag{7-4}$

The moment of inertia of an area with respect to an axis is the sum of the moments of inertia of the elements that comprise the area. This statement provides a means of approximating I_x, I_y, or J when exact formulas are not available:

$$I_x = \sum y^2(\Delta A)$$
$$I_y = \sum x^2(\Delta A) \tag{7-5}$$

The smaller the elements selected, the more accurate will be the approximation.

EXAMPLE 1: Determine the approximate value of I_x of the area shown in Fig. 7-2(a).

Solution: The area is divided into 10 strips parallel to the x-axis as shown in Fig. 7-2(a). All elements of area within each strip are assumed to be the same distance from the axis; the error in the computation lies in this assumption.

The area of the first strip, 3 sq in., is at a mean distance of 7.5 in. from the x-axis. The moment of inertia $(I_x)_1$ is

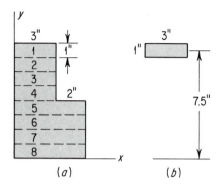

Figure 7-2

$$(I_x)_1 = \text{(area)} \times \text{(distance)}^2 = 3(1)(7.5)^2 = 168.75 \text{ in.}^4$$

The moments of inertia of the remaining strips are computed in a similar manner.

$$(I_x)_2 = 3(1)(6.5)^2 = 126.75 \text{ in.}^4$$
$$(I_x)_3 = 3(1)(5.5)^2 = 90.75 \text{ in.}^4$$
$$(I_x)_4 = 3(1)(4.5)^2 = 60.75 \text{ in.}^4$$
$$(I_x)_5 = 5(1)(3.5)^2 = 61.25 \text{ in.}^4$$
$$(I_x)_6 = 5(1)(2.5)^2 = 31.25 \text{ in.}^4$$
$$(I_x)_7 = 5(1)(1.5)^2 = 11.25 \text{ in.}^4$$
$$(I_x)_8 = 5(1)(0.5)^2 = 1.25 \text{ in.}^4$$

The moment of inertia I_x is the sum of the individual values

$$I_x = \sum (I_x)_n$$
$$I_x = 168.75 + 126.75 + 90.75 + 60.75 + 61.25 + 31.25 + 11.25 + 1.25$$
$$I_x = 552 \text{ in.}^4$$

An exact answer, obtained by methods that will be described later, gives a value of $I_x = 554.7 \text{ in.}^4$ The error in the approximation is less than 0.5 per cent.

The method of finding moments of inertia illustrated in Example 1 has particular value when the area of concern has an irregular shape. Formulas for the more common shapes, such as rectangles, triangles, and circles, are readily available in handbooks that deal with properties of areas.

7-2 Parallel Axis Theorem

The centroidal moment of inertia, that which is computed with reference to an axis through the centroid, can be used to find the value of moments of inertia about other parallel axes. Consider the area shown in Fig. 7-3, where the y-axis is assumed to pass through the centroid O. The moment of

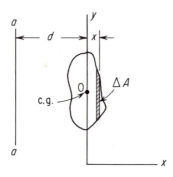

Figure 7-3

inertia of an element of area ΔA about axis a-a parallel to the y-axis is $(x + d)^2 \, \Delta A$, and the moment of inertia of the entire area about axis a-a is

$$I_{a-a} = \sum (x + d)^2 \, \Delta A$$

Expanding the bracketed term gives

$$I_{a-a} = \sum (x^2 + 2xd + d^2)\Delta A$$
$$= \sum x^2(\Delta A) + \sum 2xd(\Delta A) + \sum d^2 \Delta A$$
$$= \bar{I} + 2d\bar{x}A + Ad^2$$

The term $\bar{x}$ in the expression is zero, since the y-axis lies on the centroid of the area: the equation reduces to

$$I_{a-a} = \bar{I} + Ad^2 \tag{7-6}$$

where $\bar{I}$ is the moment of inertia with respect to the centroidal axis and Ad^2 is the product of the area and the square of its distance to the parallel axis. This expression is called the *parallel-axis theorem*.

EXAMPLE 2: The area of 25 sq in. shown in Fig. 7-4 has a centroidal moment of inertia of $\bar{I}_y = 60$ in.[4] Determine the moment of inertia with respect to line a-a, 10 in. to the left of the y-axis.

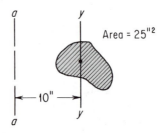

Figure 7-4

Solution: The moment of inertia is given by Eq. (7-6).

$$I_{a-a} = \bar{I}_y + Ad^2$$
$$= 60 + 25(10)^2 = 2560 \text{ in.}^4$$

7-3 Radius of Gyration

The *radius of gyration* is a number, symbolized by the letter k, which, when squared and multiplied by the area, equals the moment of inertia of that area.

$$I = Ak^2 \qquad (7-7)$$

As with the moment of inertia, it is difficult to attach a physical meaning to the radius of gyration of an area; it is simply a term that is frequently used in mechanics.

EXAMPLE 3: The 12-in. by 5-in. I beam shown in Fig. 7-5 weighs 35 lb per lineal ft. $\bar{I}_x$ and $\bar{I}_y$ are 227.0 in.^{4} and 10.0 in.^{4} respectively, and the cross-sectional area of the beam is 10.2 sq in. Determine the radius of gyration with respect to the x-axis and with respect to the y-axis.

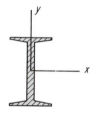

Figure 7-5

Solution: The data are substituted into Eq. (7-7).

$$k = \sqrt{\frac{\bar{I}}{A}}$$

$$k_x = \sqrt{\frac{227}{10.2}} = 4.72 \text{ in.}$$

$$k_y = \sqrt{\frac{10}{10.2}} = 0.99 \text{ in.}$$

7-4 Composite Areas

Formulas can be derived which give the exact value of the moment of inertia for the various geometric shapes. Several of these are listed in Table 7-1; others, particularly those for structural sections, can be found in handbooks which list the properties of geometric sections.

Table 7-1. Moments of Inertia of Areas

Square:	$I_x = I_y = \frac{1}{12}a^4$
Rectangle:	$I_x = \frac{1}{12}bh^3$ $I_y = \frac{1}{12}hb^3$
Triangle:	$I_x = \frac{1}{36}bh^3$
Circle:	$I_x = I_y = \frac{1}{4}\pi r^4$ $J_o = \frac{1}{2}\pi r^4$
Quarter circle:	$I_x = I_y = \frac{1}{16}\pi r^4$

When an area is composed of two or more simple areas, its moment of inertia is the sum of the individual values all with respect to a common reference line. The parallel-axis theorem is employed in this type of problem to transfer the individual inertias to a common reference line.

The examples that follow illustrate both the use of the table and the method of determining the moment of inertia of composite areas.

EXAMPLE 4: Determine the values of the centroidal moments of inertia $\bar{I}_x$ and $\bar{I}_y$ for the area shown in Fig. 7-6.

Solution: The moment of inertia of the composite area is the sum of the moments of inertia of its parts, each found with reference to the centroidal axis of the com-

Structural Shapes

		Size (in.)	Weight per ft (lb)	Area (in.²)	w (in.)	d (in.)	$\bar{x}$ (in.)	I_x (in.⁴)	I_y (in.⁴)
Wide flange beam:		12×12	120	35.31	12.32	13.12	0	1072	345
		10×10	100	29.43	10.35	11.12	0	625	207
Channel:		12×3	30	8.79	3.17	12.00	0.68	161.2	5.2
		6×2	13	3.81	2.16	6.00	0.52	17.3	1.1
Equal leg angle:		$6 \times 6 \times \frac{1}{2}$	19.6	5.75	6	6	1.68	19.9	19.9
		$3 \times 3 \times \frac{3}{8}$	7.2	2.11	3	3	0.89	1.8	1.8

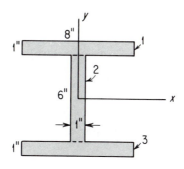

Figure 7-6

posite area before addition. In this example, the centroid is found simply by inspection.

The y-axis passes through the centroid of each of the three rectangles. The formulas from Table 7-1 can be used directly to find $\bar{I}_y$.

$$\bar{I}_y = (\bar{I}_y)_1 + (\bar{I}_y)_2 + (\bar{I}_y)_3$$

Hence

$$I_y = \tfrac{1}{12}(1)(8)^3 + \tfrac{1}{12}(6)(1)^3 + \tfrac{1}{12}(1)8^3$$
$$= 42.67 + 0.5 + 42.67$$
$$= 85.84 \text{ in.}^4$$

To find I_x, the parallel-axis theorem must be employed to transfer the inertias of areas A_1 and A_3 to the centroidal x-axis.

$$(I_x)_1 = \bar{I}_1 + Ad^2$$
$$= \tfrac{1}{12}(8)(1)^3 + 8(1)(3.5)^2$$
$$= 0.67 + 98.00$$
$$= 98.67 \text{ in.}^4$$

$(I_x)_1$ and $(I_x)_3$ are numerically equal to one another, since the areas A_1 and A_3 are the same distance from the centroidal x-axis.

$(I_x)_2$ can be found directly by formula,

where
$$(I_x)_2 = \tfrac{1}{12}bh^3 = \tfrac{1}{12}(1)(6)^3$$
$$(I_x)_2 = 18.00 \text{ in.}^4$$

The moment of inertia of the composite area about the centroidal x-axis is, then, the sum of the three component inertias.

$$(\bar{I}_x) = (I_x)_1 + (I_x)_2 + (I_x)_3$$
$$= 98.67 + 18.00 + 98.76$$
$$= 215.34 \text{ in.}^4$$

EXAMPLE 5: A rectangular plate 8 in. high by 6 in. wide has a hole 4 in. in diameter drilled through as shown in Fig. 7-7. Determine the moment of inertia of the plate with reference to axis a-a.

Solution: The missing area is assumed to contribute a negative moment of inertia to the section.

The moment of inertia of the rectangle, transferred to axis a-a, is

$$(I_{a-a})_1 = \tfrac{1}{12}bh^3 + Ad^2$$
$$= \tfrac{1}{12}(6)(8)^3 + 6(8)(4)^2$$
$$= 256 + 768 = 1024 \text{ in.}^4$$

The moment of inertia of the circle, also transferred to axis a-a, is

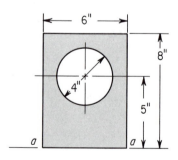

Figure 7-7

$$(I_{a-a})_2 = \frac{\pi r^4}{4} + Ad^2$$

$$= \frac{\pi(2)^4}{4} + \pi(2)^2(5)^2$$

$$= 12.56 + 314.00$$

$$= 326.56 \text{ in.}^4$$

The moment of inertia of the composite section is

$$I_{a-a} = (I_{a-a})_1 - (I_{a-a})_2$$

$$= 1024 - 326.56$$

$$= 697.44 \text{ in.}^4$$

7-5 Moment of Inertia of Bodies

The mathematical expression $\sum (\Delta W/g)r^2$ is frequently encountered in problems concerned with rotation of solid bodies. The term [weight]/[gravity] is called the *mass* of the body and for this reason the mathematical expression is referred to as the *mass moment of inertia*. The equation indicates a process of summation of the products of the individual elements of mass and the square of their distance from the particular reference axis.

The term W/g is called a *slug* and has the dimensions

$$\frac{W}{g} = \frac{\text{lb}}{\text{ft/sec}^2}$$

Mass moment of inertia has the dimensions, therefore, of slug ft^2 or ft lb sec^2

$$I = \frac{W}{g}r^2 = \frac{\text{lb}}{\text{ft/sec}^2} \times \text{ft}^2 = \text{ft lb sec}^2$$

7-6 Radius of Gyration of Bodies

The *radius of gyration* with reference to an inertia axis is a number k, which, when squared and multiplied by the mass of a solid body, is equal to the moment of inertia of the body with reference to the same inertia axis.

$$I = mk^2 = \frac{W}{g}k^2 \qquad (7\text{-}8)$$

A physical meaning can be attached to mass radius of gyration; it is the radius of a ring of concentrated mass which has the same moment of inertia as the body in question.

EXAMPLE 6: A solid body weighing 64.4 lb has a mass moment of inertia about a particular axis of 10 slug ft^2. Determine the radius of gyration of the body about the same inertia axis.

Solution: The mass, based upon the accepted value of gravity, 32.2 ft per sec², is

$$m = \frac{W}{g} = \frac{64.4}{32.2} = 2 \text{ slugs}$$

Substitution of the data into Eq. (7-8) gives the value of the radius of gyration:

$$k = \sqrt{\frac{I}{m}} = \sqrt{\frac{10 \text{ slug ft}^2}{2 \text{ slug}}} = 2.24 \text{ ft}$$

7-7 Transfer of Inertia Axis

Just as with areas, the mass moments of inertia can be transferred from a centroidal axis to some prescribed axis by the parallel-axis theorem. In this instance, mass replaces area in the formula.

$$I = \bar{I} + md^2 \qquad (7\text{-}9)$$

This equation, together with the values of centroidal moments of inertia of solid bodies given in Table 7-2, permits finding the moment of inertia of a body about a variety of axes. The example that follows illustrates the method.

EXAMPLE 7: Determine the mass moment of inertia of a solid right-circular cylinder of radius r about an axis parallel to the longitudinal axis and at a distance r from it.

Solution: Table 7-2 gives the value of the centroidal moment of inertia as

$$\bar{I} = \tfrac{1}{2} mr^2$$

By the parallel-axis theorem, the desired value of I is

$$I = \bar{I} + md^2$$
$$= \tfrac{1}{2}mr^2 + mr^2 = \tfrac{3}{2}mr^2$$

7-8 Moment of Inertia of Composite Bodies

The mass moments of inertia of several geometric bodies are given in Table 7-2. These, together with the parallel-axis theorem, are used to compute the mass moments of inertia of composite bodies in a manner similar to that employed for composite areas. Moments of inertia of the components are computed with reference to a common axis and then added; the absence of a portion of the body merely indicates a negative contribution to the moment of inertia of the system.

EXAMPLE 8: Determine the moment of inertia of the brass collar shown in Fig. 7-8 about the geometric axis *o-o*. Brass has a specific weight of 0.31 lb per cu in.

Table 7-2. Moments of Inertia of Uniform Masses

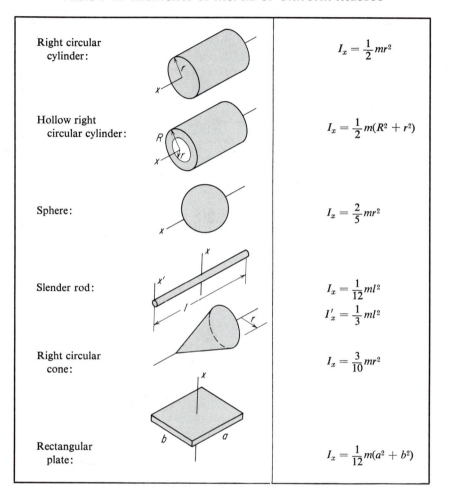

Right circular
cylinder:

$$I_x = \frac{1}{2}mr^2$$

Hollow right
circular cylinder:

$$I_x = \frac{1}{2}m(R^2 + r^2)$$

Sphere:

$$I_x = \frac{2}{5}mr^2$$

Slender rod:

$$I_x = \frac{1}{12}ml^2$$

$$I'_x = \frac{1}{3}ml^2$$

Right circular
cone:

$$I_x = \frac{3}{10}mr^2$$

Rectangular
plate:

$$I_x = \frac{1}{12}m(a^2 + b^2)$$

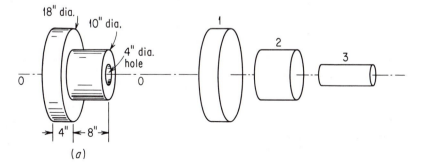

18" dia.

10" dia.

4" dia.
hole

4" 8"

(a)

Figure 7-8

Solution: Three cylindrical elements, each with a different mass, make up the collar: the section 18 in. in diameter, the section 10 in. in diameter, and the hole 4 in. in diameter.

Before the formulas from Table 7-2 can be used, the mass in slugs of each element must be computed.

Where:
$$m = \frac{W}{g} = \frac{[\text{volume}] \times [\text{specific weight}]}{\text{gravity}}$$

a. The 18-in. cylinder:

$$m_1 = \frac{[\pi(9)^2 4] \text{ in.}^3 [0.31] \text{ lb/in.}^3}{32.2 \text{ ft/sec}^2} = 9.79 \text{ slugs}$$

from Table 7-2 $I = \frac{1}{2}mr^2$

$$I_1 = \frac{1}{2}(9.79)(\tfrac{9}{12})^2 = 2.75 \text{ slug ft}^2$$

b. The 10-in. cylinder:

$$m_2 = \frac{[\pi(5)^2 8] \times [0.31]}{32.2} = 6.05 \text{ slugs}$$

$$I_2 = \frac{1}{2}mr^2 = \frac{1}{2}(6.05)(\tfrac{5}{12})^2 = 0.53 \text{ slug ft}^2$$

c. The 4-in.-diameter hole:

$$m_3 = \frac{[\pi(2)^2 12] \times [0.31]}{32.2} = 1.45 \text{ slugs}$$

$$I_3 = \frac{1}{2}mr^2 = \frac{1}{2}(1.45)(\tfrac{2}{12})^2 = 0.02 \text{ slug ft}^2$$

The moment of inertia of the composite section is equal to the sum of the moments of inertia of its elements:

$$I = I_1 + I_2 - I_3$$
$$= 2.75 + 0.53 - 0.02 = 3.26 \text{ slug ft}^2$$

QUESTIONS, PROBLEMS, AND ANSWERS

7-1. Determine I_x, I_y, and J of 1 sq. in. of area located in the x-y plane at $x = 5$ in. $y = 10$ in.

Ans. 100 in.[4]; 25 in.[4]; 125 in.[4].

7-2. Determine I_x, I_y, and J of an element of area of 2 sq. in. located in the x-y plane at 6, −3.

Ans. 18 in.[4]; 72 in.[4]; 90 in.[4].

7-3. One sq. in. of area in the x-y plane has a polar moment of inertia of 169 in.[4] and a moment of inertia with respect to the x-axis of 25 in.[4] Find I_y.

Ans. 144 in[4].

7-4. Determine by approximation the second moment I_x about the base of the rectangular area shown. Divide the area into 6 equal strips for the computation.

Ans. 643.5 in.⁴.

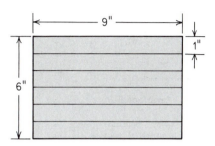

Problem 7-4

7-5. The exact answer for I_x in Prob. 7-4 is 648 in⁴. What is the per cent of error in the approximation?

Ans. 0.694 per cent.

7-6 to 7-8 Determine by approximation the value of I_x about the horizontal base line of the area shown. Use 1-in.-high horizontal strips of area.

Ans. **7-6.** *Hint:* Exact value is 205.3 in.⁴.

7-9. Use 1-in. strips to approximate the value of I_x and I_y of the area illustrated.

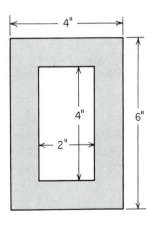

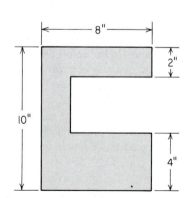

Problem 7-6   *Problem 7-7*

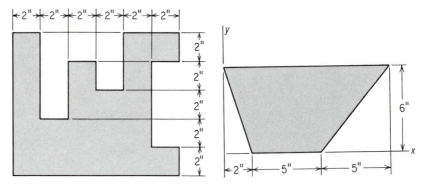

Problem 7-8   Problem 7-9

7-10. If the centroidal moments of inertia of the 10-sq-in. area illustrated are $\bar{I}_x = 25$ in.^{4} and $\bar{I}_y = 40$ in.^{4}, determine the moments of inertia about the x'-axis and y'-axis.

Ans. 665 in.^{4}; 290 in.^{4}.

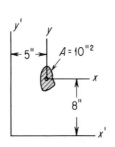

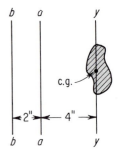

Problem 7-10   Problem 7-11

7-11. In the figure, the moments of inertia of the area about axes $a\text{-}a$ and $b\text{-}b$ are 250 in.^{4} and 450 in.^{4} respectively. Determine the centroidal moment of inertia $\bar{I}_y$ and the quantity of area A in the element.

Ans. 90 in.^{4}; 10 in.^{2}.

7-12. The centroidal moment of inertia of the element of area shown is $\bar{I}_y = 50$ in.^{4} and the moment of inertia about a parallel axis $a\text{-}a$ is $I_{a\text{-}a} = 400$ in.^{4} Determine the amount of area in the element.

Ans. 3.5 in.^{2}.

7-13. The centroidal moment of inertia of the 10-sq-in. area illustrated is $\bar{I}_x = 40$ in.^{4} Determine the distance d for which $I_{a\text{-}a} = 120$ in.^{4}

Ans. 2.83 in.

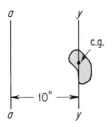

Problem 7-12

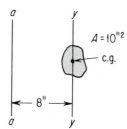

Problem 7-13

7-14. Determine the second moment of an area of 40 sq in. if the radius of gyration of the area is $k = 8$ in.

Ans. 2560 in.^{4}.

7-15. Find the centroidal radius of gyration of an area $A = 15$ in.^{2} with respect to the x-axis if the area has a moment of inertia of $I_x = 60$ in.^{4}

Ans. 2 in.

7-16. Find the polar radius of gyration of the area described in Prob. 7-2.

Ans. 6.71 in.

7-17. Determine the radius of gyration with reference to the x-axis of the area described in Prob. 7-1.

Ans. 10 in.

7-18. As shown, the radius of gyration of the 10-sq-in. area about axis a-a is $k = 11$ in. Determine the centroidal moment of inertia I_y.

Ans. 570 in.^{4}

Problem 7-18

Note: Refer to tables in the appendix for exact values of moments of inertia where required in the problems that follow.

7-19. A rectangle is 4 in. wide by 12 in. high. Find its moment of inertia with respect to a line drawn along the short side.

Ans. 2304 in.⁴.

7-20. Determine the centroidal polar moment of inertia of a rectangle 5 in. wide by 7 in. high.

Ans. 216 in.⁴.

7-21. Find the polar moment of inertia of a 10-in. by 12-in. rectangle with respect to a line perpendicular to the plane of the rectangle and passing through a corner.

Ans. 9760 in.⁴.

7-22. The rectangular plate section illustrated has a square hole cut from it. Determine the moment of inertia of the plate with respect to the centroidal x-axis and with respect to axis a-a.

Ans. 392 in.⁴; 1840 in.⁴.

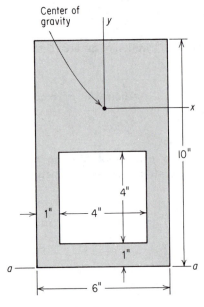

Problem 7-22

7-23. Find the radius of gyration of the plate described in Prob. 7-22 about an axis perpendicular to the plate at the center of gravity.

Ans. 3.54 in.

7-24. Locate the centroid of the area shown and find the moment of inertia with respect to the $\bar{x}$-axis.

Ans. 315 in.⁴.

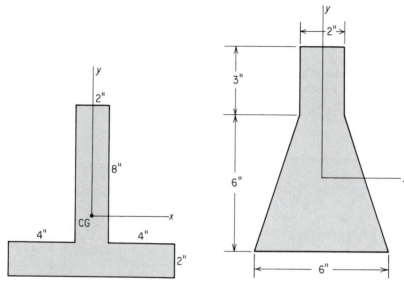

Problem 7-24   **Problem 7-25**

7-25. Locate the centroid of the area illustrated and find the moment of inertia with respect to the $\bar{x}$-axis.

Ans. 190.5 in.^{4}.

7-26. Determine the polar moment of inertia of the area shown with respect to its centroidal axis.

Ans. 118 in.^{4}.

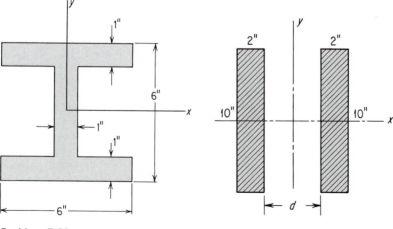

Problem 7-26   **Problem 7-27**

7-27. Find the distance d between the two rectangular areas illustrated in order that $\bar{I}_x = \bar{I}_y$.

Ans. 3.66 in.

7-28. Three 2-in. by 10-in. timbers are nailed together to form a tee-beam as shown. Locate the center of gravity of the cross-section and determine $\bar{I}_x$ and $\bar{I}_y$.

Ans. 820 in.^{4}; 340 in.^{4}.

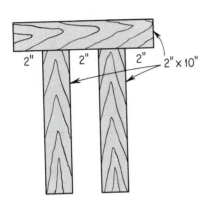 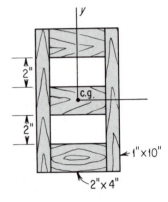

Problem 7-28   *Problem 7-29*

7-29. A box-beam is fabricated from two 1-in. by 10-in. and three 2-in. by 4-in. planks. Determine the centroidal moments of inertia $\bar{I}_x$ and $\bar{I}_y$ of the cross-section shown.

Ans. 431 in.^{4}; 159 in.^{4}.

7-30. Find the moment of inertia $\bar{I}_y$ of the hexagonal area shown.

Ans. 701.5 in.^{4}.

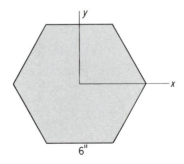

Problem 7-30

7-31. Two 12-in. by 120-lb/ft-wide flange beams are welded to form the section shown. Determine the centroidal moments of inertia and the centroidal radius of gyration.

Ans. $I_x = 5180$ in.^{4}; $k_x = 8.60$ in.

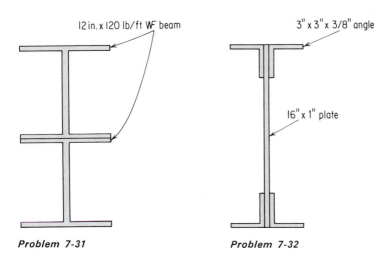

Problem 7-31   Problem 7-32

7-32. An *I* beam is fabricated by welding 4 angles to a plate as shown. Determine the centroidal moments of inertia of the section.

Ans. $I_x = 775$ in.^{4}; $I_y = 24.84$ in.^{4}.

7-33. A box-beam is made by welding a 10-in. by $\frac{1}{2}$-in. cover plate to two 12-in. by 3-in. channels as shown. Find $\bar{I}_x$, $\bar{I}_y$, and $\bar{J}$ of their combined area.

Ans. 713 in.^{4}; 422 in.^{4}; 1135 in.^{4}.

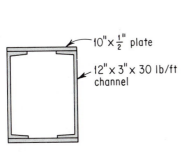

Problem 7-33

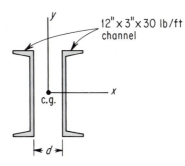

Problem 7-34

7-34. Find the spacing d between the two 12-in. by 3-in. channels illustrated in order that $I_x = I_y$.

Ans. 7.06 in.

7-35. Determine the moments of inertia of the area shown with respect to the x- and y-axis.

Ans. 7814 in.^{4}; 1161 in.^{4}.

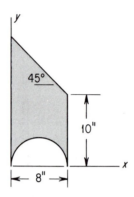

Problem 7-35

7-36. Find the radius of gyration of a 4-slug mass that has a moment of inertia of 16 slug ft^2.

Ans. 2 ft.

7-37. Determine the mass moment of inertia of a 64.4-lb right-circular cylinder about axis a-a. The cylinder, shown in the figure, has a diameter of 1 ft.

Ans. 0.75 slug ft^2.

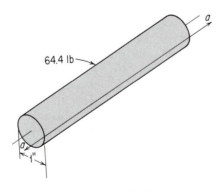

Problem 7-37

7-38. Find the mass moment of inertia about a diametral axis of an aluminum sphere with a diameter of 12 in. The specific weight of aluminum is 160 lb per cu ft.

Ans. 0.26 slug ft².

7-39. A pendulum consisting of two aluminum spheres welded to a 20-lb uniform slender rod is shown. Determine the mass moment of inertia of the pendulum about axis *O-O*. The specific weight of aluminum is 160 lb per cu ft.

Ans. 33.7 slug ft².

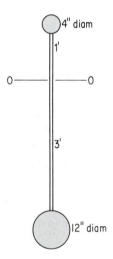

Problem 7-39

7-40. A 10-in. by 8-in. by 1-in. plate weighs 0.30 lb per cu in. Find the mass moment of inertia of the plate about an axis perpendicular to the face of the plate through one corner.

Ans. 0.283 slug ft².

7-41. Determine the moment of inertia of a 128.8-lb right-circular disk about an axis perpendicular to the centroidal longitudinal axis. The disk has a diameter of 18 in. Its thickness may be considered negligible.

Ans. 0.563 slug ft².

7-42. A slender rod is 4 ft long and weighs 32.2 lb. Compute the value of the mass moment of inertia of the rod about an axis on the rod that is perpendicular to it and 1 ft from its centroid.

Ans. 2.33 slug ft².

7-43. Find the centroidal radius of gyration of the hollow steel cylinder shown. The specific weight of steel is 0.28 lb per cu in.

Ans. 4.48 in.

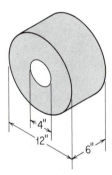

Problem 7-43

7-44. A 6-in.-square bar weighing 0.30 lb per cu in, has two 3-in.-diameter holes drilled through as shown. Determine the mass moment of inertia about a longitudinal centroidal axis *x-x*.

Ans. 0.15 slug ft². *Hint:* Use a handbook to determine the moment of inertia of the short cylinder.

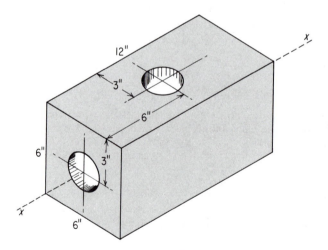

Problem 7-44

7-45. The rectangular plate, made of a material weighing 0.25 lb per cu in., has four holes drilled through as shown. Find the mass moment of inertia of the plate about a centroidal axis normal to the plate.

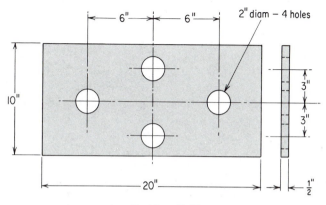

Problem 7-45

Ans. 0.22 slug ft². *Hint:* Use tables and transfer-of-axis formulas.

7-46. A pulley, made of material weighing 0.30 lb per cu in., has the cross-section shown. Assume the rim, hub, and web to be hollow cylinders and compute the mass moment of inertia of the pulley about its polar axis.

Ans. 9.33 slug ft².

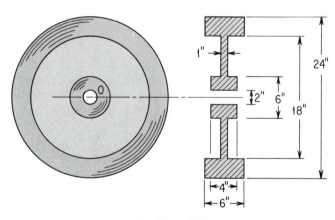

Problem 7-46

Chapter 8

CONCEPT OF STRESS

Mechanics of materials is a science which relates the physical properties of matter to the geometric properties of form. It is a science born of mechanics and developed through intuition, reasoning, and experimentation. In its many applications it provides a "first step" in the design of bodies and structures which must resist deformation as they transmit force from one point to another.

8-1 Internal Reactions

The rigid body emphasized throughout the study of engineering mechanics provided a means to an end in the analysis of action and reaction of forces. The mere fact that such a body had never existed in nature in no way influenced computations, since only forces acting on the body and not within the body were involved. Consider, for example, two bars that each support two 100-lb weights, as shown in Fig. 8-1. By rigid-body analysis the two situations are equivalent; a 200-lb reactive force is required in both cases to balance the external load. From the point of view of deformation, however, the two bars are far from equal: every particle of matter in the first bar helps, equally, to support the total weight, and every particle of matter, therefore, deforms equally. In contrast, forces in the second bar vary with position, as the free-body diagrams indicate. The deformation in the two segments would, therefore, be different.

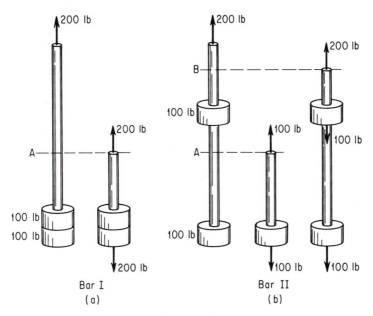

Bar I   Bar II
(a)   (b)

Figure 8-1

It will be apparent, as the theory is developed, that many of the concepts considered valid in rigid-body mechanics do not apply to deformable bodies since *forces and moments that act within bodies depend upon the locations as well as the magnitudes of the external forces and moments.* This simply means that couples cannot be transferred from place to place nor forces moved, at will, along their lines of action without changing the character of the deformation.

Internal reactions can be classified into five distinct groups: tensile or compressive forces, shear forces, bearing forces, moments, and torques. Typical force systems that illustrate each type of internal reaction are shown in Fig. 8-2. Although the illustrations depict a single reaction in each instance,

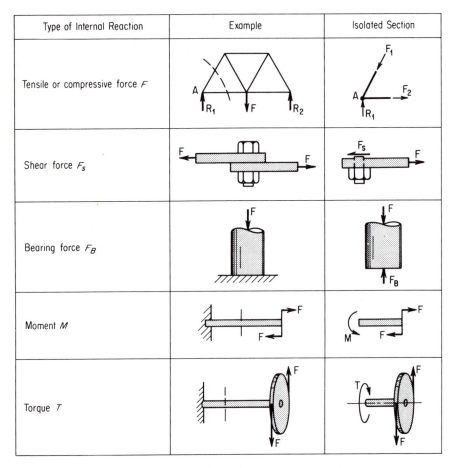

Type of Internal Reaction	Example	Isolated Section
Tensile or compressive force F		
Shear force F_s		
Bearing force F_B		
Moment M		
Torque T		

Figure 8-2

the possibility exists that these reactions can act in combination in response to more sophisticated external force systems. The solved examples that follow will serve as a guide to the method of finding internal reactions.

EXAMPLE 1. The crank shown in Fig. 8-3(a) is rigidly fastened to a support at A. Determine the internal forces, moments, and torques that act within the member on a plane normal to the axis at point B.

Solution: A free-body diagram of a portion of the crank cut at B is drawn as shown in Fig. 8-3(b). Three reactive components are necessary to maintain equilibrium: the force F_z, a moment M_y, and a torque $T_{\underline{x}}$. Numerically, these are equal to

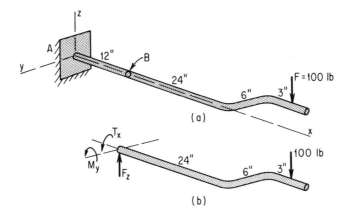

Figure 8-3

$$\sum F_z = 0$$
$$F_z - 100 = 0$$
$$F_z = 100 \text{ lb}$$
$$\sum M_y = 0$$
$$M_y - 100(24 + 3) = 0$$
$$M_y = 2700 \text{ lb in.}$$

and

$$\sum M_x = 0$$
$$T_x - 100(6) = 0$$
$$T_x = 600 \text{ lb in.}$$

EXAMPLE 2. The column of Fig. 8-4(a) supports a uniformly distributed horizontal load and a concentrated vertical load as shown. Determine the reactions within the column on a plane perpendicular to the axis at A.

Solution: Three reactions at A are exposed by the free-body diagram pictured in Fig. 8-4(b). These reactions consist of a shear force A_x, a compressive force A_y, and a moment M perpendicular to the x-y plane. The equations of static equilibrium are used to evaluate these reactions.

$$\sum F_x = 0$$
$$A_x = -1000 \text{ lb}$$
$$\sum F_y = 0$$
$$A_y = 5000 \text{ lb}$$
$$\sum M_{xy} = 0$$
$$M - 250(4)9 - 5000(3) = 0$$
$$M = 24{,}000 \text{ lb in.}$$

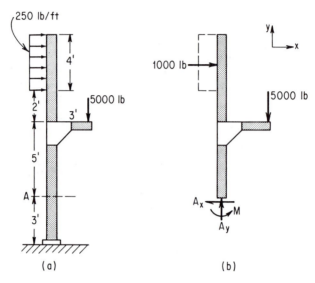

Figure 8-4

EXAMPLE 3. The post pictured in Fig. 8-5(a) supports a compressive load of
$P = 5$ kips. Find the internal force components that act normal and tangent to plane
$ABCD$, which cuts the post at an angle of 30 deg with the vertical.

Solution: The free-body diagram and the accompanying closed polygon of forces
are shown in Fig. 8-5(b). Solving the vector diagram for the desired forces gives

$$F_n = 5 \sin 30° = 5(0.5)$$
$$F_n = 2.5 \text{ kips}$$

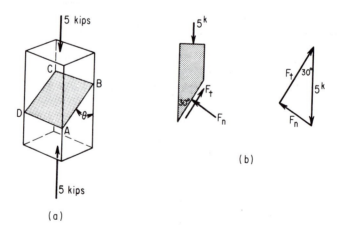

Figure 8-5

and

$$F_t = 5 \cos 30° = 5(0.866)$$
$$F_t = 4.33 \text{ kips}$$

8-2 Stress

Stress, like pressure, is a term used in mechanics of materials to describe the *intensity of a force*—the quantity of force that acts on a unit of area. To say, for instance, that a particular piece of steel can withstand a tensile stress of 60,000 psi simply means that every square inch of cross-sectional area can support 60,000 lb in tension. The inference in the preceding statement is that every portion of the area supports an equal share of the load; the stress, in other words, is *uniform*.

If, for some reason, the stress varies from point to point over a given area within a body, the stress is referred to as *non-uniform stress*. It will be shown, as the theory is developed, that there are many more cases of non-uniform stress than of uniform stress.

8-3 Axial, Shear, and Bearing Stresses

Stress can be classified in accord with the internal reaction that produces it. An axial tensile or compressive force, as shown in Fig. 8-6(a), produces a tensile or compressive stress. This type of stress can be thought of as being caused by a longitudinal "push" or "pull." Mathematically, the average value of the axial stress, represented by the Greek letter σ, is the ratio of the force to the area:

$$\sigma = \frac{F}{A} \tag{8-1}$$

Shear stress, the second type of stress illustrated, is caused by a force that acts at right angles to the axis of the member. This form of stress acts parallel to the cross-sectional area and has an average value of

$$\tau = \frac{F}{A_s} \tag{8-2}$$

where the Greek letter τ represents shear stress, F the force causing shear, and A_s the area being sheared.

The third fundamental type of stress, the bearing stress, is actually a pressure, since it represents the intensity of force between a body and its support. Like the previous two stresses, the average bearing stress is defined in terms of force and area:

$$\sigma_b = \frac{F}{A_b} \tag{8-3}$$

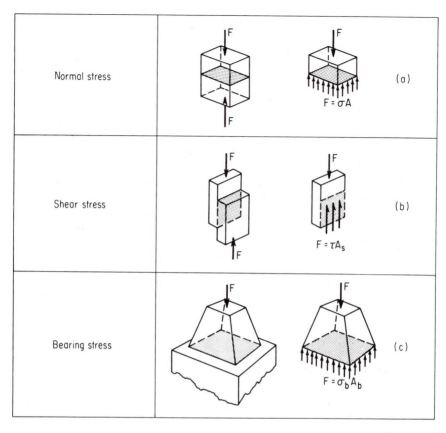

Normal stress		(a)
	$F = \sigma A$	
Shear stress		(b)
	$F = \tau A_s$	
Bearing stress		(c)
	$F = \sigma_b A_b$	

Figure 8-6

EXAMPLE 4. The clevis shown in Fig. 8-7(a) supports a load $P = 5$ tons. Determine: (a) the tensile stress in the circular section at A; (b) the tensile stress in the rectangular section at B; (c) the shear stress in the bolt.

Solution: Parts (a) and (b). The tensile stress in both the circular section and the rectangular section is given by Eq. (8-1). For the circular section:

$$\sigma = \frac{P}{A} = \frac{5(2000)}{\frac{\pi}{4}\left(\frac{3}{4}\right)^2} = 22{,}600 \text{ psi tension}$$

For the rectangular section:

$$\sigma = \frac{P}{A} = \frac{5(2000)}{0.5(2)} = 10{,}000 \text{ psi tension}$$

Part (c). The bolt is in *double shear;* two transverse areas, one to the right and the other to the left of the rectangular bar, help support the load. This is illustrated in Fig. 8-7(b). The shear stress, found through Eq. (8-2), is

$$\tau = \frac{P}{A_s} = \frac{5(2000)}{2\left(\frac{\pi}{4}\right)\left(\frac{1}{2}\right)^2} = 25{,}500 \text{ psi}$$

EXAMPLE 5. Angle clips welded to the column pictured in Fig. 8-8(a) support the loads indicated. The column, an 8-in., wide-flanged section weighing 35 lb per ft (8 WF 35), has a cross-sectional area of 10.3 sq in. Determine the axial stress at sections A, B, and C, and the bearing stress between the plate and the pedestal. Neglect, in each instance, the weight of the column.

Solution: Three free-body diagrams, each showing the force required to maintain internal equilibrium, are drawn as illustrated in Fig. 8-8(b). Equation (8-1) is used to compute the stress at each section.

At A: $\sigma_A = \dfrac{P}{A} = \dfrac{6(2000)}{10.3}$

$\qquad = 1170$ psi tension

At B: $\sigma_B = \dfrac{P}{A} = \dfrac{10(2000)}{10.3}$

$\qquad = 1940$ psi compression

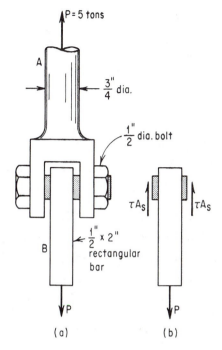

Figure 8-7

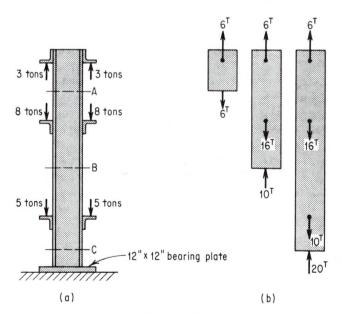

Figure 8-8

At C: $\sigma_c = \dfrac{P}{A} = \dfrac{20(2000)}{10.3} = 3880$ psi compression

The bearing stress, equivalent to the pressure between the bearing plate and the pedestal, is

$$\sigma_b = \frac{P}{A_b} = \frac{20(2000)}{12 \times 12} = 278 \text{ psi}$$

8-4 Stresses on Oblique Planes

Situations often arise in design when consideration must be given to stresses that act within a body on planes other than transverse planes. Consider a bar of cross-sectional area A, Fig. 8-9(a), to be acted upon by a tensile force P. The transverse stress, that on plane a-a, is equal to P/A. By selecting an oblique plane b-b inclined as shown in Fig. 8-9(c), a free-body can be drawn with the force P resolved into a component P_n normal to the plane and

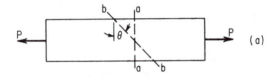

(a)

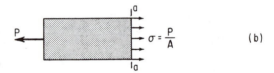

(b)

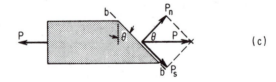

(c)

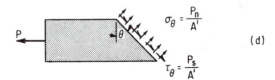

(d)

Figure 8-9

a component P_s parallel to the plane. Regardless of the inclination of the oblique plane, the vector sum of P_n and P_s must always be equal to the force P. These components, in terms of θ and P are

$$P_n = P \cos \theta$$

and

$$P_s = P \sin \theta$$

The stresses associated with these components, Fig. 8-9(d), can be found by dividing each by the magnitude of the oblique area A', which in terms of the transverse area A is

$$A' = \frac{A}{\cos \theta}$$

The resulting normal stress and shear stress are, respectively,

$$\sigma_\theta = \frac{P_n}{A'} = \frac{P \cos \theta}{\dfrac{A}{\cos \theta}} = \frac{P}{A} \cos^2 \theta \qquad (8\text{-}4)$$

and

$$\tau_\theta = \frac{P_s}{A'} = \frac{P \sin \theta}{\dfrac{A}{\cos \theta}} = \frac{P}{A} \sin \theta \cos \theta$$

Since

$$\sin \theta \cos \theta = \frac{\sin 2\theta}{2}$$

the shear stress can be written

$$\tau_\theta = \frac{P}{2A} \sin 2\theta \qquad (8\text{-}5)$$

By inspection, the normal stress is a maximum on the transverse plane, $\theta = 0$ deg, and a minimum on an axial plane, $\theta = 90$ deg.

$$\sigma_{\theta\,max} = \frac{P}{A} \cos 0° = \frac{P}{A}$$

and

$$\sigma_{\theta\,min} = \frac{P}{A} \cos 90° = 0$$

Similarly, the shear stress is a maximum at $2\theta = \pm 90$ deg or $\theta = \pm 45$ deg, and a minimum at $2\theta = 0$ deg.

$$\tau_{\theta\,max} = \frac{P}{2A} \sin (2 \times 45°) = \frac{P}{2A}$$

and

$$\tau_{\theta \min} = \frac{P}{2A} \sin (2 \times 0°) = 0$$

EXAMPLE 6. The rectangular bar of Fig. 8-10 is made of two pieces of steel solidly welded along a plane inclined at 30 deg as shown. Determine: (a) the maximum safe load P if the normal stress and the shearing stress are not to exceed 8000 psi and 4000 psi, respectively; (b) the ratio of the design load to the tensile strength of the bar based on a working stress of 25,000 psi in tension.

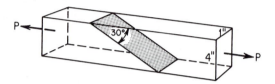

Figure 8-10

Solution: Part (a). Two values of the load P must be computed: one based on the allowable normal stress and the other on the allowable shearing stress. The smaller of these two values will be the design load P. From Eq. (8-4)

$$\sigma_\theta = \frac{P}{A} \cos^2 \theta$$

Solving for P gives

$$P = \frac{\sigma_\theta A}{\cos^2 \theta} = \frac{8000(4 \times 1)}{(\cos 60°)^2} = 128,000 \text{ lb}$$

Note: The angle θ in Eqs. (8-4) and (8-5) lies between a perpendicular drawn to the inclined surface and the axis of the bar.

Next, a value of the allowable load P' is computed on the basis of the maximum allowable shearing stress of 4000 psi; thus

$$\tau_\theta = \frac{P'}{2A} \sin 2\theta$$

$$P' = \frac{2\tau_\theta A}{\sin 2\theta}$$

where

$$P' = \frac{2\tau_\theta A}{\sin 2\theta} = \frac{2(4000)(4 \times 1)}{\sin 120°} = 37,000 \text{ lb}$$

Therefore the design load, the lesser value of P, is 37,000 lb.

Part (b). The tensile strength of the bar, based on a working stress of 25,000 psi is

$$P_{ts} = \sigma A = 25,000(4 \times 1) = 100,000 \text{ lb}$$

and the ratio of the allowable load to the strength of the bar is, therefore,

$$\frac{P'}{P_{ts}} = \frac{37,000}{100,000} = 0.37$$

This numerical value is called the *efficiency of the joint.*

8-5 Geometric Stress Concentration

An abrupt change in the geometry of a structural member, such as that caused by a hole, a notch, or a groove, results in a non-uniform stress pattern, as illustrated in Fig. 8-11. The maximum stress in each instance occurs at the boundary of the geometric discontinuity, and failure, particularly under dynamic loading, begins at these points of high localized stress.

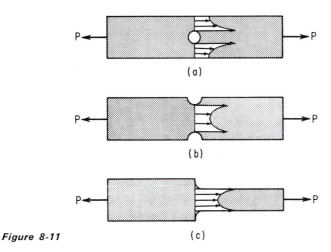

Figure 8-11

The problem of geometric stress concentration has attracted both the experimenter and the theorizer; their labors have produced data that enable one to predict with fair precision the maximum stress that will occur under a given condition. Data are usually presented in the form of curves that cover a large range of variables for a given type of geometrical discontinuity. Five typical curves[1] for members in tension are shown in Figs. 8-12, 8-13, 8-14, 8-15, and 8-16. In each a stress concentration factor k is given in terms of the critical dimensions of the member. The actual stress, then, in terms of k and the average stress P/A is

$$\sigma_{max} = k\sigma_{avg} = k\frac{P}{A} \tag{8-6}$$

EXAMPLE 7. Determine the safe axial load P for the rectangular bar shown in Fig. 8-17 if the tensile stress in the bar is not to exceed 15,000 psi.

Solution: Two regions of geometric stress concentration are present in the bar;

[1] For a complete discussion of stress concentration, see Charles Lipson. G.C. Noll, and L.S. Clock, *Stress and Strength of Manufactured Parts* (New York: McGraw-Hill Book Company, Inc., 1950.)

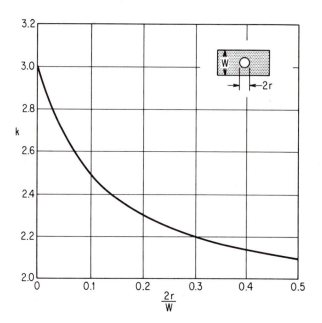

Figure 8-12

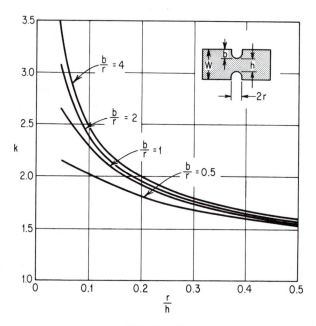

Figure 8-13

168

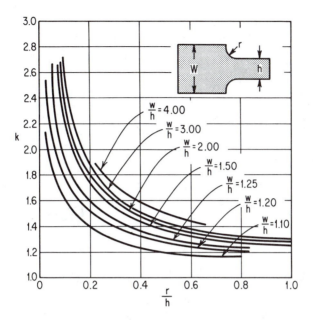

Figure 8-14

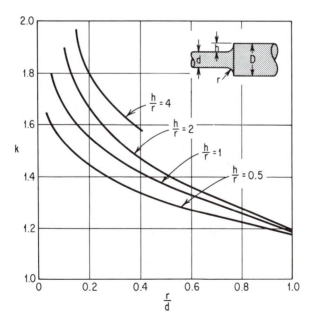

Figure 8-15

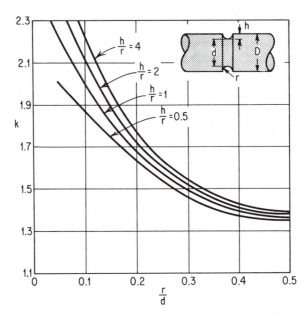

Figure 8-16

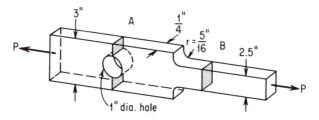

Figure 8-17

one near the hole at A and the other near the fillet at B. An allowable load must be determined for each section and the least value selected as the design load.

Section A. The average stress on a transverse plane through the hole is

$$\sigma_{\mathrm{avg}} = \frac{P}{A} = \frac{P}{(3-1)\frac{1}{4}} = 2P \text{ psi}$$

The stress concentration factor k from Fig. 8-12 is dependent upon the ratio of the diameter of the hole to the width of the plate.

$$\frac{2r}{W} = \frac{1}{3} = 0.333$$

The data—the given allowable stress, the computed average stress, and the approximate value of k from the graph—are substituted into Eq. (8-6).

$$\sigma_{\max} = k\sigma_{\mathrm{avg}}$$
$$15,000 = 2.18(2P)$$

Solving for P gives

$$P = \frac{15,000}{2(2.18)} = 3440 \text{ lb}$$

Section B. The average stress at B, the region of the fillet, is

$$\sigma_{avg} = \frac{P}{A} = \frac{P}{2.5(\frac{1}{4})} = 1.6 P$$

The ratios r/h and W/h are needed to determine k from the graph of Fig. 8-14.

$$\frac{r}{h} = \frac{\frac{5}{16}}{2.5} = 0.125$$

and

$$\frac{W}{h} = \frac{3}{2.5} = 1.2$$

The approximate value of k, read from the graph, is 1.7. Substitution of these data into Eq. (8-6) gives a second critical value of P; thus

$$\sigma_{max} = k\sigma_{avg}$$
$$15,000 = 1.7(1.6P)$$

where

$$P = \frac{15,000}{1.7(1.6)} = 5510 \text{ lb}$$

The least of the two values of P is the safe load; thus, $P = 3440$ lb, where the critical section is at the hole.

8-6 Stresses in Thin-Walled Cylindrical Vessels

As is often the case in the study of mechanics of materials, certain simplifying assumptions are made that enable one to arrive at a reasonably correct answer to what might be an exceedingly complex problem. Such is the case with thin-walled cylinders—cylinders whose wall thicknesses are less than $\frac{1}{20}$ of their diameters. The simplifying assumption in this instance is that the stress in the walls of the cylinder is uniform. Consider the closed cylindrical vessel subjected to an internal pressure p, as shown in Fig. 8-18(a). A free-body diagram of a portion of the cylinder cut about a circumferential line, Fig. 8-18(b), indicates an external pressure force P_l to act on the closed end and an internal axial tensile force T_l to act within the wall. To maintain equilibrium in the longitudinal direction, the tensile force and the pressure force must be equal.

$$T_l = P_l = p\frac{\pi D^2}{4} \tag{a}$$

The tensile force in terms of the average longitudinal stress across the severed circumferential area is

$$T_l = \sigma_l A = \sigma_l \pi Dt \tag{b}$$

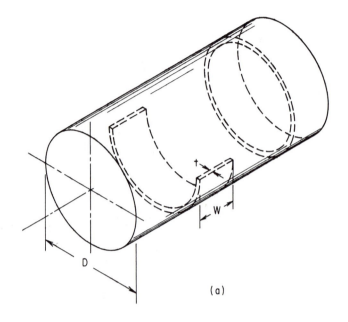

(a)

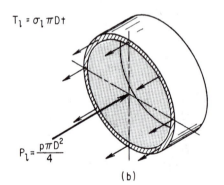

(b)

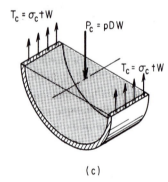

(c)

Figure 8-18

Equating Eqs. (a) and (b) gives

$$\sigma_l \pi D t = p\frac{\pi D^2}{4}$$

or

$$\sigma_l = p\frac{D}{4t} \tag{8-7}$$

To arrive at an expression for the circumferential stress, a second free-body is sketched as shown in Fig. 8-18(c). This is drawn to include the portion of the pressurized gas or liquid contained by this segment of the cylinder. The forces in the vertical direction include a pressure force P_c equal to the product of the internal pressure p and the area WD, and tensile forces on each of the two severed areas. To maintain equilibrium, then,

$$2T_c = P_c = pWD \tag{c}$$

The tensile force T_c in terms of the circumferential stress on the severed area is

$$T_c = \sigma_c A = \sigma_c Wt \tag{d}$$

Combining Eqs. (c) and (d) gives

$$2\sigma_c Wt = pWD$$

or

$$\sigma_c = \frac{pD}{2t} \tag{8-8}$$

Since the circumferential stress is twice the longitudinal stress, the cylinder would fail, burst, or explode, by splitting longitudinally.

An expression for the stress in a spherical shell subjected to an internal pressure p can be found by an approach similar to that used for the cylinder. A hemispherical free-body which includes the pressurized liquid or gas is drawn as shown in Fig. 8-19. To maintain equilibrium, the tensile force in

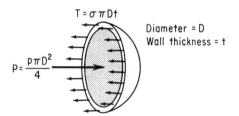

Figure 8-19

the wall must just equal the pressure force required to maintain the matter within the hemisphere.

$$T = p\frac{\pi D^2}{4} \tag{e}$$

The tensile force, in terms of the product of the stress and the severed area, is substituted in Eq. (e):

$$\sigma \pi D t = \frac{p\pi D^2}{4}$$

Simplifying gives

$$\sigma = \frac{pD}{4t} \tag{8-9}$$

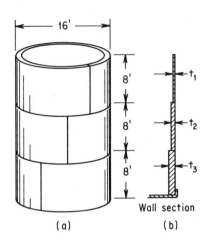

(a)

(b)

Wall section

Figure 8-20

EXAMPLE 8. A tank, 16 ft in diameter and 24 ft high, is filled to the top with a liquid weighing 160 lb per cu ft. If the tank is fabricated from three 8-ft-wide steel rings, as shown in Fig. 8-20(a), determine the proper thickness of each ring for the most economical construction. Assume that the tensile stress in the steel is not to exceed 5000 psi.

Solution: The pressure within the tank varies directly with the depth h.

$$p = \gamma h$$

where γ is the specific weight of the fluid:

$$\gamma = 160 \frac{\text{lb}}{\text{ft}^3} \times \frac{1 \text{ ft}^3}{1728 \text{ in.}^3}$$

$$= 0.0926 \text{ lb/in.}^3$$

Thus, the pressures at the 8 ft, 16 ft, and 24 ft marks are

$$p_8 = 0.0926(8 \times 12) = 8.89 \text{ psi}$$

$$p_{16} = 0.0926(16 \times 12) = 17.8 \text{ psi}$$

$$p_{24} = 0.0926(24 \times 12) = 26.7 \text{ psi}$$

The required thicknesses from Eq. (1-8) are

$$t_8 = \frac{pD}{2\sigma_c} = \frac{8.89(16 \times 12)}{2(5000)} = 0.085 \text{ in.}$$

$$t_{16} = \frac{17.8(16 \times 12)}{2(5000)} = 0.342 \text{ in.}$$

$$t_{24} = \frac{26.7(16 \times 12)}{2(5000)} = 0.769 \text{ in.}$$

8-7 Working Stress and Factor of Safety

In several of the problems that have served as examples the terms *working stress* and *allowable stress* have been used. These are values of stress that provide a margin of safety in the design. The need for this safety margin is apparent for many reasons: stress itself is seldom uniform; materials lack the homogeneous properties theoretically assigned to them; abnormal loads might occur; manufacturing processes often impart dangerous stresses within the component. These and other factors make it necessary to select working stresses substantially below those known to cause failure.

The phrase *factor of safety* is a term generally defined as the ratio of the stress necessary to produce failure to the working stress. Under this definition, a factor of safety of 3 would mean that the load could be increased 3 times before failure would occur.

QUESTIONS, PROBLEMS,
AND ANSWERS

8-1. An elastic rod of negligible weight hangs downward from a support as shown in Fig. (a). A downward force F is applied to the rod at point B and then at point C, as illustrated in the figure.

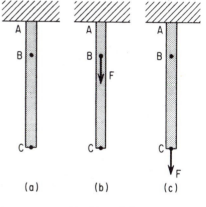

(a) Would the internal reaction at the support A to the external force be the same for either loading arrangement? (Yes or no?)

Ans. Yes.

(b) Would the internal reaction between points A and B to the external force be the same for either loading arrangement? (Yes or no?)

Ans. Yes.

(c) Would the internal reaction between points B and C be the same for either loading arrangement? (Yes or no?)

Ans. No.

(a)   (b)   (c)

Problem 8-1

(d) Answers to queries (a) and (b) were yes because an internal vertical force equal to F would be necessary to maintain equilibrium if the bar were "cut" between A and B. Why was the answer to question (c) no?

Ans. Because with the force applied at the very end, every particle within the rod must help support the force. With the force applied at B only, that portion above B supports the force.

(e) The word "elastic," a key word in the problem, implies that the rod will _____ when the force is applied.

Ans. Deform, stretch, or elongate.

(f) The total elongation or stretch will be greater with the force applied at C than at B. Why?

Ans. Because total length stretches in the first case as compared to only a portion stretching in the second.

(g) Suppose the rod is 20 in. long and A and B are 5 in. apart. How much more stretch would you expect with the force applied at C as compared to B?

Ans. Four times as much.

8-2. A bar of uniform cross section is shown. The long axis of the bar is represented by a dotted line. Cross-sectional planes within the bar are sometimes defined in terms of the angle that the plane makes with the long axis. Plane A is perpendicular to the long axis and is called a _____ plane. Plane C, parallel to the long axis, is referred to as an _____ plane. Plane B, which is neither normal nor axial, is an _____ plane.

Ans. Normal; axial; oblique.

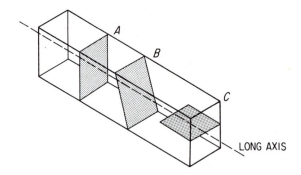

Problem 8-2

8-3. A force P is applied to an L-hook in each of three different ways, as shown. Below each figure is a free-body diagram showing the internal reactions, an axial force and a moment M. What are the magnitudes of F and M?

Ans. (a) $F = P$, $M = 0$.
     (b) $F = P$, $M = Pd$.
     (c) $F = P$, $M = 2Pd$.

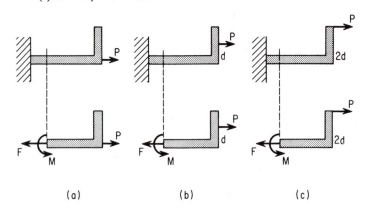

Problem 8-3

8-4. A force is applied to the hook as shown. What are the internal reactions at a section normal to the axis of the force at A; at B; at C?

Ans. A: $F \leftarrow$, $M = 0$; B: $F \leftarrow$, $M = FR$ $\searrow$; C: zero.

8-5. Determine the internal reactions at A on a plane normal to the axis of the lamp post. Each arm and lamp has a combined weight of 500 lb and a center of gravity at G. The column weighs 100 lb per ft.

Ans. 2500 lb $\uparrow$; 4330 lb ft $\nearrow$.

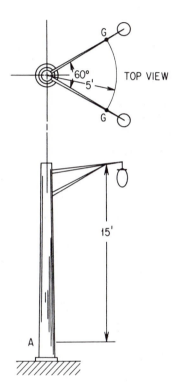

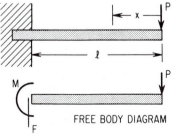

Problem 8-4   **Problem 8-5**

8-6. A cantilevered beam supports a concentrated load as shown; the beam is elastic and the support unyielding.

(a) When the load is applied, the beam will bend; the originally straight beam is now slightly curved with the radius of curvature pointing _____ .

Ans. Downward.

(b) With the load removed, the beam will return to its original position, providing that the beam is _____ .

Ans. Elastic.

Problem 8-6

(c) Two internal reactions would be necessary to maintain equilibrium in the free-body diagram. These reactions would be a(n) (upward, downward) force and a (clockwise, counterclockwise) moment.

Ans. Upward; counterclockwise.

(d) A non-yielding support implies that the slope of the beam at the support is _____.

Ans. Zero.

(e) Within the beam there is an internal counterclockwise moment and an upward force; the force is constant but the moment (increases, decreases) with x and has a magnitude of _____ for any x.

Ans. Increases; Px.

(f) The maximum moment occurs at x equal to _____. If the load were increased and the beam were to fail, we would expect the failure to occur at or near the _____.

Ans. l; support.

8-7. The simply supported beam is subjected to the loading shown. Determine the internal reactions within the beam at C.

Ans. 200 lb; 600 lb ft.

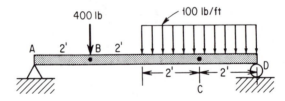

Problem 8-7

8-8. Solve Prob. 8-7 with the 100-lb-per-ft distributed load assumed to be concentrated at point C; account for any difference in the answers to the two problems.

Ans. The shear force can be found just to the right or just to the left of C, but not at C. $M = 800$ lb ft.

It is easier for the beam to carry the distributed load; with the load distributed the moment is also distributed.

8-9. Write an equation for the moment within the beam illustrated, first as a function of distance x measured from the left support, and next as a function of distance $(l - x)$ from the right support. What is the moment at midspan for both equations?

Ans. From left: $M = \dfrac{wlx}{2} - \dfrac{wx^2}{2}$; from right: $M = \dfrac{wl}{2}(l - x) - \dfrac{w}{2}(l - x)^2$;

at midspan: $M = \dfrac{wl^2}{8}$.

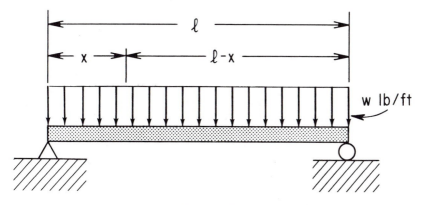

Problem 8-9

8-10. A concrete beam weighing 100 lb per lin ft is hoisted by the cable arrangement shown. Determine the internal reactions that act on a plane perpendicular to the axis of the beam at its centroid.

Ans. Force and moment are both zero.

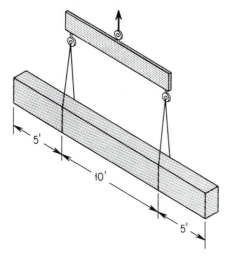

Problem 8-10

8-11. Solve Prob. 8-10, assuming that a single cable at the midpoint is used to hoist the beam.

Ans. Force just to right or just to left of midpoint equals 1000 lb. Moment at midpoint equals 5000 lb ft.

8-12. The torque arm shown supports a 1000-lb load. Determine the internal reactions within the arm: (a) as a function of x, (b) at x equal to 4 ft, (c) at a maximum value of x.

Ans. (a) Shear force = 1000 lb; torque = 2000 lb ft; moment = $1000 \, x$.
 (b) Shear force and torque as in (a); moment = 4000 lb ft.
 (c) Shear force and torque as in (a); moment = 5000 lb ft.

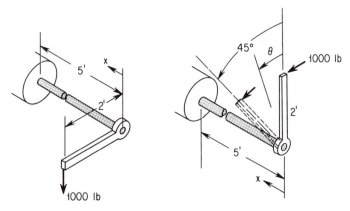

Problem 8-12   Problem 8-13

8-13. Two other positions of the torque arm of Prob. 8-12 are illustrated. Describe how the position of the arm as defined by the angle θ affects the moment and the torque within the shaft at any point.

Ans. The torque and moment are not affected as long as the force remains perpendicular to the 2-ft arm.

8-14. The vertical hanging bar shown weighs 150 lb per ft and carries a load *W*. Write an equation for the tensile force *F* in the column as a function of *y*, and sketch the curve of this function. Let the ordinate represent *F* and the abscissa *y*.

Ans. $F = W + 150(20 - y)$.

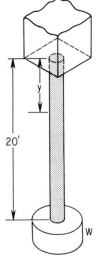

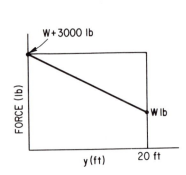

Problem 8-14   Problem 8-14 Answer

8-15. The jack-shaft shown is subjected to belt tensions. Determine the components of the internal reactions within the shaft at a section normal to the axis at C.

Ans. $C_x = 18.2$ lb; $C_y = 16.5$ lb; $M_x = 247$ lb in.; $M_y = 328$ lb in.; torque $= 320$ lb in.

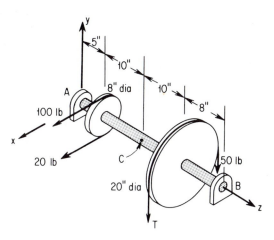

Problem 8-15

8-16. The circular bar illustrated is subjected to an axial tension force. Determine the components of force normal and tangent to a plane inclined at 45 deg to the axis of the bar.

Ans. $F_N = F_S = 707$ lb.

8-17. Determine the components of force, normal and tangential, that act on the plane inclined at 60 deg to the axis of the tie-bar AB, as shown.

Ans. $F_N = 4330$ lb; $F_T = 2500$ lb.

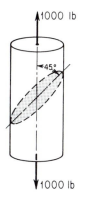

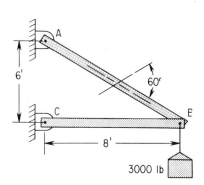

Problem 8-16   *Problem 8-17*

8-18. The plate illustrated is subjected to the actions shown. Determine the normal and tangential components of force that act on the diagonal plane.

Ans. $F_N = 520$ lb, $F_T = 360$ lb.

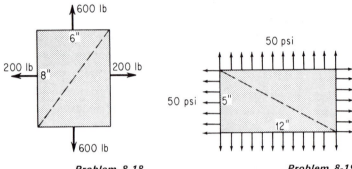

Problem 8-18   Problem 8-19

8-19. A uniform plate 5 in. thick is subjected to the edge pressure shown. Find the normal and tangential components of force that act on the diagonal plane.

Ans. $F_N = 3250$ lb, $F_T = 0$.

8-20. The truss illustrated supports a load of 5 tons. Determine the required cross-sectional area of each member if the stresses in tension and compression are not to exceed 20,000 psi and 10,000 psi, respectively.

Ans. By inspection: BH, BG, CG, DG, and DF carry zero load; $AC = CE = 0.833$ in.^{2}; $AE = 0.333$ in.^{2}.

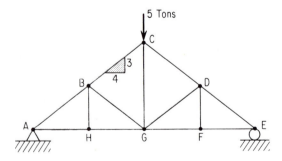

Problem 8-20

8-21. Determine the force F required to maintain equilibrium and the stresses in sections AB, BC, and CD of the bar shown.

Ans. AB: 9050 psi compression.
     BC: 30,600 psi tension.
     CD: 20,000 psi tension.

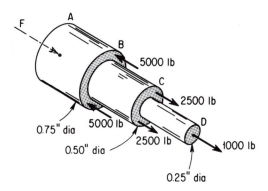

Problem 8-21

8-22. Write an equation for the tensile stress in the conic rod as a function of the distance x measured from the small end. Test your answer by letting $d = D$.

Ans. $\sigma = \dfrac{Fl^2}{\dfrac{\pi}{4}[(D - d)x + ld]^2}$; if $D = d$, $\sigma = \dfrac{F}{\dfrac{\pi d^2}{4}}$.

When the diameters are equal, the stress is constant throughout and equal to the force divided by the area.

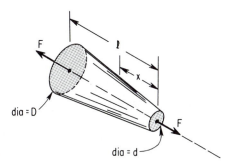

Problem 8-22

8-23. The flanged pivot shown sustains a compressive stress of 10,000 psi when it is acted upon by the force F. Determine the shearing stress between the flange and the cylinder, and the bearing stress between the flange and the support.

Ans. Shear: 20,000 psi; bearing: 2010 psi.

8-24. An adjustable ratchet bar used in machine setups is shown in the figure. Find the shearing stress and the bearing stress developed in the teeth. Assume that the teeth in contact share the load equally.

Ans. Shear: 4000 psi; bearing: 5330 psi.

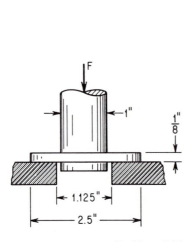

Problem 8-23   **Problem 8-24**

8-25. Find the maximum value of belt pull P and the resulting torque that can be transmitted to the 2-in.-diameter keyed shaft illustrated. The $\frac{1}{2}$-in.-square key is 3 in. long and has a limiting shear stress of 10,000 psi.

Ans. 2100 lb.

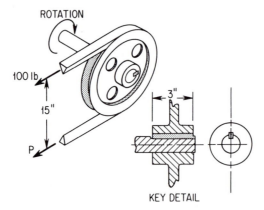

KEY DETAIL   *Problem 8-25*

8-26. The following questions apply to Prob. 8-25.

(a) The key is subjected to side pressures as well as shearing forces. These pressures are called _____ stresses.

Ans. Bearing.

(b) If the diameter of the pulley were doubled, what changes would be required in the key? Assume the belt pulls remain the same.

Ans. Either select a material with at least twice the allowable shear stress, or increase the size of the key, or do both. Doubling the length of the key will not help if the width of the hub remains at 3 in.

(c) Why is it sometimes advantageous to make the key the weakest link in the system?

Ans. Because if failure occurs, it will be localized in an easily replaceable and inexpensive component. Shafts and pulleys are expensive; keys are cheap.

(d) List some typical examples of the key used as a "weak link."

Ans. The "propeller-to-shaft" assembly of an outboard motor.

8-27. The geometry of a punched hole is shown in the figure. Determine the maximum force that must be exerted on a punch to shear this hole in 0.020-in.-thick steel stock. The stock has a shear failure stress of 30,000 psi. What is the maximum compressive stress in the punch during the stamping?

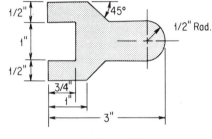

Ans. 6291 lb; 1854 psi.

Problem 8-27

8-28. Two materials are being considered for a machine component, a circular tension member 15 in. long that must sustain an axial tensile load of 10,000 lb. One material is a "super-strength" titanium alloy and the other is plain carbon steel. The working stress for the former is 100,000 psi and for the latter 30,000 psi. Compare the weights of the two members if the titanium alloy and the steel have specific weights of 0.165 lb per cu in. and 0.285 lb per cu in., respectively.

Ans. $W_t = 0.248$ lb; $W_s = 1.425$ lb.

8-29. A short column is made by welding an 8 WF 31 beam to two 10 WF 45 beams, as illustrated. If the maximum allowable compressive stress in the beams is 10,000 psi, determine (a) the maximum value of P_1 if $P_2 = 100$ kips; (b) the maximum value of P_1 if $P_2 = 20$ kips. See appendix for the properties of these beams.

Ans. (a) 64.8 kips; (b) 91.2 kips.

8-30. The column shown is fabricated by inserting one tube into another for a distance d and then brazing the two to make a homogeneous joint. The maximum allowable shear stress in the joint is 1000 psi, and the maximum compressive stress in each tube is 5000 psi. Find the inside diameter of the smaller tube, the outside diameter of the larger tube, and the distance d.

Ans. 1.65 in.; 2.30 in.; 0.80 in.

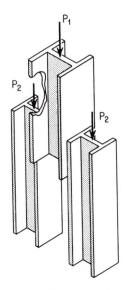

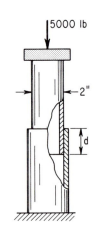

Problem 8-29   **Problem 8-30**

8-31. The uniform weight $W = 10$ kips is supported by three tension members as shown. The stress in C is twice the stress in A and one-half the stress in B; member B carries 50 per cent of the load. Find the cross-sectional area of each member if the maximum stress that any member can withstand is 5000 psi. Assume members A and C support equal loads.

Ans. A: 2.0 in.^{2}; B: 1 in.^{2}; C: 1 in.^{2}.

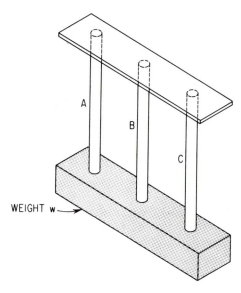

Problem 8-31

8-32. Axial stresses in members A and B of the figure are 1200 psi compression and 5000 psi tension, respectively. Determine the load P and the shearing stress in the pin at C.

Ans. 4600 lb; 271 psi.

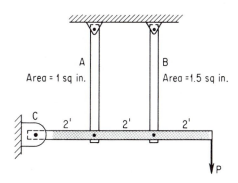

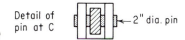

Detail of
pin at C

Problem 8-32

8-33. Select the most economical (lightest) wide-flanged beams A and B that will serve as short columns for the structure shown if the compressive stress in the column must not exceed 12,000 psi. Assume the horizontal member is rigid (see Appendix C for data on wide-flanged sections).

Ans. A: 10 WF 33; *B:* 16 WF 58.

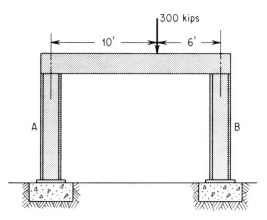

Problem 8-33

8-34. A short cylindrical post with a diameter of 4 in. is known to withstand compressive and shearing stresses of 12,000 psi and 5000 psi, respectively. What maximum axial load can be placed on the post?

Ans. Shear governs; 126 kips.

8-35. A square tension member, 2 in. on edge, is known to have a shearing stress component of 5000 psi on a plane inclined at 60 deg to the axis of the bar. Find the tensile load acting on the member and the normal stress on the inclined plane.

Ans. 46.2 kips; 8660 psi.

8-36. One-quarter-inch birch dowels are used to secure the frame shown. What maximum load *P* can be applied to the frame if the dowels have a limiting shear stress of 1200 psi? Neglect frictional effects.

Ans. 333 lb.

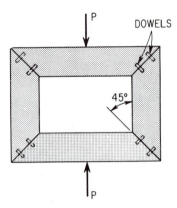

Problem 8-36

8-37 Determine the maximum permissible static load *P* that may be applied to the
to　tension member if the tensile stress must not exceed 15,000 psi. What is the
8-41. per cent loss in strength due to stress concentration?

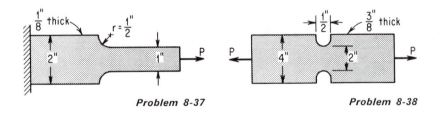

Problem 8-37　　　　　　　　　　　　　　*Problem 8-38*

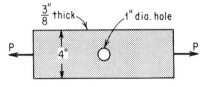

Problem 8-39

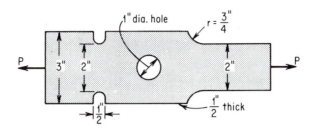

Problem 8-40

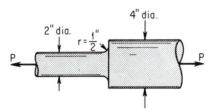

Problem 8-41

Ans.
8-37. 1340 lb; 28.5 per cent.
8-38. 4890 lb; 56.5 per cent.
8-39. 7500 lb; 56.5 per cent.
8-40. 6670 lb; 55.6 per cent.
8-41. 29,400 lb; 37.6 per cent.

8-42. Sections *A* and *B* have the same stress; determine the proper fillet radius *r*.
Ans. 0.2 in.

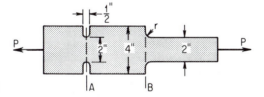

Problem 8-42

8-43. A cylindrical pressure vessel 2 ft in diameter is to be designed to operate at a pressure of 500 psi. How thick must the wall be if the tensile stress must not exceed 15,000 psi?
Ans. 0.4 in.

8-44. A 4-ft-diameter cylindrical pressure vessel is fabricated from 12 gage (0.1046 in.) steel. Determine the maximum permissible pressure within the vessel if the tensile stress must not exceed 15,000 psi.
Ans. 65.4 psi.

8-45. Large cylindrical tanks that are sealed in warm weather have buckled and collapsed when the temperature has dropped just a few degrees. Expalin the failure and discuss methods that will prevent this failure.

    Ans. A drop in temperature caused a drop in internal pressure. With a greater external pressure the walls of the vessel are placed in compression with the resultant buckling. To prevent this type of failure, vent the tank.

8-46. Argue against the case for rectangular pressure vessels.

8-47. The figure shows the dimensional details of a 100-ton hydraulic jack. Determine the minimum wall thickness t if the cylinder is fabricated from high-strength steel with a permissible working stress of 40,000 psi in tension.

    Ans. 0.4 in.

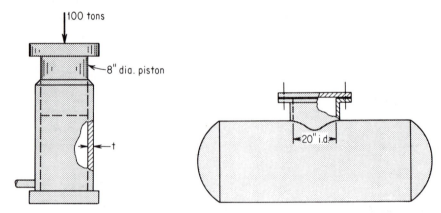

Problem 8-47   **Problem 8-48**

8-48. An inspection cover on a pressure vessel is secured with twenty $\frac{1}{4}$-in.-diameter bolts, as shown. Find the maximum stress in the bolts if the vessel is pressurized to 150 psi. The cross-sectional area at the thread root of each bolt is 0.126 sq in. If the fasteners are under-stressed, why not use fewer bolts?

    Ans. 18,700 psi. First of all, the assumption is made that the bolts are tightened just enough to maintain the pressure and all the stress is caused only by the pressure forces. The only way to accomplish this is by tightening with a torque wrench, an unlikely process. Secondly, the cover plate would probably buckle if fewer bolts were used.

8-49. A vertical standing oil tank made of $\frac{1}{2}$ in. plate is 30 ft in diameter and 60 ft high. Determine the height of oil (specific gravity $= 0.84$) that will cause a circumferential stress of 5000 psi.

    Ans. 38.2 ft.

8-50. A spherical pressure vessel with an inside diameter of 100 in. is made of material having an allowable stress in tension of 10,000 psi. Determine the thickness of the wall if the vessel is to sustain a pressure of 200 psi.

Ans. 0.50 in.

8-51. A spherical vessel is made by bolting two hemispherical sections together, as shown. How many 1-in.-diameter bolts are required to secure the two halves if the internal pressure is 20 psi? Use a working stress of 15,000 psi and a cross-sectional area of 0.551 sq in. for the bolts.

Ans. 110 bolts.

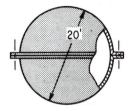

Problem 8-51

Chapter **9**

CONCEPT OF STRAIN

Most materials of construction deform under the action of loads according to a set pattern. They behave first elastically, then, as the load increases, plastically, until failure occurs. Although many designs can tolerate a certain amount of plastic deformation, emphasis in this chapter will be on the elastic character of materials.

9-1 Deformation and Strain

The terms *deformation* and *strain* have much the same meaning; they both are a measure of a change in a physical dimension. The first represents a *total change*, and the second a *unit change*. Consider, for example, the rectangular bar illustrated in Fig. 9-1, which has the dimensions of t, w, and l. Under the action of an axial tensile load P the bar stretches to a length of l'; accompanying the stretch will be a shortening of the thickness and width to t' and w'. Deformations, in terms of an x-, y-, z-coordinate system, are

192

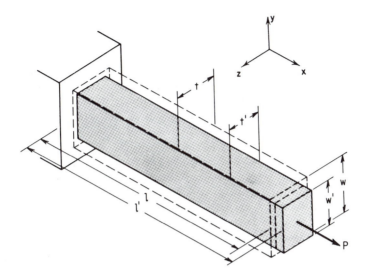

Figure 9-1

$$\delta_x = l' - l$$
$$\delta_y = w' - w \qquad (9\text{-}1)$$
$$\delta_z = t' - t$$

Each of the three terms carries the units of the particular dimension, and each represents a *total* or *net change*. The strain ϵ, a measure of the unit change δ/L along each of the coordinate axes, is defined as

$$\epsilon_x = \frac{l' - l}{l}$$

$$\epsilon_y = \frac{w' - w}{w} \qquad (9\text{-}2)$$

$$\epsilon_z = \frac{t' - t}{t}$$

The equations clearly show that strain is a dimensionless number that represents a change in length per unit length; a strain of 0.1 can equally mean a change of 0.1 in. per inch of length, 0.1 ft per foot, or 0.1 yd per yard. It is also important to note that deformation and strain can be either positive or negative, depending on whether the particular dimension increases or decreases.

EXAMPLE 1. A rod 10 ft long deforms 0.024 in. because of the action of an axial tensile force. Determine the accompanying axial strain.

Solution: The strain given by Eq. (9-2) is

$$\epsilon = \frac{\delta}{L} = \frac{0.024}{10(12)} = 0.0002 \text{ in./in.}$$

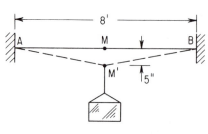

Figure 9-2

EXAMPLE 2. The midpoint M on the stretched horizontal wire drops to M' when weight W is suspended as shown in Fig. 9-2. Find the strain in the wire.

Solution: The change in length of wire is

$$\Delta MB = M'B - MB$$
$$= \sqrt{(4 \times 12)^2 + 5^2} - (4 \times 12)$$
$$= \sqrt{2329} - 48 = 0.26 \text{ in.}$$

The strain, by Eq. (9-2), is

$$\epsilon = \frac{\delta}{L} = \frac{0.26}{4 \times 12} = 0.00542 \text{ in./in.}$$

9-2 Elasticity: The Relationship Between Stress and Strain

There is little doubt that the most important discovery in the science of mechanics of materials was that pertaining to the elastic character of materials. This discovery, made by an English scientist, Robert Hooke, in 1678, mathematically relates stress to strain. The relationship, known as Hooke's law, states that *in elastic materials stress and strain are proportional.* Hooke approached the problem experimentally; he applied weights to "springy bodies" and measured their deformations. Although techniques have improved, this experiment is still performed hundreds upon hundreds of times to determine the elastic and plastic properties of materials. Universal testing machines, similar to the one shown in Fig. 9-3, are now employed to apply precise loads at precise rates to standardized tensile or compressive specimens. A variety of devices for measuring and recording strain can be attached to the specimen to make an accurate plot of the variation of stress with strain. Typical of these strain-measuring devices, which are called *extensometers,* is the linear differential transformer shown attached to a specimen in Fig. 9-4. As the specimen stretches, the activating knife edge moves, which in turn moves the core of a differential transformer. This movement creates an alternating voltage which can be amplified to run recording instrumentation. Modern extensometers have the ability to measure strains to an accuracy of a millionth of an inch—a far cry from Robert Hooke's yardstick. Figure

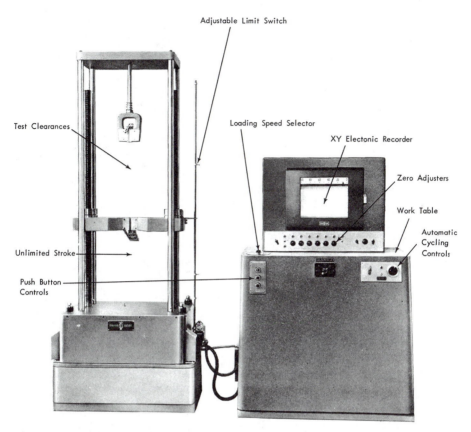

Figure 9-3. *Universal testing machine. Courtesy Tinius Olsen Testing Machine Company.*

9-5, a photograph of a mild steel tensile specimen, illustrates the type of failure typical of this material.

All materials deform differently under the action of tensile loads; some materials, such as mild steel, show a true proportionality between stress and strain up to a point, whereas others, like aluminum and copper, only approximate a proportionality. Typical plots of stress versus strain for several materials are shown in Fig. 9-6. Regardless of how the deformation progresses with load, a great deal of information can be obtained from the stress-strain diagram. Consider the curve shown in Fig. 9-7; stress in the units of pounds per square inch, computed on the basis of the original area, is represented as the ordinate in the curve, and strain in the units of inches per

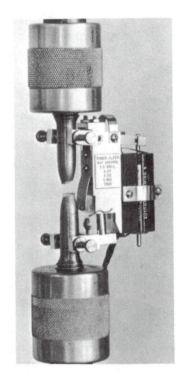

Figure 9-4. *Differential transformer extensometer which automatically separates when its measuring range has been exceeded. Courtesy Tinius Olsen Testing Company.*

inch is the abscissa. As can be seen, the strain in this particular specimen increases uniformly with stress up to 66,000 psi. Beyond this value of stress, called the *proportional limit*, the material deforms at a non-uniform rate. The maximum stress obtained in the plot, 116,000 psi in this example, is called the *ultimate stress*, or in some instances the *ultimate strength*, or the *tensile strength*. The final point in the curve represents the *rupture stress* or *rupture strength* of the material. It must be remembered that the stress plotted in the curve is computed on the basis of the original cross-sectional area, which, by the time the specimen fails, may be reduced appreciably. For this reason the rupture strength obtained from the curve is a fictitious number, and little emphasis is placed on its value.

The slope of the elastic portion of the stress-strain curve, the modulus line in Fig. 9-7, is called the *modulus of elasticity*, and the constant that represents the slope, Young's modulus E.

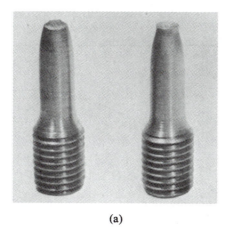

(a)

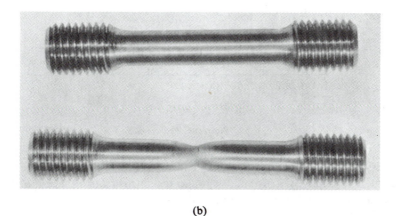

(b)

Figure 9-5. *Ductile steel tensile specimen showing the physical features of a tensile specimen before and after break.*

$$\frac{\text{Stress}}{\text{Strain}} = \text{Constant}$$

Symbolically

$$\frac{\sigma}{\epsilon} = E \qquad (9\text{-}3)$$

Another point of concern in the stress-strain diagram is the *yield strength*, or *yield point*, which is usually defined in terms of a specified amount of permanent deformation.

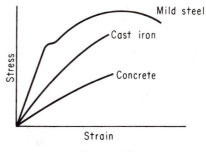

Figure 9-6

A yield strength at 0.2 per cent offset, as illustrated in Fig. 9-7, is that value

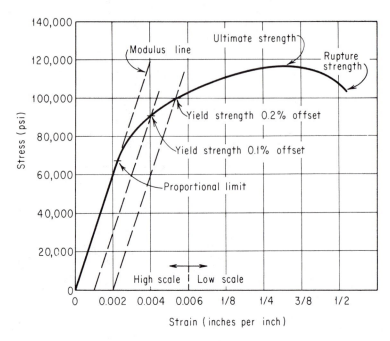

Figure 9-7

of stress which would cause a permanent deformation of 0.002 in. per in. with the load removed. This point is found by constructing a line parallel to the modulus line, offset by the specified amount as shown.

EXAMPLE 3. A tensile test on a steel specimen produced the stress-strain diagram illustrated in Fig. 9-7. Prior to testing, the specimen had a diameter of 0.505 in., and after rupture the diameter at the break was measured and found to be 0.425 in. Before testing, small punch marks 2 in. apart were made on the specimen, and, by placing the two portions of the specimen together after the test, the gage marks were found to be $2\frac{3}{8}$ in. apart. Determine the following: (a) modulus of elasticity, (b) proportional limit, (c) ultimate strength, (d) rupture strength based on the original area, (e) rupture strength based on the actual area, (f) yield strength at 0.02 per cent offset, (g) per cent elongation, and (h) per cent reduction in area.

Solution: (a) The slope of the modulus line, the ratio of a particular stress to the corresponding strain, is equal to the modulus of elasticity.

$$E = \frac{30,000 \text{ psi.}}{0.001 \text{ in./in.}} = 30 \times 10^6 \text{ psi}$$

(b) The proportional limit, that point where stress and strain cease to be proportional, is approximated from the graph; thus

$$\sigma_{pl} = 66,000 \text{ psi}$$

(c) The ultimate strength, the maximum point on the curve, is approximately

$$\sigma_{ult} = 116{,}000 \text{ psi}$$

(d) Since stress in the diagram is computed on the basis of the original cross-sectional area, the rupture strength is read directly from the curve

$$\sigma_{rup} = 103{,}000 \text{ psi}$$

(e) The load at failure P_f, computed in terms of the original area, is

$$P_f = \sigma A = 103{,}000\frac{\pi}{4}(0.505)^2 = 20{,}600 \text{ lb}$$

This load divided by the actual area gives the true rupture strength:

$$\sigma_{rup'} = \frac{P}{A_{actual}} = \frac{20{,}600}{\pi(0.425)^2/4} = 145{,}000 \text{ psi}$$

(f) The yield strength at 0.02 per cent offset is found by constructing a line parallel to the modulus line, offset by 0.002 in. per in. at the zero stress point, as shown. The yield strength read from the graph is approximately 100,000 psi.

(g) and (h) The per cent elongation and per cent reduction in area, both measures of ductility, are, respectively,

$$\text{Per cent elongation} = \frac{2.375 - 2}{2} \times 100 = 18.8 \text{ per cent}$$

$$\text{Per cent reduction in area} = \frac{\pi(0.505)^2/4 - \pi(0.425)^2/4}{\pi(0.505)^2/4} \times 100 = 9.5 \text{ per cent}$$

Unit stress and unit strain, as Hooke defined the terms, refer to the change in load expressed in pounds per square inch and the change in deformation in actual inches per inch. To correctly evaluate materials according to Hooke's Law, strain should be measured in inches per inch, and not be determined from assumed nor calculated distances between the jaws of the testing machine. Yet a large amount of physical test data has been collected in recent years based on assumed or calculated conditions of unit strain.

To circumvent this problem, testing machines have been built, and are being used, in which unit strain is calculated on the basis of crosshead motion per unit of time. However, curves produced in this manner contain many inaccuracies, ranging from errors due to grip slippage, variables in the specimen loading train, and machine deformation. . . to non-uniform length and non-uniform cross-section of exposed specimen between grips. All these basic mechanical operational errors are contained in the curves, and become an inextricable part of all data derived from them.

One of the chief shortcomings of the stress-time method is the wide range of modulus values that can be obtained by varying the crosshead speed and specimen length. For example, notice the divergent modulus values obtained for a plastic sample using different combinations of the two principal variables in Fig. 9-8. Such erratic results can raise havoc with any research or quality

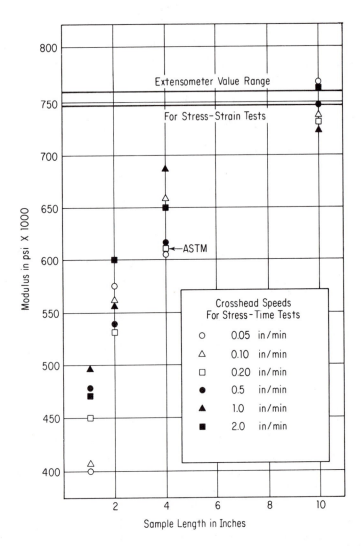

Figure 9-8. *Contrast the inconsistent modulus values obtained for a typical plastic film using the stress-time method with the uniform results secured with an extensometer—regardless of sample length or crosshead speed. Courtesy of Tinius Olsen Testing Machine Company.*

control program—especially since the optimum combination of crosshead speed and sample length is different for different materials.

Contrast these disputable stress-time calculations with the consistently accurate results secured when an extensometer is attached to the specimen. Here actual strain in inches per inch is measured between gauge points—completely independent of crosshead speed and sample length.

In evaluating new materials, how can reliable physical test data be

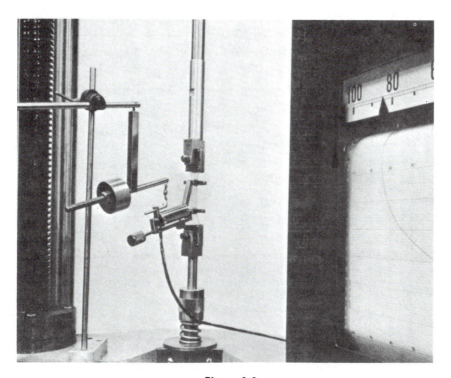

Figure 9-9

obtained unless an accurate stress-strain diagram is produced using strain instrumentation?

By employing the simple expedient of a counterbalance stand, which offsets the tare weight of the instrument, an extensometer can be attached to virtually any material without impairing it (Fig. 9-9). Thus, actual strain can be measured for such comparatively fragile specimens as films, yarns, rubber, thin metal sheets, foils, paper, fine wire, etc. Specially engineered clamping devices are available for a wide range of materials—each designed to grip gently but firmly without damaging the specimen or influencing test results.

The fallacy of the assumed and/or calculated strain has become readily apparent with the development of the modern universal testing machine. With this universal testing machine, which incorporates an XY recorder, it is possible to make and compare both stress-strain (extensometer) and stress-elongation (time base) curves.

The two sets of comparative curves (both made with identical brass specimens) are shown in Fig. 9-10. The graph to the left is a stress-strain curve, using an extensometer to record the strain, while the graph to the right is a stress-elongation (time) curve with strain calculated on the basis of specimen length between the jaws of the testing machine.

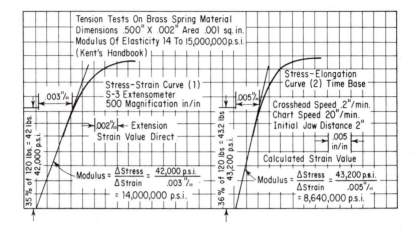

Figure 9-10. *These two curves show the divergent modulus values obtained for a brass specimen. Note that the value obtained with an extensometer matches published data, while the stress-time value is over 38% less. Courtesy Tinius Olsen Testing Machine Company.*

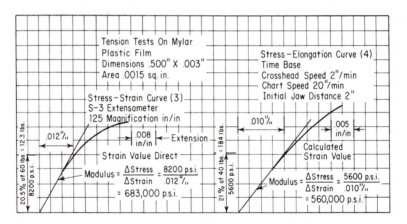

Figure 9-11. *Two identical plastic samples were tested using the stress-strain method (left) and the stress-time method (right). Courtesy Tinius Olsen Testing Machine Company.*

Examination of the two curves clearly shows the variances between stress-strain and stress-elongation (time) calculations. Particular attention is called to the data derived from the stress-strain curve for brass, which coincides with authenticated engineering data. Similar evaluation of the load-stress-elongation (time) curve prompts the question: "Why are the results different?" It is because of the errors in the stress-elongation or time method previously described. These errors are eliminated completely when the stress-strain extensometer method is used.

Curves produced by means of the stress-strain and stress-elongation (time) methods with identical plastic specimens are shown in Fig. 9-11. A definite discrepancy exists. Based on this evidence and countless other comparative tests, on which data would you base your material evaluation—stress-strain or stress-elongation (time)? The answer is obvious.

9-3 The Equations of Elasticity

Hooke's law provides a means of predicting with reasonable accuracy the elastic deformation in an axial loaded member. The mathematical expression for the law, in terms of the definitions of stress and strain, is

$$\frac{\sigma}{\epsilon} = \frac{\dfrac{P}{A}}{\dfrac{\delta}{L}} = E$$

Rearranging terms and solving for δ gives

$$\delta = \frac{PL}{EA} \tag{9-4}$$

Equation (9-4) expresses the important relationship among the variables: deformation δ, axial load P, length L, cross-sectional area A, and modulus of elasticity E.

It must be remembered that Eq. (9-4) is valid *only* for axial loaded members of constant cross-sectional area which are not stressed beyond their proportional limit. Care must be exercised in the selection of appropriate dimensions for each of the terms. Most tables, like those in the appendix, give values of the modulus of elasticity, E, in pounds per square inch. To be dimensionally correct, therefore, the cross-sectional area A must be expressed in square inches and the load P in *pounds:* deformation δ will then assume the units assigned to the length L.

EXAMPLE 4. A cylindrical steel bar which has a length of 15 in. is subjected to a tensile force of 5000 lb. Determine the required diameter of the bar if the stress is not to exceed 20,000 psi or the total elongation 0.005 in. $E = 30 \times 10^6$ psi.

Solution: Two restrictions are stated in the problem; the first places an upper limit on the stress, and the second places an upper limit on the deformation. To satisfy the first condition, the area must be

$$A = \frac{F}{\sigma} = \frac{5000}{20,000} = 0.25 \text{ sq in.}$$

The area, in terms of the deformation δ, by Eq. (9-4) is

$$A = \frac{FL}{\delta E} = \frac{5000(15)}{0.005(30 \times 10^6)} = 0.50 \text{ sq in.}$$

To satisfy both conditions, the larger of the two areas must be used. The required diameter, then, is given by

$$A = \frac{\pi D^2}{4}$$

$$D = \sqrt{\frac{4A}{\pi}} = \sqrt{\frac{4(0.5)}{\pi}} = 0.798 \text{ in.}$$

EXAMPLE 5. The roof truss shown in Fig. 9-12 is simply supported at A and E, and it in turn supports the panel loads as illustrated. Member GH is an annealed carbon steel tube with an outside diameter of 2 in. and a wall thickness of $\frac{1}{8}$ in. Determine the deformation and stress in member GH and the factor of safety in the design based on the ultimate tensile strength. $E = 30 \times 10^6$ psi.

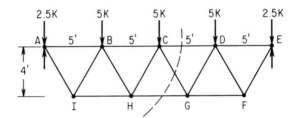

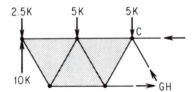

Figure 9-12

Solution: Since the truss is symmetrical, the reactions A and E are each equal to one-half the total external load.

$$R_A = R_E = \frac{2.5 + 5 + 5 + 5 + 2.5}{2} = 10 \text{ kips}$$

The force in member GH is found by sectioning the truss and taking moments about joint C in the free-body diagram shown in Fig. 9-8(b).

$$\sum M_C = 0$$

$$4GH + 5(5) + 2.5(10) - 10(10) = 0$$

$$GH = \tfrac{50}{4} = 12.5 \text{ kips (tension)}$$

The cross-sectional area of member GH is next computed.

$$A = \frac{\pi}{4}(2^2 - 1.75^2) = 0.736 \text{ in.}^2$$

Substitution of numerical data into Eq. (9.4) gives the desired deformation.

$$\delta = \frac{PL}{EA} = \frac{12,500(5 \times 12)}{30 \times 10^6(0.736)} = 0.034 \text{ in.}$$

The second quantity desired, the stress, is given by

$$\sigma = \frac{P}{A} = \frac{12,500}{0.736} = 17,000 \text{ psi}$$

This value of stress is considerably less than the yield strength of 38,000 psi given in the appendix, thus validating the use of Hooke's law in the prior step.

Lastly, the safety factor in the design based on the tensile strength from the appendix is

$$\text{S.F.} = \frac{\text{tensile strength}}{\text{working stress}}$$

$$= \frac{65,000}{17,000} = 3.82$$

EXAMPLE 6. A rigid beam of negligible weight is supported in a horizontal position by two rods, as shown in Fig. 9-13(a). Where, with respect to end A, should the force P act if the beam is to remain horizontal?

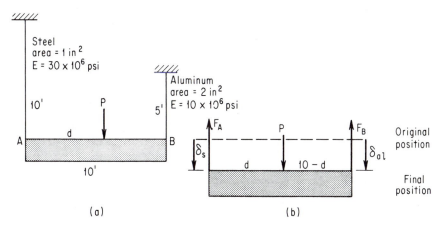

(a)   (b)

Figure 9-13

Solution: The beam will remain horizontal only if both rods deform equally.

$$\delta_s = \delta_{al}$$

$$\left(\frac{FL}{EA}\right)_s = \left(\frac{FL}{EA}\right)_{al}$$

$$\frac{F_A(10)12}{30(10)^6(1)} = \frac{F_B(5)12}{10(10)^6(2)}$$

$$F_A = \tfrac{3}{4} F_B$$

This result is combined with the equations of static equilibrium to give the desired answer.

$$\sum F_y = 0$$

$$F_A + F_B = P$$

$$F_B(\tfrac{3}{4} + 1) = P$$

$$F_B = \tfrac{4}{7}P$$

$$\sum M_A = 0$$

$$Pd = \tfrac{4}{7}P(10)$$

$$d = \tfrac{40}{7} = 5.71 \text{ ft to right of } A$$

9-4 Poisson's Ratio

Experimentation has shown that an axial elongation is always accompanied by a lateral contraction, and for a given material strained within its elastic region the ratio of lateral strain to axial strain is a constant.

$$\mu = \frac{\text{lateral strain}}{\text{axial strain}} \qquad (9\text{-}5)$$

Discovered in the early nineteenth century by the French mathematician Poisson, the constant μ, called *Poisson's ratio*, varies between 0 and $\tfrac{1}{2}$ for all materials. Precise measurements indicate that for most metals it lies between $\tfrac{1}{4}$ and $\tfrac{1}{3}$.

EXAMPLE 7. A load of 40,000 lb applied to a brass cylinder 15 in. long and 4 in. in diameter caused the length to decrease 0.0032 in. and the diameter to increase 0.00022 in. Find the modulus of elasticity E and Poisson's ratio μ of the brass.

Solution: The modulus of elasticity, the ratio of axial stress to axial strain, is

$$E = \frac{PL}{\delta A} = \frac{40,000(15)}{0.0032\,\pi(2)^2} = 15 \times 10^6 \text{ psi}$$

Poisson's ratio, the ratio of lateral strain to axial strain, is

$$\mu = \frac{\text{lateral strain}}{\text{axial strain}} = \frac{(0.00022)/4}{(0.0032)/15} = 0.26$$

9-5 Thermal Strain

Most materials expand when heated and contract when cooled. Careful measurements have shown that the ratio of strain E to temperature change ΔT is a constant.

$$\alpha = \frac{\text{strain}}{\text{change in temperature}} = \frac{\delta/L}{\Delta T}$$

Solving this equation for the deformation gives

$$\delta = \alpha L \Delta T \qquad (9\text{-}6)$$

where α is called the *thermal expansion coefficient*. In American engineering practice the customary units of this constant are inches per inch per degree

Fahrenheit. Steel, for example, has a thermal expansion coefficient of 6.5×10^{-6} in./in./°F, which simply means that for every degree Fahrenheit temperature change, the length, width, or breadth will change by 0.0000065 in. per in.

Problems in which both thermal deformation and stress deformation are involved are best treated by the method of *superposition:* by considering separately the deformations in the structure caused by thermal expansion or contraction and the deformations caused by axial forces. A geometric picture is obtained which will relate the strains, and this relationship, together with equations of static equilibrium, is used to solve the problem.

EXAMPLE 8. A manganese bronze rod is fastened securely to support A, as shown in Fig. 9-14(a). The free end can move 0.01 in. before contacting the solid support B. Determine (a) the temperature increase necessary to cause the free end to just touch B, and (b) the stress in the rod if its temperature increases 50°F.

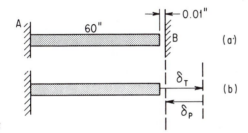

Figure 9-14

Solution: The following data are obtained from the Appendix:

$$\alpha = 11.2 \times 10^{-6} \text{ in./in./°F}$$
$$E = 15 \times 10^6 \text{ psi}$$

Part (a). The change in length as a function of temperature is

$$\delta_T = \alpha L \Delta T$$
$$0.01 = 11.2 \times 10^{-6}(60)\Delta T$$

Solving for the temperature change gives

$$\Delta T = \frac{0.01}{11.2 \times 10^{-6}(60)} = 14.9°F$$

Part (b). With wall B removed, Fig. 9-14(b), the bar could expand freely a distance equal to δ_T. To return the bar to its constrained position, the wall must exert a force sufficiently large to deform the bar an amount δ_P. From the geometry of the picture

$$\delta_T = \delta_P + 0.01$$
$$\alpha L \Delta T = \frac{PL}{EA} + 0.01 = \frac{\sigma L}{E} + 0.01$$

Substitution of data gives

$$11.2 \times 10^{-6}(60)50 = \frac{60\sigma}{15 \times 10^6} + 0.01$$

$$4\sigma = 11.2(60)50 - 10^4$$

$$\sigma = 5900 \text{ psi}$$

EXAMPLE 9. A 25 ton weight is supported by three columns, two steel and one manganese bronze, as shown in Fig. 9-15(a). Determine the temperature change necessary to just relieve the bronze column of all stress. The columns have equal cross-sectional areas of 2 sq in. and are initially the same length.

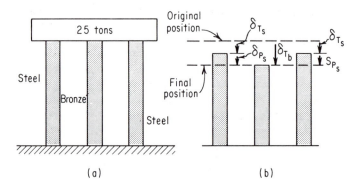

Figure 9-15

Solution: The following data are obtained from the appendix:

$$\alpha_s = 6.5 \times 10^{-6} \text{ in./in./}^\circ\text{F}$$

$$\alpha_b = 11.2 \times 10^{-6} \text{ in./in./}^\circ\text{F}$$

$$E_s = 30 \times 10^6 \text{ psi}$$

$$E_b = 15 \times 10^6 \text{ psi}$$

A sketch is drawn as shown in Fig. 9-15(b). With the weight removed and the temperature assumed to drop, the two steel columns contract a distance δ_T. and the bronze column a distance δ_{T_b}. To be realistic, the latter deformation is sketched as a greater distance, since the coefficient of expansion of bronze is nearly twice that of steel. The weight is then replaced, and the two steel columns are allowed to deform a distance δ_{P_s}; this replacement positions all three columns at the same level. Since the steel alone supports the weight of 25 tons, each steel column supports half of the total load, or 25,000 lb.

$$\delta_{T_s} + \delta_{P_s} = \delta_{T_b}$$

$$(\alpha L \Delta T)_s + \left(\frac{PL}{EA}\right)_s = (\alpha L \Delta T)_b$$

Substitution of data gives

$$(6.5 \times 10^{-6})L\Delta T + \frac{25,000L}{30 \times 10^6(2)} = (11.2 \times 10^{-6})L\Delta T$$

Every term in the equation is a linear function of L; therefore,

$$\Delta T(10^{-6})(11.2 - 6.5) = \frac{25,000}{30 \times 10^6(2)}$$

$$\Delta T = \frac{25,000}{30(2)(4.7)} = 88.7°F \text{ (drop)}$$

EXAMPLE 10. A steel bar and an aluminum bar, each secured to a rigid support, are fastened at their free ends by a 1-in.-diameter pin, as shown in Fig. 9-16(a). Determine the shearing stress in the pin if the temperature drops 50°F. The bars are initially free of stress.

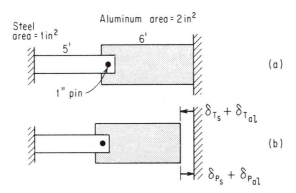

Figure 9-16

Solution: The following data are taken from the appendix:

$$\alpha_s = 6.5 \times 10^{-6} \text{ in./in./°F}$$

$$\alpha_{al} = 13.1 \times 10^{-6} \text{ in./in./°F}$$

$$E_s = 30 \times 10^6 \text{ psi}$$

$$E_{al} = 10 \times 10^6 \text{ psi}$$

With one constraint removed, the right-hand wall in this instance, the joined members would contract an amount equal to $\delta_{T_s} + \delta_{T_{al}}$, as the temperature drops. To return the members to their original length, the wall must exert a force that will cause the combination to stretch $\delta_{P_s} + \delta_{P_{al}}$.

$$\delta_{P_s} + \delta_{P_{al}} = \delta_{T_s} + \delta_{T_{al}}$$

$$\left(\frac{PL}{EA}\right)_s + \left(\frac{PL}{EA}\right)_{al} = (\alpha L\Delta T)_s + (\alpha L\Delta T)_{al}$$

The force P is common to both members; factoring and substituting data gives

$$P\left[\frac{5(12)}{30 \times 10^6(1)} + \frac{6(12)}{10 \times 10^6(2)}\right] = 6.5(10^{-6})(5 \times 12)50 + 13.1(10^{-6})(6 \times 12)50$$

$$P(\tfrac{5}{30} + \tfrac{6}{20}) = [6.5(5) + 13.1(6)]50$$
$$0.467P = 5560$$
$$P = 11,900 \text{ lb}$$

Finally, the shear stress in the pin is

$$\tau = \frac{P}{A} = \frac{11,900}{\pi(0.5)^2} = 15,200 \text{ psi}$$

9-6 Rigidity

When shear forces act on a body, Fig. 9-17, the deformation that occurs is due to a sliding action between adjoining layers of matter, and as a result of this action the shape of the body, rather than its volume, changes. The measure of shearing strain is the ratio of the deformation δ_s to the length L:

$$\epsilon_s = \frac{\delta_s}{L}$$

Since δ_s is small, the strain can be expressed in terms of the angular displacement γ in radians:

$$\epsilon_s = \frac{\delta_s}{L} = \gamma \tag{9-7}$$

Hooke's law relates the proportionality between shearing stress and shearing strain in terms of the constant G, called the *modulus of rigidity*.

$$\frac{\tau}{\gamma} = \frac{FL}{\delta_s A_s} = G \tag{9-8}$$

An important theoretical equation shows the three elastic constants E, G, and μ to be dependent upon one another for homogeneous materials.

$$G = \frac{E}{2(1 + \mu)} \tag{9-9}$$

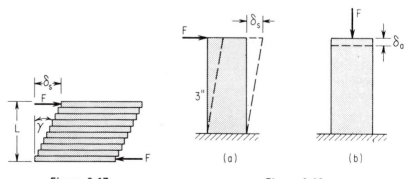

Figure 9-17   Figure 9-18

EXAMPLE 11. A brass specimen having a square cross-section 2 in. on edge and a height of 3 in. is subjected to a force of 20,000 lb, first in shear and then in compression, as shown in Fig. 9-18. The following data are observed:

$$\delta_s = 0.00234 \text{ in.}$$

$$\delta_a = 0.00088 \text{ in.}$$

Determine the modulus of elasticity, the modulus of rigidity, and Poisson's ratio.

Solution: Hooke's law is used to find the first two constants E and G.

$$E = \frac{FL}{\delta A} = \frac{20,000(3)}{0.00088(2 \times 2)} = 17 \times 10^6 \text{ psi}$$

and

$$G = \frac{FL}{\delta_s A_s} = \frac{20,000(3)}{0.00234(2 \times 2)} = 6.4 \times 10^6 \text{ psi}$$

Solving Eq. (9-9) for Poisson's ratio gives

$$G = \frac{E}{2(1 + \mu)}$$

$$\mu = \frac{E}{2G} - 1 = \frac{17 \times 10^6}{2(6.4 \times 10^6)} - 1 = 1.33 - 1 = 0.33$$

QUESTIONS, PROBLEMS, AND ANSWERS

9-1. Explain why *strain* is a dimensionless term.

Ans. Strain is a ratio of change of length to original length; to avoid confusion strain is sometimes expressed in inches per inch, or feet per foot. In either instance, however, the magnitudes are equal.

9-2. What is implied by positive or negative strain?

Ans. Positive strain implies an elongation; negative strain implies that dimensions are diminishing.

9-3. A cylindrical bar L feet long deforms l inches when an axial tensile force is applied. What is the strain in the bar?

Ans. $l/12L$.

9-4. The maximum permissible strain in a particular elevator cable is 0.0005 in./in. What is the elongation in 100 ft of this cable?

Ans. 0.60 in.

9-5. Precise measurements on a bar 20 in. long, 3 in. wide, and $\frac{1}{2}$ in. thick, subjected to an axial tensile force, show its length to increase by 0.024 in. and its width and thickness to decrease by 0.0009 in. and 0.00015 in., respectively. What are the three components of strain in the bar?

Ans. $\epsilon_L = 0.0012$ in./in.; $\epsilon_W = -0.0003$ in./in.; $\epsilon_T = -0.0003$ in./in.

9-6. A brass cube 5 in. on edge is surrounded by a pressure force that strains each side 0.0004 in./in. What is the change in volume of the cube?

Ans. 0.150 in.³ decrease.

9-7. Cable *BD* is fastened to the midpoint of a stretched horizontal wire, as shown. With the weight suspended from the end of the cable, *B* moves to *B'* and *D* to *D'*. What is the displacement *DD'* if the strain in the horizontal wire is 0.005 in./in. and in the vertical cable 0.010 in./in.?

Ans. 7.21 in.

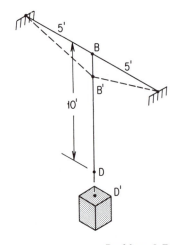

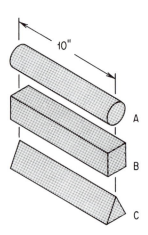

Problem 9-7   **Problem 9-8**

9-8. The three rods shown are each axially strained 0.001 in. per in. What is the axial deformation in each? What part does the cross-sectional geometry play in the calculation?

Ans. $\delta_A = \delta_B = \delta_C = 0.01$ in. None, as long as the cross-section remains constant. If the cross-section varies, the original statement describing equal strain throughout would not be realistic.

9-9. Due to the action of the load the strain in the stepped shaft is inversly proportional to the cross-sectional area. If this strain is equal to 0.01 inch per inch in section *A*, what is the total deformation of the shaft?

Ans. 0.713 in.

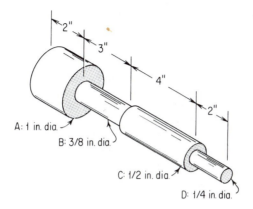

Problem 9-9

9-10. A stress-strain diagram for an alloy tensile specimen is shown. The following data are observed:

Initial diameter = 0.505 in.
Diameter at break = 0.408 in.
Initial gage length = 2 in.
Final gage length = $2\frac{7}{8}$ in.

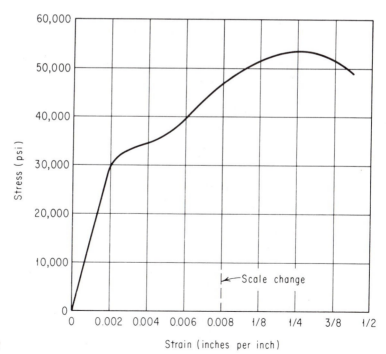

Problem 9-10   Strain (inches per inch)

Find: (a) the modulus of elasticity, (b) the proportional limit, (c) the yield strength at 0.2 per cent offset, (d) the per cent elongation, (e) the per cent reduction in area, (f) the indicated rupture strength, and (g) the true rupture strength based on area at break.

9-11. The following data were obtained during a tensile test on a mild steel specimen having an initial diameter of 0.505 in. At failure the reduced diameter of the specimen was 0.305 in. Plot the data and determine (a) the modulus of elasticity, (b) the proportional limit, (c) the yield strength at 0.2 per cent offset, (d) the ultimate strength, (e) the per cent reduction in area, (f) the per cent elongation, (g) the indicated strength at rupture, (h) the true rupture strength.

Axial load (lb)	Elongation in a 2 in. length (in.)	Axial load (lb)	Elongation in a 2 in. length (in.)
0	0		
1640	0.00050	8040	0.00938
3140	0.00100	8060	0.0125
4580	0.00150	9460	0.050
6000	0.00200	12,000	0.125
7440	0.00250	13,260	0.225
8000	0.00300	13,580	0.325
7980	0.00375	13,460	0.475
7900	0.00500	13,220	0.535
8040	0.00624	9860	0.625

9-12. A 1-in.-diameter manganese bronze test specimen was subjected to an axial tensile load, and the following data were observed:

Gage length	10 in.
Final gage length	12.25 in.
Load at proportional limit	18,500 lb
Elongation at proportional limit	0.016 in.
Maximum load	55,000 lb
Load at rupture	42,000 lb
Diameter at rupture	0.845 in.

Find: (a) the modulus of elasticity, (b) the proportional limit, (c) the ultimate strength, (d) the per cent elongation, (e) the per cent reduction in area, (f) the indicated rupture strength, and (g) the true rupture strength.

9-13. Interest has been focused recently on the tensile properties of alloys at elevated temperatures. The indicated data were observed during a tensile test on aluminum alloy.

Prepare a single graph of four curves that will show (a) the ultimate strength, (b) the yield strength, (c) the per cent elongation, and (d) the safe

Temperature (°F)	Ultimate strength (psi)	Yield strength (0.2 per cent offset psi)	Elongation (per cent)
75	35,000	29,000	12
212	34,000	29,000	13
300	28,000	25,000	22
400	21,000	15,000	35
500	14,000	7500	55
600	7500	5000	80
700	5000	3000	90

working stress based upon 40 per cent of the ultimate strength, all as functions of temperature.

9-14. A bronze bar having a cross-sectional area of 1 sq in. is suspended as shown. What is the total elongation of the bar? $E_b = 17 \times 10^6$ psi.

Ans. 0.107 in.

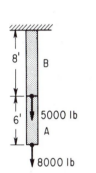

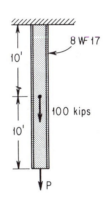

Problem 9-14   **Problem 9-15**

9-15. An 8 WF 17 steel beam supports the two loads as shown. What is the force and the deflection of the free end if the maximum stress in the beam must not exceed 30,000 psi?

Ans. 50 kips, 0.160 in.

9-16. The aluminum bar illustrated has a uniform cross-sectional area of 0.50 in.²

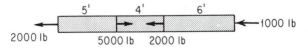

Problem 9-16

What is the total change in length of the bar if it is subjected to the axial forces shown? $E_{al} = 10 \times 10^6$ psi.

Ans. −0.0192 in.

9-17. The short steel tube shown is used as a post. Determine the allowable load that can be applied if the compressive stress and deformation are not to exceed 20,000 psi and 0.005 in., respectively. $E_s = 30 \times 10^6$ psi. Is it possible to select a wall thickness that would allow both maximum stress and maximum deformation to be satisfied simultaneously? Explain.

Ans. 11,100 lb. Not likely; since the ratio of stress to deformation is equal to the ratio of modulus to length, the given length and given material will not satisfy these conditions.

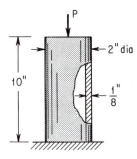

Problem 9-17

9-18. A concrete cylinder 6 in. in diameter and 5 ft high is to support an axial compressive load *P*. What is the maximum value of the load if the normal stress is not to exceed 1200 psi, the shearing stress 540 psi, and the deformation 0.03 in? $E_c = 2 \times 10^6$ psi.

Ans. 28,300 lb.

9-19. A uniform weight *W* is supported by two bars as shown. If the ends of the bars are initially at the same level and the weight is to remain horizontal, what is the required cross-sectional area of the bronze bar? $E_b = 17 \times 10^6$ psi; $E_s = 30 \times 10^6$ psi.

Ans. 3.53 in.^{2}.

9-20. What is the stress in sections *AB* and *BC* if the top support yields 0.005 in. when the load is applied? The support at *C* remains rigid and the bar has a cross-sectional area of $\frac{1}{2}$ sq in. The bar is steel.

Ans. *AB*: 14,750 psi tension; *BC*: 25,250 psi compression.

9-21. A bronze cylinder and an aluminum tube jointly support a 20-kip load, as shown. What is the stress in each material if prior to loading the cylinder and tube are the same length? $E_{al} = 10 \times 10^6$ psi; $E_b = 17 \times 10^6$ psi.

Ans. Aluminum: 5410 psi; bronze: 9240 psi; both compression.

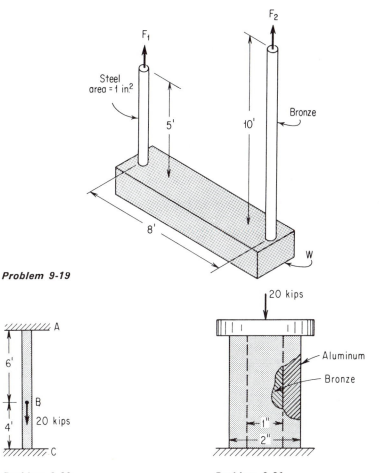

Problem 9-19

Problem 9-20

Problem 9-21

9-22. In the ordinary design of machine and structural components, the anticipated stress and strain are kept well below the yield point. Is it necessary to "play safe"?

Ans. Not always; if the loads are static and some permanent strain is permissible, higher stresses than those within the elastic region can be used. When materials become scarce, the technique by necessity will become more popular.

9-23. The column illustrated consists of a number of steel reinforcing rods imbedded in a concrete cylinder. Determine the cross-sectional area of the concrete and the steel and the stresses in each if the column deflects 0.01 in. while carrying a load of 100,000 lb. $E_s = 30 \times 10^6$ psi; $E_c = 2 \times 10^6$ psi.

Ans. Concrete: 146 in.^{2}, 167 psi; steel: 30.6 in.^{2}, 2500 psi.

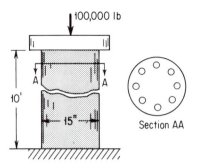

Problem 9-23

9-24. In the figure, four steel tie-rods secure the end plates to the aluminum tube. The ends of the rods are threaded with a standard $\frac{1}{4}$-20 die, and each nut on the right plate is tightened one-half turn from the snug position. What is the stress in the rods and in the tube? The thickness of the end plates may be considered negligible. $E_s = 30 \times 10^6$ psi; $E_a = 10 \times 10^6$ psi.

Ans. Steel: 26,100 psi tension; aluminum: 1320 psi compression.

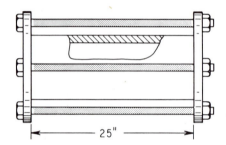
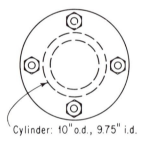

Problem 9-24

9-25. In the figure, determine the greatest permissible weight W that can be supported; the stresses in the steel and in the aluminum are not to exceed 20 ksi and 8 ksi, respectively. Assume the pinned beam AB is rigid and weightless. $E_s = 30 \times 10^6$ psi; $E_a = 10 \times 10^6$ psi.

Ans. 10 kips.

9-26. In the figure, three short columns, each with a cross-sectional area of 1 sq in., support a rigid and weightless beam. Where, with respect to A, should a force P be placed if the beam is to remain horizontal? $E_s = 30 \times 10^6$ psi; $E_b = 15 \times 10^6$ psi; $E_a = 10 \times 10^6$ psi.

Ans. 6.82 ft to right of A.

9-27. What is the decrease in diameter of a solid circular brass bar 2 in. in diameter that is subjected to an axial tensile force of 15,000 lb? $E_b = 15 \times 10^6$ psi; $\mu_b = 0.26$.

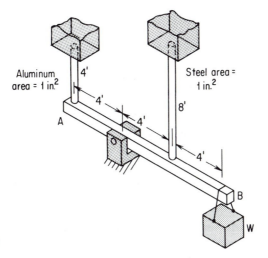

Aluminum area = 1 in.^{2}

Steel area = 1 in.^{2}

Problem 9-25

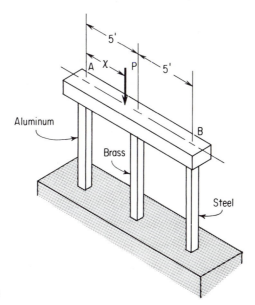

Aluminum

Brass

Steel

Problem 9-26

Ans. 1.65×10^{-4} in.

9-28. An axially loaded bar 2 in. by 2 in. and 60 in. long becomes 0.0006 in. narrower and 0.072 in. longer after loading. What is Poisson's ratio of this material?

Ans. 0.25.

9-29. The determination of Poisson's ratio requires precise measurements. What are some of the variables that would lead to an experimentally incorrect value?

Ans. The rate at which the load is applied; the dynamic value of Poisson's ratio would conceivably be different from the static value. Stress concentration, even microscopic, would also interfere with measurements.

9-30. A cylindrical concrete test specimen 5 in. in diameter and 20 in. long is subjected to an axial compressive load of 270 kips. Precise measurements indicate the length decreases by 0.138 in. and the circumference increases by 0.0157 in. What are the modulus of elasticity and Poisson's ratio of this material?

Ans. 2×10^6 psi; 0.145.

9-31. An axial force of 8000 lb was required to reach the proportional limit of an alloy specimen 0.505 in. in diameter. Careful measurements indicated that the length increased 0.0125 in. in a gage length of 10 in. and the diameter decreased by 0.000156 in. What are the proportional limit, the modulus of elasticity, and Poisson's ratio of this alloy?

Ans. $\sigma_{PL} = 40,000$ psi; $E = 32 \times 10^6$ psi; $\mu = 0.25$.

9-32. The circular aluminum bar shown fits snugly in a steel collar when the bar is subjected to an axial compressive force of 10,000 lb. How much larger should the diameter of the collar be? $E_a = 10 \times 10^6$ psi; $\mu_a = 0.33$.

Ans. 4.2×10^{-4} in.

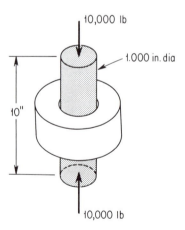

10,000 lb

1.000 in. dia

10"

10,000 lb   *Problem 9-32*

9-33. Problem solving by *superposition* is a powerful mathematical tool. To illustrate a typical application consider the following example. One edge of a square plate is maintained at 100°F while the other three are kept at 0°F. What is the temperature in the center of the plate?

Ans. Imagine that a square plate has a constant temperatue of 100°F throughout and that it consists of four plates stacked one on top of the other.

Each of the four plates is similar to the one in the problem; one edge is held at 100°F while the other three are at 0°F. The center temperatures are the same in all four and are equal to T. Thus $4T$ is equal to 100°F and T is equal to 25°F.

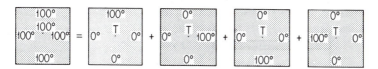

Problem 9-33

9-34. A steel bridge spans a channel 2000 ft wide. What change in length would be caused by the change in summer and winter temperatures if the range is $-20°F$ to $100°F$? $\alpha_s = 6.5 \times 10^{-6}$ in./in./°F.

Ans. 18.7 in.

9-35. A circular hole 5.000 in. in diameter is cut in a sheet of aluminum at 0°F. By how much will the area of the hole change if the aluminum is heated to $100°F$? $\alpha_a = 12.5 \times 10^{-6}$ in./in./°F.

Ans. 0.1 in.² increase.

9-36. What should be the separation between successive 40 ft steel rails to allow for expansion if they are laid when the temperature is 30°F and the maximum temperature reached is 120°F? $\alpha_s = 6.5 \times 10^{-6}$ in./in./°F.

Ans. 0.281 in.

9-37. The steel bar illustrated is held by the rigid supports at A and B. What is the stress in the bar if the temperature drops 100°F? Assume the bar to be initially free of stress. $\alpha_s = 6.5 \times 10^{-6}$ in./in./°F. Does it seem reasonable that the answer would be the same for any length?

Ans. 19,500 psi. Yes, since unit values of deformation are used in the calculation and the length is the same for both thermal expansion and stress determinations.

9-38. An aluminum rod and a steel rod are fastened to non-yielding supports A

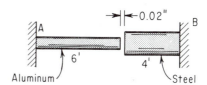

and *B*. What temperature change is required to close the gap? $\alpha_s = 6.5 \times 10^{-6}$ in./in./°F; $\alpha_{al} = 12.5 \times 10^{-6}$ in./in./°F.

Ans. 16.5°F.

9-39. The composite steel and brass member shown is held, with no initial axial stress, by the rigid supports at *A* and *B*. What is the stress in each material if the temperature drops 100°F? $E_s = 30 \times 10^6$ psi; $E_b = 15 \times 10^6$ psi; $\alpha_s = 6.5 \times 10^{-6}$ in./in./°F; $\alpha_b = 10.4 \times 10^{-6}$ in./in./°F.

Ans. Steel: 27,300 psi tension; brass: 13,650 psi tension.

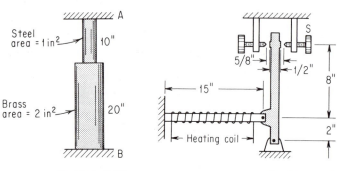

Problem 9-39   **Problem 9-40**

9-40. The essential features of a temperature-actuated relay are shown. The sensing device is a copper rod 15 in. long that is heated by a resistance coil. Through what temperature increment will the switch *S* be in the "open" position illustrated? $\alpha_c = 9.3 \times 10^{-6}$ in./in./°F.

Ans. ± 89.5°F.

9-41. A temperature-measuring device consists of an aluminum bar *AB* pinned through a series of gears to a light pointer. What scale reading, in inches, would correspond to an incremental temperature change of 100°F? $\alpha_a = 12.5 \times 10^{-6}$ in./in./°F.

Ans. ± 0.375 in.

9-42. The important features of a *differential dilatometer*, a device employed to measure the difference in linear expansion coefficients, are shown. A light ray is deflected by mirror *M*, as block *B* rotates due to the differential expansion of the steel and aluminum rods. What equal change in the temperature of both rods would cause a scale reading of $\frac{1}{4}$ in.? $\alpha_s = 6.5 \times 10^{-6}$ in./in./°F; $\alpha_a = 12.5 \times 10^{-6}$ in./in./°F.

Ans. 86.8°F.

9-43. A material with unknown expansion characteristics is substituted for the aluminum bar in Prob. 9-42. What is the coefficient of expansion of this

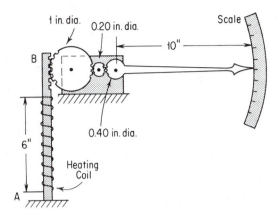

Problem 9-41

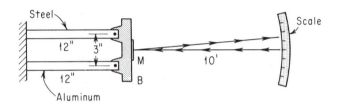

Problem 9-42

material if a differential scale reading of $\frac{1}{2}$ in. is recorded in the direction shown when both the material and the steel are heated to 200°F?

Ans. 11.7 × 10⁻⁶ in./in./°F.

9-44. Three rods of equal length support the 10,000-lb weight as shown. What temperature change in the system would relieve the brass rod of all load? $E_s = 30 \times 10^6$ psi; $E_b = 15 \times 10^6$ psi; $\alpha_s = 6.5 \times 10^{-6}$ in./in./°F; $\alpha_b = 10.4 \times 10^{-6}$ in./in./°F.

Ans. 85.5°F decrease.

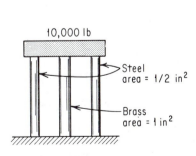

Problem 9-44

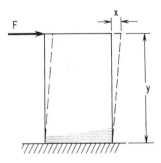

Problem 9-45

9-45. The block illustrated is subjected to a shear force that causes the deformation shown. What is the shearing strain?

Ans. x/y.

9-46. The following data apply to the block of Prob. 9-45: $x = 0.0005$ in., $y = 5.00$ in., material = steel, cross-sectional area = 2 sq in. What is the force?

Ans. 2400 lb.

9-47. The steering wheel transmits a torque of 20 lb ft. The spokes are made of aluminum and have the uniform cross-section shown. What is the shear deformation in each spoke? List any assumptions that must be made. $G_a = 3.8 \times 10^6$ psi.

Ans. 11.8×10^{-5} in.

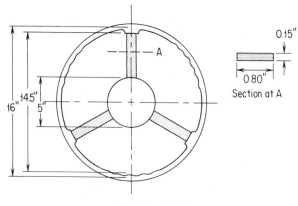

Problem 9-47

9-48. Two steel blocks separated by a rigid dowel are deformed equally, as shown. What is the force P if one block has a cross-sectional area of 1 sq in. and the other a cross-sectional area of $\frac{1}{2}$ sq in.? $G = 12 \times 10^6$ psi.

Ans. 36,000 lb.

9-49. What is Poisson's ratio of manganese bronze based on values of E and G given in the appendix?

Ans. 0.339.

9-50. A particular metal has a Poisson's ratio of 0.28 and a modulus of elasticity of 10.3×10^6 psi. What is its modulus of rigidity?

Ans. 4×10^6 psi.

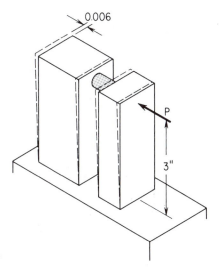

Problem 9-48

9-51. What is the ratio of the shear modulus to the elastic modulus if Poisson's ratio is 0.25?

Ans. 0.4.

Chapter **10**

TORSION

The previous chapters have considered the stress-strain relationships for axial loaded members—members acted upon by either tensile or compressive forces. A second mode of loading, the pure torque, will be studied next.

10-1 Torsional Stress and Strain

Consider a circular shaft fixed to a rigid support, as shown in Fig. 10-1(a). The shaded plane represents a surface within the shaft. Experimental evidence has shown that, although the couple T will warp the plane as it twists the shaft, circular sections within the bar will remain circular. A thin slice of the shaft having a length ΔL, Fig. 10-1(b), is isolated as a free-body, as shown in Fig. 10-1(c). It can be seen that point a on the circumference of this element rotates to a', and point b within the element rotates to b' as the slice twists through an angle $\Delta\theta$. A general expression for the shearing strain γ_r at a

226

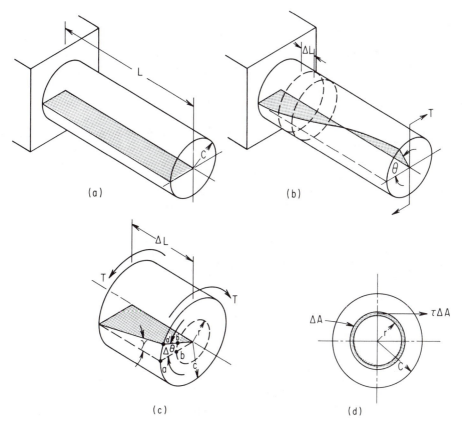

(a)

(b)

(c)

(d)

Figure 10-1

radius r within the body can be written in terms of the geometry of the rotation.

$$\gamma_r = \frac{\overline{bb'}}{\Delta L} = \frac{r\Delta\theta}{\Delta L} \qquad (a)$$

Since all elements of the shaft twist equally, the ratio $\Delta\theta/\Delta L$ is constant and can be replaced by its equivalent, θ/L:

$$\gamma_r = \frac{r\theta}{L} \qquad (b)$$

A second relationship is obtained by considering the internal reaction to the torque T. Internal moments are created, Fig. 10-1(d), which must statically balance this torque.

$$T = \sum \tau_r \Delta A \cdot r \qquad (c)$$

Equations (b) and (c) are related through Hooke's law:

$$G = \frac{\tau}{\gamma}$$

Thus,

$$T = \sum G\gamma \cdot \Delta A \cdot r = \sum \frac{Gr\theta}{L} \cdot \Delta A \cdot r$$

$$T = \frac{G\theta}{L} \sum r^2 \Delta A \tag{d}$$

The term $\sum r^2 \Delta A$ is the polar moment of inertia of the section, thus

$$T = \frac{G\theta J}{L}$$

When the terms are rearranged, the angle of twist of the shaft can be expressed as a function of torque, length, rigidity, and polar moment of inertia.

$$\theta = \frac{TL}{GJ} \tag{10-1}$$

An expression for maximum shearing stress is next obtained by regrouping the terms in the previous derivation.

$$\tau_{max} = G\gamma = G \cdot \frac{c\theta}{L} = G \cdot \frac{c}{L} \cdot \frac{TL}{GJ}$$

$$\tau_{max} = \frac{Tc}{J} \tag{10-2}$$

By replacing J by its appropriate value, detailed equations may be obtained for solid and hollow sections. For solid circular sections:

$$J = \frac{\pi d^4}{32}$$

$$\theta = \frac{TL}{G} \frac{32}{\pi d^4} \tag{10-3}$$

$$\tau_{max} = \frac{Td/2}{\pi d^4/32} = \frac{16T}{\pi d^3} \tag{10-4}$$

For hollow circular sections:

$$J = \frac{\pi}{32}(d_0^4 - d_i^4)$$

$$\theta = \frac{TL}{G} \frac{32}{\pi(d_0^4 - d_i^4)} \tag{10-5}$$

$$\tau_{max} = \frac{16Td_0}{\pi(d_0^4 - d_i^4)} \tag{10-6}$$

The examples that follow illustrate the use of the equations of torsional stress and strain. Of particular interest are those problems which are statically indeterminate.

EXAMPLE 1. A core drill for earth strata study consists of a hollow circular steel shaft 50 ft long with a 2 in. outside diameter and a 1.5-in. inside diameter. If a twisting moment of 10,000 lb in. acts on the shaft, find the maximum shearing stress and the angle of twist. The modulus of rigidity of steel is $G = 12 \times 10^6$ psi.

Solution: To compute the stress, numerical values are substituted into Eq. (10-6).

$$\tau_{max} = \frac{16Td_o}{\pi(d_o^4 - d_i^4)} = \frac{16(10,000)(2)}{\pi(2^4 - 1.5^4)} = 9310 \text{ psi}$$

The angle of twist is computed by substituting numerical values into Eq. (10-5).

$$\theta = \frac{TL}{G}\frac{32}{\pi(d_o^4 - d_i^4)} = \frac{(10,000)(50 \times 12)32}{12 \times 10^6\pi(2^4 - 1.5^4)}$$
$$= 0.466 \text{ rad} = 0.466 \times 57.3 = 26.7 \text{ deg}$$

EXAMPLE 2. Find the safe torque that may be applied to a solid steel shaft 4 in. in diameter. The working stress in shear is 5000 psi, and the allowable angle of twist per foot of length is $\frac{1}{10}$ deg.

Solution: Two values of torque must be computed: one based on stress and the other on deformation. The smaller of the two will be the safe torque. By Eq. (10-4),

$$\tau_{max} = \frac{16T}{\pi d^3}$$

Where

$$T = \frac{\tau_{max}\pi d^3}{16} = \frac{5000\pi(4)^3}{16} = 62,800 \text{ lb in.}$$

By Eq. (10-3),

$$\theta = \frac{TL}{G}\frac{32}{\pi d^4}$$

Solving for T gives

$$T = \frac{\theta G}{L}\frac{\pi d^4}{32}$$
$$= \left(\frac{1}{10} \times \frac{\pi}{180}\right)\left(\frac{12 \times 10^6}{12}\right)\left(\frac{\pi 4^4}{32}\right)$$
$$= 43,800 \text{ lb in.}$$

The lesser of the two values, $T = 43,800$ lb in., is the permissible torque for this shaft.

EXAMPLE 3. The composite shaft shown in Fig. 10-2 is held between rigid supports at A and C. Find the torque in section AB and the torque in section BC. $G_s = 12 \times 10^6$ psi; $G_b = 6.4 \times 10^6$ psi.

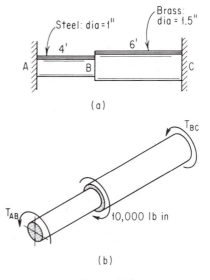

(a)

(b)

Figure 10-2

Solution: This problem is *statically indeterminate*. Only one equation of equilibrium $\sum M = 0$ can be written, and in this equation are the unknowns T_A and T_B. The sum of these two torques is the applied torque.

$$T_A + T_B = 10,000 \tag{a}$$

Since the torque acts at the junction of the two shafts, the twists in the two sections are equal. This equality provides a second equation involving T_A and T_B.

$$\theta_{AB} = \theta_{BC}$$

$$\left(\frac{TL}{G}\frac{32}{\pi d^4}\right)_{AB} = \left(\frac{TL}{G}\frac{32}{\pi d^4}\right)_{BC}$$

Like terms are canceled and numerical data substituted:

$$\frac{T_{AB}(4 \times 12)}{(12 \times 10^6)(1)^4} = \frac{T_{BC}(6 \times 12)}{(6.4 \times 10^6)(1.5)^4}$$

$$T_{AB} = 0.556\, T_{BC} \tag{b}$$

Equations (a) and (b) are then combined:

$$T_{BC}(0.556 + 1) = 10,000$$

$$T_{BC} = 6427 \text{ lb in.}$$

and

$$T_{AB} = (0.556)6427 = 3573 \text{ lb in.}$$

EXAMPLE 4. Two steel shafts are gear-connected, as shown in Fig. 10-3. Find the angle of twist of end D if shaft CD is subjected to a twisting moment of 10,000 lb in. $G = 12 \times 10^6$ psi.

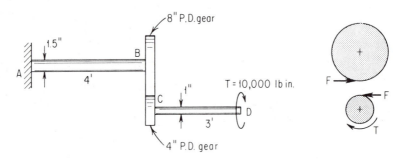

Figure 10-3

Solution: The torque that acts in shaft AB is twice as great as that in shaft CD. This is disclosed by the free-body diagram of the two gears. The mutual contact force F can be computed in terms of the applied torque:

$$F_r = T_{CD}$$

$$F(2) = 10,000 \text{ lb in.}$$

$$F = 5000 \text{ lb}$$

The moment produced by this force in the larger gear is double, since it acts on a lever arm double in length.

$$T_{AB} = F'_r = 5000(4) = 20,000 \text{ lb in.}$$

A study of the geometry of motion also indicates the angle of rotation of gear C to be twice that of gear B; thus

$$\theta_C = 2\theta_B$$

The total rotation at D, then, is the angle of twist in shaft CD added to twice the angle of twist of shaft AB.

$$\theta_D = \theta_{CD} + 2\theta_{AB}$$

By Eq. (10-3),

$$\theta_D = \left(\frac{TL}{G} \times \frac{32}{\pi d^4}\right)_{CD} + 2\left(\frac{TL}{G} \times \frac{32}{\pi d^4}\right)_{AB}$$

$$\theta_D = \frac{10,000(3 \times 12)32}{(12 \times 10^6)\pi(1)^4} + 2\frac{(20,000)(4 \times 12)32}{(12 \times 10^6)\pi(1.5)^2}$$

$$= 0.628 \text{ rad or } 36 \text{ deg}$$

10-2 Power Transmission

There are many applications in which shafts must be designed to transfer power to various machines. Computations which involve the determination of proper shaft sizes can be easily made, since power and torque are related by the kinetic equation:

$$P = T\omega$$

In this relationship power (P) has the units of pound feet per second, torque (T) the units of pound feet, and angular velocity (ω) the units of radians per second. The equation can be rewritten in terms of horsepower (hp) and revolutions per minute (n), since

$$1 \text{ hp} = 550 \times 12 \times \frac{\text{lb in.}}{\text{sec}}$$

and

$$\omega = \frac{2\pi n}{60}$$

thus

$$\text{hp} = \frac{Tn}{550(12)} \times \frac{2\pi}{60} = \frac{Tn}{63,000} \tag{10-7}$$

EXAMPLE 5. Design a solid steel shaft to transmit 250 hp at 1800 rpm. The maximum allowable shearing stress is 5000 psi.

Solution: By Eq. (10-7),

$$T = \frac{63,000(\text{hp})}{n} = \frac{63,000(250)}{1800} = 8750 \text{ lb in.}$$

The relationship between shearing stress and torque is given by Eq. (10-4).

$$\tau = \frac{16T}{\pi d^3}$$

Solving for the diameter d gives

$$d = \sqrt[3]{\frac{16T}{\pi \tau}} = \sqrt[3]{\frac{16(8750)}{\pi(5000)}} = 2.07 \text{ in.}$$

EXAMPLE 6. A solid 2-in.-diameter steel line-shaft, Fig. 10-4(a), is driven by a 50-hp motor at a speed of 315 rpm. Power take-offs are situated at A and C. Determine the angle of twist of gear C relative to gear A. Assume $G_s = 12 \times 10^6$ psi.

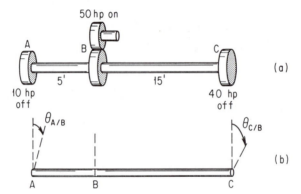

Figure 10-4

Solution: The torque in segments AB and BC is computed by Eq. (10-7):

$$T_{AB} = \frac{63,000(\text{hp})}{n} = \frac{63,000(10)}{315} = 2000 \text{ lb in.}$$

$$T_{BC} = \frac{63,000(\text{hp})}{n} = \frac{63,000(40)}{315} = 8000 \text{ lb in.}$$

To determine the twist of gear C relative to gear A (written as $\theta_{C/A}$), imagine B to be fixed, Fig. 10-4(b), and compute the twists of gears A and C. Since these gears rotate in the same direction relative to B, their rotation relative to each other is the difference $\theta_{C/B} - \theta_{A/B}$:

$$\theta_{C/A} = \theta_{C/B} - \theta_{A/B}$$

By Eq. (10-3).

$$\theta_{C/B} = \frac{32}{\pi d^4} \frac{TL}{G} = \frac{32(8000)(15 \times 12)}{\pi(2)^4(12 \times 10^6)} = 0.0764 \text{ rad}$$
$$= 0.0764 \times 57.3 = 4.38 \text{ deg}$$

and

$$\theta_{A/B} = \frac{32}{\pi d^4} \times \frac{TL}{G} = \frac{32(2000)(5 \times 12)}{\pi(2)^4(12 \times 10^6)} = 0.00637 \text{ rad}$$
$$= 0.00637 \times 57.3 = 0.365 \text{ deg}$$

Therefore,

$$\theta_{C/A} = 4.38 - 0.365 = 4.015 \text{ deg}$$

10-3 Torsion Bars

There are countless applications in the field of design in which torsion members can be employed as springs; the *torsion bar* is one of the many examples. The *spring constant*, or *spring rate*, as it is sometimes called, is the ratio of torque to twist expressed in pound inches per radian (lb in./rad). Since the twist in radians is in itself a function of torque, the spring constant k is equal to

$$k = \frac{T}{\theta} = \frac{T}{\dfrac{TL}{GJ}} = \frac{GJ}{L} \tag{10-8}$$

For solid sections and for hollow sections k has the specific value of

$$k_s = \frac{\pi d^4}{32} \frac{G}{L} \tag{10-9}$$

and

$$k_h = \frac{\pi(d_o^4 - d_i^4)}{32} \frac{G}{L} \tag{10-10}$$

Two or more torsion bars can be coupled to form a spring more flexible than any of those combined, a *series combination*, or stiffer than any of those combined, a *parallel combination*. Consider three torsion bars connected end to end, as shown in Fig. 10-5. The internal reaction in each segment is the applied torque T, and the angle of twist of D relative to A is the sum

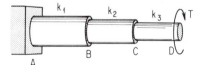 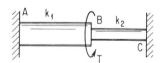

Figure 10-5   Figure 10-6

$$\theta_{D/A} = \theta_{B/A} + \theta_{C/B} + \theta_{D/C}$$

Since θ for any segment is T/k, it follows that

$$\theta_{D/A} = \frac{T}{k_1} + \frac{T}{k_2} + \frac{T}{k_3}$$

Dividing by T gives

$$\frac{\theta_{D/A}}{T} = \frac{1}{k_1} + \frac{1}{k_2} + \frac{1}{k_3}$$

The reciprocal of this expression is the equivalent constant k, of the system

$$k_e = \frac{T}{\theta_{D/A}} = \cfrac{1}{\cfrac{1}{k_1} + \cfrac{1}{k_2} + \cfrac{1}{k_3}} \tag{10-11}$$

Thus, *torsion bars are in series if the same torque acts in each. The equivalent spring constant of a series system is equal to the reciprocal of the sum of the reciprocals of the individual spring constants.*

EXAMPLE 7. In Fig. 10-5, let k_1, k_2, and k_3 be 6×10^6 lb in./rad, 3×10^6 lb in./rad, and 2×10^6 lb in./rad, respectively. Find the equivalent spring constant of the system and the angle of twist $\theta_{D/A}$ if $T = 10,000$ lb in.

Solution: By Eq. (10-11),

$$k_e = \cfrac{1}{\cfrac{1}{k_1} + \cfrac{1}{k_2} + \cfrac{1}{k_3}} = \cfrac{1}{\cfrac{1}{6 \times 10^6} + \cfrac{1}{3 \times 10^6} + \cfrac{1}{2 \times 10^6}} = \cfrac{10^6}{\cfrac{1}{6} + \cfrac{1}{3} + \cfrac{1}{2}}$$

$$= 10^6 \text{ lb in./rad}$$

and

$$\theta_{D/A} = \frac{T}{k_e} = \frac{10,000}{10^6} = 0.01 \text{ rad}$$

$$= 0.573 \text{ deg.}$$

Torsion bars are *parallel-connected* when their angular displacements are equal. The connected bars illustrated in Fig. 10-6 represent a parallel system; the bars share the torque while twisting equally. The applied torque T is the sum

$$T = T_{AB} + T_{BC}$$

The torques T_{AB} and T_{BC} can be expressed in terms of the spring constants k_1 and k_2:

$$T = k_1\theta + k_2\theta$$

Dividing by the angle of twist gives

$$\frac{T}{\theta} = k_1 + k_2$$

where T/θ is the equivalent spring constant k_e.

$$k_e = k_1 + k_2 \tag{10-12}$$

Formally stated: *Torsion bars are in parallel if they have equal angles of twist and an applied torque apportioned between them; the equivalent spring constant of such a system is equal to the sum of the individual spring constants.*

EXAMPLE 8. In Fig. 10-6, let $k_1 = 2 \times 10^6$ lb in./rad and $k_2 = 3 \times 10^6$ lb in./rad. Find the equivalent spring constant for the system and the angle of twist at B. Let the torque T equal 20,000 lb in.

Solution: By Eq. (10-12),

$$k_e = k_1 + k_2 = (2 + 3)10^6 = 5 \times 10^6 \text{ lb in./rad}$$

$$\theta_B = \frac{T}{k_e} = \frac{20,000}{5 \times 10^6} = 0.004 \text{ rad} = 0.229 \text{ deg}$$

10-4 Helical Springs

The torsion theory can be expanded to include a rather interesting and important application, the design of close-coiled helical springs. One of the primary functions of a spring, of course, is to store or release energy; this it must do without becoming overstressed. Consider the tension spring acted upon by a force P as shown in Fig. 10-7(a). It is presumed that the spring has been made by winding a helical coil of a solid circular wire of diameter d on a mandril of diameter D_m. The mean radius R of the coil, a term that appears in the derivation that follows, is

$$R = \frac{D_m + d}{2}$$

A free-body diagram of the upper portion of the spring, Fig. 10-7(b), indicates that both a shear force P and a torque $P \times R$ are required at the cut section to maintain equilibrium. Each of these two internal reactions results in a shearing stress. The first, $\tau_1 = P/A$, is distributed uniformly over the area as shown in Fig. 10-7(c), whereas the second, $\tau_2 = Tc/J$, varies uniformly with distance from the center of the wire, Fig. 10-7(d). By superimposing the stress patterns, Fig. 10-7(e), it can be seen that the greatest stress in the spring occurs at the innermost portion of the coil; it is here that the stresses τ_1 and τ_2 add directly to give τ_{max}.

$$\tau_{max} = \tau_1 + \tau_2 = \frac{P}{A} + \frac{Tc}{J}$$

$$= \frac{4P}{\pi d^2} + \frac{16PR}{\pi d^3} = \frac{16PR}{\pi d^3}\left(1 + \frac{d}{4R}\right) \tag{10-13}$$

When the wire diameter is small compared to the mean radius, the term $d/4R$ can be neglected. In this case the stress in the spring is caused principally by the torsional load. In heavy coil springs, however, the term $d/4R$ *cannot* be ignored. In this instance the direct stress is appreciable.

Considerable liberty was taken in the derivation with the expression Tc/J. In theory this term applies only to straight circular bars; here it is used to describe the shearing stress in a curved wire. The derivation, therefore, is in error and particularly so when the springs are heavy and closely coiled. A corrected empirical equation which gives a true picture of the maximum stress states that

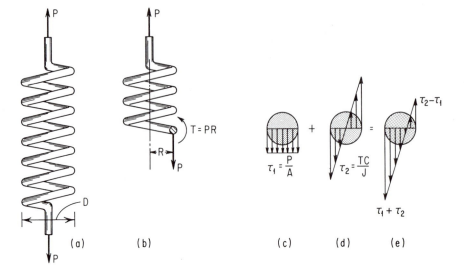

Figure 10-7

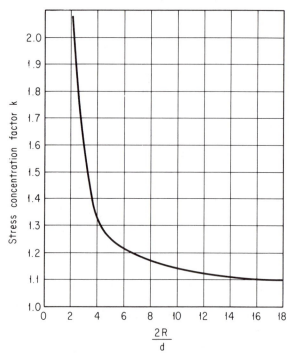

Figure 10-8

$$\tau_{max} = K\left(\frac{16PR}{\pi d^3}\right) \tag{10-14}$$

Where K, *the stress concentration factor*,[1] is a function of the ratio $2R/d$. The value of K can be obtained from the graph of Fig. 10-8.

The deflection of a coil spring is caused, for the most part, by twisting rather than by direct shear. To obtain a relationship for the deflection, one can imagine that the spring, consisting of n coils of wire, is straightened into a shaft $2\pi Rn$ in length, as shown in Fig. 10-9. The deflection δ caused by the rotation of the shaft is approximately

$$\delta = R\theta$$

and, since $\theta = TL/GJ$,

$$\delta = R \times \frac{TL}{GJ} = \frac{R \times PR \times 2\pi Rn}{G \times \pi d^4/32}$$

$$= \frac{64PR^3n}{Gd^4} \tag{10-15}$$

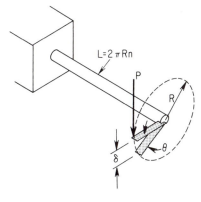

Figure 10-9

A spring constant can now be defined in terms of force and deflection.

$$k = \frac{P}{\delta} = \frac{Gd^4}{64R^3n} \tag{10-16}$$

A *series spring system* is one in which an equal force acts in each spring, and the total deflection is the sum of the individual deflections. For the system shown in Fig. 10-10(a), the total deflection is

$$\delta = \delta_1 + \delta_2 + \delta_3$$

where

$$\delta = \frac{F}{k_1} + \frac{F}{k_2} + \frac{F}{k_3}$$

Dividing both sides of the equation by F gives

$$\frac{\delta}{F} = \frac{1}{k_1} + \frac{1}{k_2} + \frac{1}{k_3}$$

Since δ/F is the reciprocal of an equivalent spring whose constant is k_e, it follows that

[1] The constant K is known as *Wahl's correction factor*. For a complete discussion of spring design, see A. M. Wahl, *Mechanical Springs* (Cleveland, Ohio: Penton Publishing Co., 1944).

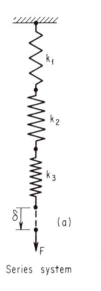

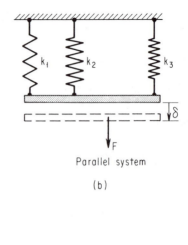

Parallel system

(b)

Series system

Figure 10-10

$$k_e = \cfrac{1}{\cfrac{1}{k_1} + \cfrac{1}{k_2} + \cfrac{1}{k_3}} \qquad (10\text{-}17)$$

In a *parallel system* the springs deform equally as they each share a portion of the load. The equivalent constant in this case is the sum of the individual constants. To prove this, consider the system shown in Fig. 10-10(b). If the bar is to remain horizontal, each spring will support a share of the load.

$$F = F_1 + F_2 + F_3 = k_1\delta + k_2\delta + k_3\delta$$

Dividing this equation by the deflection δ gives an expression for the equivalent spring constant.

$$\frac{F}{\delta} = k_e = k_1 + k_2 + k_3 \qquad (10\text{-}18)$$

In many instances the *equivalent spring* concept is a helpful aid in setting up and solving complicated spring problems.

EXAMPLE 9. A close-coiled helical spring, made of 10 coils of solid 1-in.-diameter steel wire, has a mean coil diameter of 4 in. Determine the maximum stress, corrected for stress concentration, and the elongation of the spring if it supports an axial load of 1200 lb.

Solution: The concentration factor K is found from the graph of Fig. 10-8.

$$\frac{2R}{d} = \frac{2(2)}{1} = 4$$

$$K = 1.33$$

To compute the stress, numerical data are substituted into Eq. (10-14).

$$\tau_{max} = K\left(\frac{16PR}{\pi d^3}\right) = \frac{1.33(16)1200(2)}{\pi(1)^3} = 16,300 \text{ psi}$$

The deflection, by Eq. (10-15), is

$$\delta = \frac{64PR^3n}{Gd^4} = \frac{64(1200)(2)^3(10)}{12 \times 10^6(1)^4} = 0.512 \text{ in.}$$

EXAMPLE 10. Four identical steel springs are joined together, as shown in Fig. 10-11. Each spring has 10 coils of 0.2-in.-diameter wire wound to a mean radius of 2 in. Find the elongation δ if a 10 lb force F is applied as shown. Assume the bar to remain horizontal.

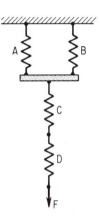

Solution: The constant of each of the springs, by Eq. (10-16), is

$$k = \frac{Gd^4}{64R^3n} = \frac{12 \times 10^6(0.2)^4}{64(2)^310} = 3.75 \text{ lb per in.}$$

The parallel combination of springs A and B is series connected to springs C and D. The equivalent spring constant is

$$k_e = \frac{1}{\dfrac{1}{k+k}+\dfrac{1}{k}+\dfrac{1}{k}} = \frac{1}{\dfrac{1}{2k}+\dfrac{2}{k}} = \frac{2}{5}k = 0.4k$$

Figure 10-11

Thus, the system behaves as a single spring having a constant of 0.4 × 3.75. The deflection is simply F/k_e.

$$\delta = \frac{F}{k_e} = \frac{10}{0.4(3.75)} = 6.67 \text{ in.}$$

10-5 Shaft Couplings

There are many ways in which power can be transmitted directly from one shaft to another; the flanged and bolted coupling shown in Fig. 10-12 illustrates one very practical method. To analyze the strength requirements of the coupling, several assumptions must be made. First, it is assumed that

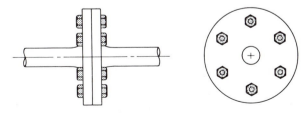

Figure 10-12

torque is transmitted by the bolts alone, and second, that the stress within the bolts is uniform. If more than one circle of bolts is used, it is also assumed that the stress in any circle varies directly with the radius.

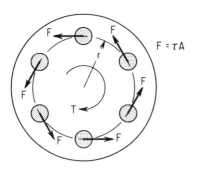

To illustrate, consider a free-body diagram of one face of a coupling, Fig. 10-13. To retain equilibrium, each of the n bolts must help to balance the external torque T.

$$T = nFr \qquad (a)$$

Since each bolt is in direct shear, the force F is

$$F = \tau A \qquad (b)$$

Figure 10-13

where A represents the cross-sectional area of a single bolt. Combining Eqs. (a) and (b) and solving for the stress gives

$$T = n\tau Ar$$

$$\tau = \frac{T}{rAn} \qquad (10\text{-}19)$$

If the bolts in a coupling are arranged in two or more concentric circles, the shearing stress decreases proportionally with the radius. In Fig. 10-14, n_1

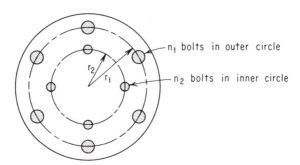

—n_1 bolts in outer circle

—n_2 bolts in inner circle

Figure 10-14

bolts are arranged in circle r_1, and n_2 bolts in circle r_2. To maintain equilibrium, the internal and external torques must balance.

$$T = (rnA\tau)_1 + (rnA\tau)_2 \qquad (a)$$

where

$$\frac{\tau_1}{r_1} = \frac{\tau_2}{r_2} \qquad (b)$$

Equations (a) and (b) can be combined to give an expression for the maximum stress τ_1.

$$\tau_1 = \frac{T}{(rnA)_1 + \frac{r_2}{r_1}(rnA)_2} \qquad (10\text{-}20)$$

It is interesting to note that if r_2 is set equal to zero, and the inner circle of bolts is thus eliminated, Eqs. (10-19) and (10-20) are identical.

Although Eq. (10-20) can prove to be a useful relationship, it is often simpler to substitute data directly into Eqs. (a) and (b).

EXAMPLE 11. Ten $\frac{1}{2}$-in.-diameter bolts are arranged in two concentric circles in a flanged coupling similar to that shown in Fig. 10-14. Find the maximum horsepower that can be transmitted by the coupling if the shaft speed is 315 rpm and the maximum permissible shearing stress is 5000 psi. $r_1 = 8$ in. and $r_2 = 4$ in.

Solution: Shearing stress is directly proportional to the radii; hence, for the inner circle,

$$\tau_2 = \frac{r_2}{r_1}\tau_1 = \frac{4}{8}(5000) = 2500 \text{ psi}$$

To maintain equilibrium the internal and external moments must balance.

$$T = (rnA\tau)_1 + (rnA\tau)_2$$

$$= 8(6)\frac{\pi}{4}\left(\frac{1}{2}\right)^2 5000 + 4(4)\frac{\pi}{4}\left(\frac{1}{2}\right)^2 2500$$

$$= 55{,}000 \text{ lb in.}$$

The permissible horsepower, given by Eq. (10-7), is

$$\text{hp} = \frac{Tn}{63{,}000} = \frac{55{,}000(315)}{63{,}000} = 275 \text{ hp}$$

QUESTIONS, PROBLEMS, AND ANSWERS

10-1. Many assumptions are made in the derivation of formulas that predict the behavior of materials under the action of forces and torques. How valid are the equations?

Ans. If the equations and formulas can be validated experimentally, the assumptions can be considered to be correct. The experienced designer knows the limitation of the formulas, however.

10-2. A 2-in.-diameter steel shaft is subjected to a torque that produces a maximum shearing stress of 20,000 psi. Determine the angle of twist in 50 ft of this shaft. $G_s = 12 \times 10^6$ psi.

Ans. 57.3 deg.

10-3. Determine the maximum shearing stress developed in a hollow shaft that is subjected to a torque of 10,000 lb in. The shaft has an outside diameter of 2 in. and an inside diameter of 1.8 in.

Ans. 18,500 psi.

10-4. Compare the strengths of a solid shaft and a hollow shaft. Both have an outside diameter of D, and the hollow shaft has an inside diameter of $D/2$. Comment on the economics of solid versus hollow shafts.

Ans. Hollow shaft is $\frac{15}{16}$ as strong as solid shaft.

10-5. A solid steel shaft 2 in. in diameter and 10 ft long is acted upon by a torque that produces a shearing stress of 4000 psi within the bar at a radius of $\frac{1}{2}$ in. Find the torque T and the angle of twist of the shaft. $G_s = 12 \times 10^6$ psi.

Ans. 4.58 deg.

10-6. What is the minimum diameter of a solid steel shaft that will not twist through more than 2 deg in 5 ft of length when subjected to a torque of 10,000 lb in.? $G_s = 12 \times 10^6$ psi.

Ans. 1.95 in.

10-7. Is it possible in a torsional design to simultaneously satisfy a desired angle of twist and a limiting stress?

Ans. Not likely if length and diameter cannot be changed, since the elastic properties of the material are involved. While a nearly infinite number of materials may be available, the one with the exact modulus probably does not exist.

10-8. Determine the total angle of twist of the stepped shaft illustrated.

Ans. $\dfrac{66TL}{\pi Gd^4}$

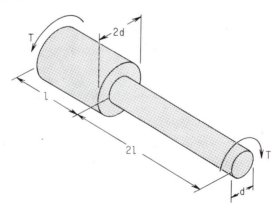

Problem 10-8

10-9. The composite shaft shown is rigidly supported at A and C. If the maximum permissible shearing stress in the brass is 8000 psi, and in the aluminum

4000 psi, what ratio of lengths a/b allows each shaft to be stressed to its limit when subjected to the torque T? $G_b = 6.4 \times 10^6$ psi; $G_a = 3.8 \times 10^6$ psi. *Ans. $a/b = 0.842$.*

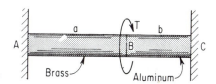

Problem 10-9

10-10. A solid steel shaft is inserted in a hollow brass tube as shown. The shaft and the tube are securely fastened at each end. Find the angle of twist of the coupled system if the applied torque is $T = 1000$ lb in. $G_b = 6.4 \times 10^6$ psi; $G_s = 12 \times 10^6$ psi.

Ans. 6.82 deg.

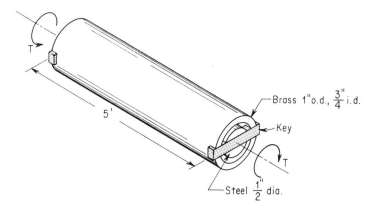

Problem 10-10

10-11. Mechanical elements, like electrical elements, can be coupled in series or in parallel. For example, equal voltage across a set of resistors is characteristic of a parallel network. Equal torques in coupled shafts would define a parallel mechanical network. The shafts of Prob. 10-10 are _____ coupled.

Ans. Series.

10-12. Proceeding with the analogy of Prob. 10-11, electrical transformers are multipliers. List two "mechanical" transformers.

Ans. The gear and the lever.

10-13. In the figure, a torque of 2000 lb in. is applied to end B of the geared 1-in.-diameter shafts. Determine the torque at A necessary to maintain equilibrium and the angle of twist of end B with respect to end A. Assume that the shafts are supported by bearings to prevent bending. $G_b = 6.4 \times 10^6$ psi; $G_s = 12 \times 10^6$ psi.

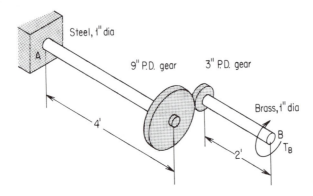

Problem 10-13

Ans. 6000 lb in.; 46.4 deg.

10-14. Shafts *A*, *B*, and *C* in the figure are interconnected by three identical bevel gears. Find the torques $T_{A'}$ and $T_{B'}$ supplied by the respective supports if a torque of 25,000 lb in. acts as shown. The three shafts are steel and are supported to prevent bending.

Ans. $T_A = 24{,}000$ lb in.; $T_B = 1000$ lb in.

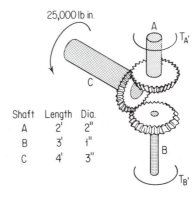

Shaft	Length	Dia.
A	2'	2"
B	3'	1"
C	4'	3"

Problem 10-14

10-15. What is the maximum horsepower a solid steel shaft 2 in. in diameter can transmit at 400 rpm if the maximum permissible shearing stress is 10,000 psi?

Ans. 100 hp.

10-16. A solid steel shaft is to transmit 100 hp at 500 rpm. What is the necessary diameter if it is not to twist more than 2 deg per foot or be stressed more than 10,000 psi in shear? $G_s = 12 \times 10^6$ psi.

Ans. 1.86 in.

10-17. What horsepower can a $\frac{1}{2}$-in.-diameter steel shaft transmit at a speed of 10,000 rpm if the working stress in shear is 8000 psi? $G_s = 12 \times 10^6$ psi.

Ans. 31.2 hp.

10-18. A motor delivers 50 hp at 630 rpm to one end of a steel line-shaft 2 in. in diameter and 10 ft long. The power drives two machines: one at the midpoint of the shaft consuming 30 hp and the other at the extreme end consuming 20 hp. Determine the maximum shearing stress in the shaft and the relative angle of twist between the ends of the shaft.

Ans. 3180 psi; 1.28 deg.

10-19. In the figure, a steel line-shaft is driven at 315 rpm by a 50-hp motor at C. What are the minimum diameters of the three shaft segments if the shearing stress in any segment is not to exceed 10,000 psi?

Ans. $d_{AB} = 1.01$ in.; $d_{BC} = 1.53$ in.; $d_{CD} = 1.15$ in. Use nominal values.

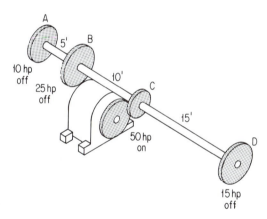

Problem 10-19

10-20. What is the torsional spring constant of a 1-in.-diameter steel shaft 10 ft long? $G_s = 12 \times 10^6$ psi.

Ans. 9.81×10^3 lb in./rad.

10-21. Find the length of a 2-in.-diameter brass shaft with a torsional spring constant of 4×10^6 lb in./rad. $G_b = 6.4 \times 10^6$ psi.

Ans. 2.51 in.

10-22. Compute a factor that will convert kip feet per degree to pound inches per radian.

Ans. 1 lb in./rad $= 1.45 \times 10^6$ kip ft/deg.

10-23. Find the length of a 2-in.-diameter steel shaft that has the same spring constant as a 4-in.-diameter brass shaft 20 ft long. $G_s = 12 \times 10^6$ psi; $G_b = 6.4 \times 10^6$ psi.

Ans. 2.34 ft.

10-24. Determine the equivalent spring constants of the systems shown and the
to angles of twist at the point of application of the torque if for each case
10-28. $T = 10{,}000$ lb in. In each instance indicate whether the spring elements are in series or parallel.

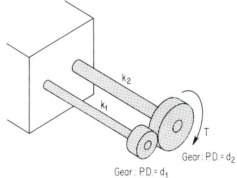

Problem 10-24

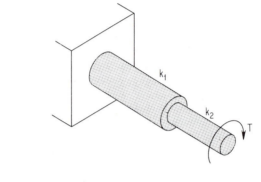

Gear: P.D.= d_1 *Problem 10-25*

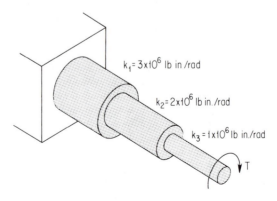

Problem 10-26

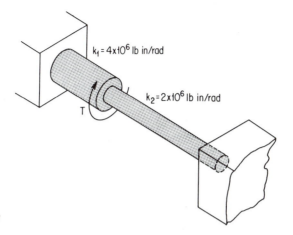

$k_1 = 4 \times 10^6$ lb in/rad

$k_2 = 2 \times 10^6$ lb in/rad

T

Problem 10-27

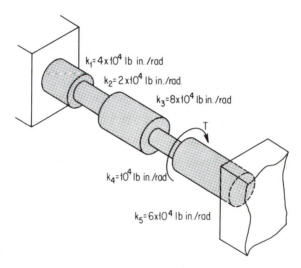

$k_1 = 4 \times 10^4$ lb in./rad

$k_2 = 2 \times 10^4$ lb in./rad.

$k_3 = 8 \times 10^4$ lb in./rad

T

$k_4 = 10^4$ lb in./rad

$k_5 = 6 \times 10^4$ lb in./rad

Problem 10-28

Ans. (10-24) $\theta = T(k_1 + k_2)/k_1 k_2$; series.
Ans. (10-25) $\theta = T/(4k_1 + k_2)$; modified parallel.
Ans. (10-26) 1.05 deg; series.
Ans. (10-27) 0.096 deg; parallel.
Ans. (10-28) 8.77 deg; series-parallel combination.

10-29. A torsion bar with a spring constant k is cut into n equal lengths. What is the spring constant of each portion?

Ans. nk.

10-30. List at least four typical engineering applications of the torsional spring.

Ans. The torsion-bar automotive suspension and. . . .

Problems 10-31 through 10-36 are problems in helical spring design in which certain data are given. Determine the unknown quantities for each problem. *Note:* Where the stress is one of the included quantities, use Eq. (10-13).

	P (lb)	R (in.)	d (in.)	n	G (psi $\times 10^6$)	τ_{max} (psi)	δ (in.)	k (lb/in.)
10-31.	100	3	1	10	12			
10-32.	10		$\frac{1}{2}$	10	6.4		2	
10-33.		3	$\frac{3}{4}$		12	25,000		100
10-34.	50	$\frac{1}{2}$	$\frac{1}{8}$		8			10
10-35.	31.4	1			12	20,000	2	
10-36.	31.4			20	6.4	10,000		40

Ans. (10-31) 1660 psi; 0.144 in.; 694 lb/in.
      (10-32) 5 in.; 2090 psi; 5 lb/in.
      (10-33) 649 lb; 22 coils; 6.49 in.
      (10-34) 24.4 coils; 69,300 psi; 5 in.
      (10-35) 0.2 in.; 19.1 coils; 15.7 lb/in.
      (10-36) 0.7 in.; 0.22 in.; 0.785 in.

10-37. A 10-turn helical compressive spring with a $\frac{1}{2}$-in. steel wire and a 4-in. outside coil diameter is used to exert an axial force on a clutch plate. The free length of the spring is $8\frac{1}{2}$ in., and the compressed length is $5\frac{1}{2}$ in. Determine the maximum stress in the wire and the pressure force exerted against the clutch plate. Note: Correct for stress concentration.

Ans. 55,700 psi.; 656 lb.

10-38. A helical spring whose constant is 400 lb per in. is cut into quarters, and the four pieces are then combined in parallel. What is the equivalent spring constant of the parallel combination?

Ans. 6400 lb/in.

10-39. Two leaf springs are separated by a coil spring as illustrated. If the constant k is the same for each spring, what is the equivalent spring constant of the system?

Ans. $3k/2$.

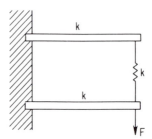

F   *Problem 10-39*

10-40. A simulated automotive suspension is shown. Locate the center of gravity as measured by x if the automobile whose weight is W is to remain horizontal.

Ans. 0.483 *l.*

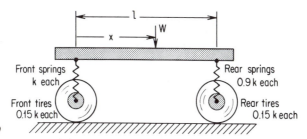

Front springs k each
Rear springs 0.9 k each
Front tires 0.15 k each
Rear tires 0.15 k each

Problem 10-40

10-41. The rigid bar AB, shown, weighs 400 lb and supports a load $P = 1000$ lb. If the free length of the springs are equal prior to loading, where should the load P be placed if the bar is to remain horizontal?

Ans. 6.4 ft to right of A.

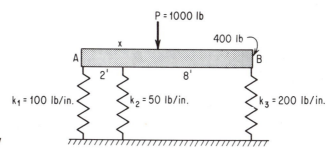

P = 1000 lb

400 lb

A   x   B

2'   8'

$k_1 = 100$ lb/in. $k_2 = 50$ lb/in. $k_3 = 200$ lb/in.

Problem 10-41

10-42. Two springs are joined with no prestress to a lever. Find the reactions at the supports A and B if a force of 100 lb acts as shown.

Ans. $R_A = 200$ lb; $R_B = 100$ lb.

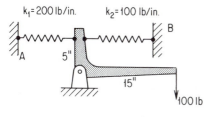

$k_1 = 200$ lb/in. $k_2 = 100$ lb/in.

B

A   5"

15"

100 lb

Problem 10-42

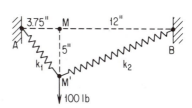

3.75"   M   12"

A   5"   B

k_1 k_2

M'

100 lb

Problem 10-43

10-43. In the figure, two springs are supported taut, but unstressed, between A and B. The junction M moves 5 in. downward when a 100-lb force is applied slowly, as shown. Find the constant of each spring.

Ans. $k_1 = 38.1$ lb/in.; $k_2 = 61.9$ lb/in.

10-44. How many $\frac{3}{8}$-in. bolts arranged in a 5-in.-diameter bolt circle would be required to transmit 100 hp at 315 rpm through a flanged coupling? The maximum permissible shearing stress is 10,000 psi.

Ans. 8 bolts.

10-45. The coupling shown consists of two sprockets held together by a flexible roller chain. The chain has 32 pins, each with a diameter of $\frac{1}{8}$ in. located in a circle of 6-in. diameter. What is the theoretical horsepower rating of this coupling at a speed of 630 rpm? The shearing stress in the pins is not to exceed 10,000 psi.

Ans. 118 hp.

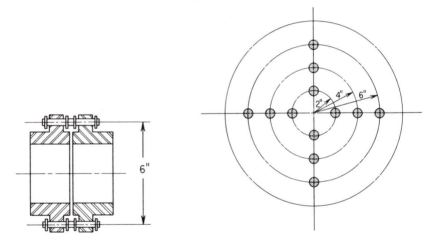

Problem 10-45   **Problem 10-46**

10-46. What is the maximum shearing stress developed in the bolts of the coupling shown? The coupling transmits 400 hp at 630 rpm, and all bolts have diameters of $\frac{3}{8}$ in.

Ans. 9710 psi.

10-47. How many bolts are required in a coupling that joins two 2-in.-diameter shafts? A torque that acts on the shafts produces a shearing stress of 12,000 psi in each. The allowable shearing stress in the bolts is 10,000 psi, the diameter of the bolt circle is 8 in., and the bolts are $\frac{1}{4}$ in. in diameter.

Ans. 10 bolts.

Chapter 11

SHEAR AND
MOMENT IN BEAMS

Structural members capable of sustaining loads normal to their axes are called *beams*. The term "beam" generally brings to mind large and massive objects employed principally in the building trades. Although this is as good a description as any, there are many types of beams that are used in machines and controls whose very function depends on their ability to bend without becoming overstressed; the leaf-spring is a typical example.

11-1 Classification of Beams and Loads

There are six fundamental methods of supporting beams; these are illustrated in Fig. 11-1. A beam that rests on two supports in a manner which offers, at most, one restraint along the beam axis is called a *simple beam*. A *cantilever beam* is held at one end only by a support which is capable of sustaining both moments and forces. As the name implies, an *overhanging*

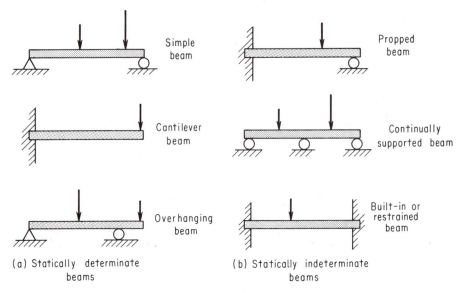

(a) Statically determinate   (b) Statically indeterminate
          beams   beams

Figure 11-1

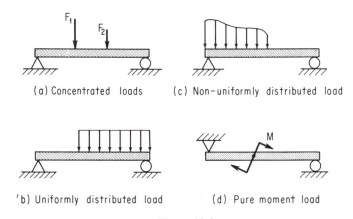

(a) Concentrated loads   (c) Non-uniformly distributed load

'b) Uniformly distributed load   (d) Pure moment load

Figure 11-2

beam is one that extends beyond its supports at either or both ends. If the external forces and moments that support the beam can be found by the equations of statics alone, the beam is *statically determinate,* whereas beams that have more supports than are necessary to maintain equilibrium are *statically indeterminate.* Beams that are *propped, continually supported,* or *built-in* are examples of this latter class.

There are four fundamental types of loads which can act on a beam; these are illustrated in Fig. 11-2. The *concentrated load,* the simplest, acts at a single point on the beam, and the *distributed load*, which may be *uniform* or *non-uniform*, acts over a given length of the beam. A fourth type of loading occurs when the beam is subjected to a *pure moment*. Although these loads are illustrated as acting singly, the possibility exists, of course, that they can act in any combination on a single beam.

11-2 Shear and Moment Diagrams

Stresses and deflections in beams are functions of the inernal reactions, forces, and moments, and for this reason it is convenient to "map" these internal reactions and form diagrams that give a complete picture of the magnitudes and directions of the forces and moments that act throughout the beam. These graphs are called *shear* and *moment diagrams.*

Consider the simply supported beam shown in Fig. 11-3(a). The reactions at the supports, computed through the equations of static equilibrium, are Pb/l and Pa/l. If the beam is cut by a transverse plane at a distance x from the left end, a force V and a moment M are required to keep the severed section in equilibirum. These two internal reactions, shown in Fig. 11-3(b), are the *shear* V and the *moment* M at x. A force summation $\sum F_y = 0$ and a moment summation $\sum M = 0$ on the left-hand portion of the free-body will show V and M to be

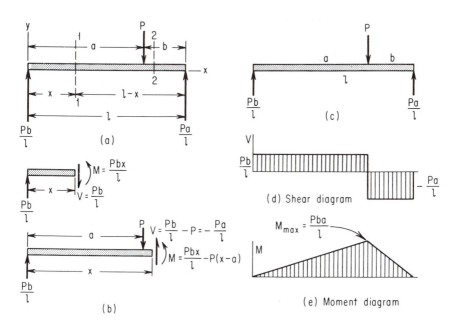

Figure 11-3

$$V = \frac{Pb}{l}$$

and

$$M = \frac{Pb}{l}x$$

The shear remains unchanged between the limits of $x = 0$ and $x = a$, whereas the moment between these limits increases uniformly with the distance x. At $x = a$ the moment is a maximum and has a value of

$$M = \frac{Pba}{l}$$

If a second free-body is drawn, this time cut by a transverse plane to the right of load P, the shear and moment are

$$V = \frac{Pb}{l} - P = \frac{P(b - l)}{l} = -\frac{Pa}{l}$$

and

$$M = \frac{Pb}{l}x - P(x - a) = \frac{Pa}{l}(l - x)$$

A study of the moment equation for the region $a < x < l$ indicates that as x increases, M decreases, and that at $x = l$ the moment is zero.

The graphs, or "maps" of the shear and the moment for all values of x between 0 and l are called the *shear* and *moment diagrams*. A uniform and consistent sign convention must be used in plotting these diagrams: *External forces that act upward on the left-hand free-body cause positive shear: positive moments are caused by forces and moments that tend to bend the beam in a way to cause its radius of curvature to be upward.* This sign convention is illustrated in Fig. 11-4.

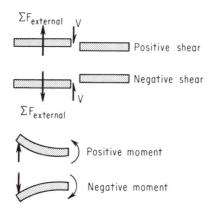

Figure 11-4

The examples that follow will further illustrate the method of plotting shear and moment diagrams. It will be noted that the free-body diagram, sketched between segments of the beam where changes in loading occur, is a tremendous aid in writing mathematical relationships for shear and moment.

EXAMPLE 1. Draw the shear and moment diagrams for the simply supported beam shown in Fig. 11-5(a) and determine the maximum values of V and M.

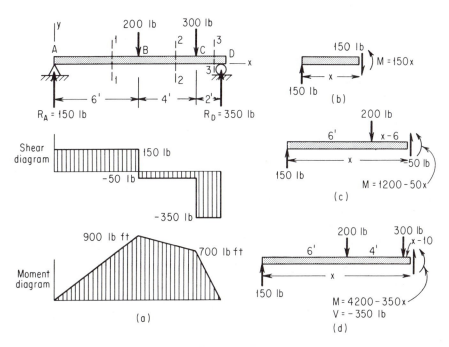

Figure 11-5

Solution: The reactions at the supports A and D are first computed:

$$\sum M_A = 0$$
$$12R_D - 6(200) - 10(300) = 0$$
$$R_D = 350 \text{ lb}$$

and

$$\sum M_D = 0$$
$$12R_A - 2(300) - 6(200) = 0$$
$$R_A = 150 \text{ lb}$$

Check:

$$\sum F_y = 0$$
$$200 + 300 - 350 - 150 = 0$$

To plot the shear and the moment, free-body diagrams of three sections of the beam are required; these are illustrated in Figs. 11-5(b), (c), and (d).

The first free-body results when a transverse cut is made at section 1-1, between points A and B, at a distance x from A. The shear in this section is constant and equal to

$$V_{AB} = +150 \text{ lb}$$

whereas the moment increases uniformly as x increases:

$$M_{AB} = +150x \text{ lb ft}$$

At $x = 6$ ft, the moment is $150(6) = 900$ lb ft.

The second free-body diagram, Fig. 11-5(c), is representative of a segment of the beam cut at section 2-2 between points B and C. The shear in this region is negative, since the sum of the external forces is downward.

$$V_{BC} = 150 - 200 = -50 \text{ lb}$$

The moment at the cut, an internal reaction required to maintain equilibrium, is

$$\begin{aligned} M_{BC} &= 150x - 200(x - 6) \\ &= 1200 - 50x \end{aligned}$$

Thus, in the portion of the beam lying between $x = 6$ ft and $x = 10$ ft, the moment decreases uniformly as x becomes greater; the limiting value of the moment given by this expression occurs at $x = 10$ ft.

$$M = 1200 - 50(10) = 700 \text{ lb ft}$$

The third and final free-body, that of a section to the left of section 3-3, shows the shear and moment to be

$$V_{CD} = 150 - 200 - 300 = -350 \text{ lb}$$

and

$$\begin{aligned} M_{CD} &= 150x - 200(x - 6) - 300(x - 10) \\ &= 4200 - 350x \end{aligned}$$

The shear is negative and constant in this segment, whereas the moment decreases with x; at $x = 12$ ft, the moment is zero.

$$M = 4200 - 350(12) = 0$$

When sketched, the shear and moment diagrams present a picture of the entire shear force and moment distribution in the beam. The greatest numerical values of V and M are

$$V = -350 \text{ lb}$$
$$M = +900 \text{ lb ft}$$

EXAMPLE 2. Draw the shear and moment diagrams for the cantilever beam shown in Fig. 11-6(a).

Solution: The shear in a section of the beam between points A and B is found by inspection to be

$$V_{AB} = -200 \text{ lb}$$

and between points B and C

$$V_{BC} = -700 \text{ lb}$$

Equations for the moment in the beam as a function of x, measured to the right from A, are

$$M_{AB} = -200x$$

and

$$M_{BC} = -200x - 500(x - 4)$$
$$= 2000 - 700x$$

The values of the moment at the *critical points* B and C are found by substitution; thus at $x = 4$ ft

$$M = -200(4) = -800 \text{ lb ft}$$

and at $x = 8$ ft

$$M = 2000 - 700(8) = -3600 \text{ lb ft}$$

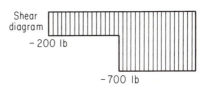

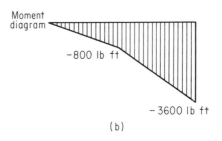

(b)

Figure 11-6

When plotted, the shear and moment diagrams appear as shown in Fig. 11-6(b).

The general procedure for drawing shear and moment diagrams outlined in the previous paragraphs applies to any loading arrangement; it follows, however, that these diagrams become more complex as the loading becomes more involved. The examples that follow will illustrate the drawing of shear and moment diagrams for the more complex loading arrangements.

EXAMPLE 3. The cantilever beam in Fig. 11-7(a) is subjected to both a couple and a concentrated load as shown. Draw the shear and moment diagrams for the beam.

Solution: Two free-body diagrams are required, one which will describe the internal reactions in a section of the beam between points A and B, and another between points B and C. In the first free-body, Fig. 11-7(b), the shear is zero and the moment is constant and equal to $+100$ lb ft. The second free-body, Fig. 11-7(c), indicates the internal reactions to consist of both shear and moment; the former is constant, whereas the latter decreases with x.

$$V_{BC} = -200 \text{ lb}$$

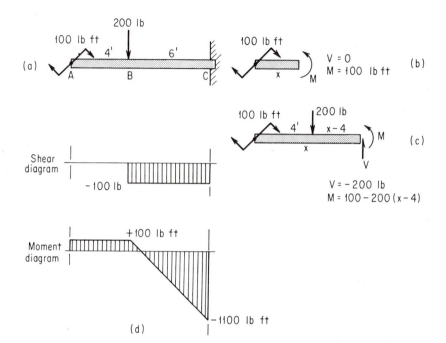

Figure 11-7

and

$$M_{BC} = 100 - 200(x - 4)$$
$$= 900 - 200x$$

and at $x = 10$ ft

$$M = 900 - 200(10) = -1100 \text{ lb ft}$$

The diagrams that descriptively picture the shear and moment in this beam are shown in Fig. 11-7(d).

EXAMPLE 4. Draw the shear and moment diagrams for the beam shown in Fig. 11-8(a) and determine the maximum value of the moment.

Solution: The reactions are determined by a moment summation, first about A and then about C:

$$\sum M_A = 0$$
$$10R_C - 6(100)3 = 0$$
$$R_C = 180 \text{ lb}$$

and

$$\sum M_C = 0$$
$$10R_A - 6(100)7 = 0$$
$$R_A = 420 \text{ lb}$$

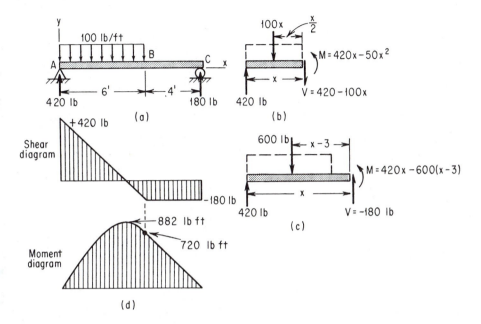

Figure 11-8

Check:

$$\Sigma \, F_y = 0$$
$$6(100) - 420 - 180 = 0$$

A free-body diagram is next drawn, Fig. 11-8(b), for a portion of the beam having a length of x, where x assumes any value between zero and 6 ft. The external forces that act on this free-body are the upward reactions of 420 lb at A and the downward weight force of $100x$ lb. The vertical shear at the transverse cut is, therefore,

$$V_{AB} = 420 - 100x$$

By this equation, the curve of the shear is shown to be a straight line sloping downward to the right; at a value of $x = 0$, the shear is $+420$ lb, and at a value of $x = 6$ ft, the shear is -180 lb.

The same two external forces that produced the vertical shear also tend to bend the beam. The sum of the moments of these two forces is equal to the internal moment at the cut:

$$M_{AB} = 420x - 100x\left(\frac{x}{2}\right) = 420x - 50x^2$$

Since the magnitude of the moment is a function of both x and x^2, the curve that represents the moment is a parabola. Careful plotting of this parabola will show *the maximum moment to occur at a point in the beam where the shear is zero.* This statement, which is true for any type of loading, places the maximum moment at $x = 4.2$ ft; thus

$$V_{AB} = 420 - 100x = 0$$
$$x = 4.2 \text{ ft}$$

The magnitude of the moment, evaluated at three points: $x = 0$, $x = 4.2$, and $x = 6$, is all that need be computed to adequately sketch the moment diagram between the limits of $x = 0$ and $x = 6$ ft.

At $x = 0$:　　$M = 420(0) - 50(0)^2 = 0$

At $x = 4.2$:　$M = 420(4.2) - 50(4.2)^2 = 882 \text{ lb ft}$

At $x = 6$:　　$M = 420(6) - 50(6)^2 = 720 \text{ lb ft}$

A free-body diagram for a portion of the beam to the left of a transverse cut between B and C, Fig. 11-8(c), indicates the shear to be constant and negative:

$$V_{BC} = 420 - 600 = -180 \text{ lb}$$

and the moment to be

$$M_{BC} = 420x - 600(x - 3) = 1800 - 180x$$

Thus, the moment in the beam between the limits of $x = 6$ ft and $x = +10$ ft decreases uniformly to zero from an initial value of 720 lb ft.

11-3 Relationships Between Load, Shear, and Moment

There are several approaches one can use to construct shear and moment diagrams; the previous section illustrates the purely mathematical method. Equations are derived for the shear and moment in the beam as a function of a distance. On the basis of these equations, V and M are plotted from point to point in the beam. While this is surely a direct approach, there are several concepts that can be employed to reduce the problem to one of simple geometry.

Consider the purely arbitrary loading arrangement shown in Fig. 11-9(a). A non-uniform load is selected simply to illustrate a general situation. A section of the beam of infinitesimal length Δx is isolated, and a free-body diagram is drawn as shown in Fig. 11-9(b). On the left-hand face, the shear and moment are assumed to be simply V and M. At the midpoint of the element, a portion of load $w\Delta x$ acts downward as shown. To maintain equilibrium, the shear and the moment on the extreme right face differ from those on the left by small incremental values ΔV and ΔM.

The equations of static equilibrium require the resultant force and resultant moment both to be zero; thus

$$\sum F_y = 0$$
$$V + w\Delta x - (V + \Delta V) = 0$$
$$\Delta V = w\Delta x \qquad\qquad \text{(a)}$$

and

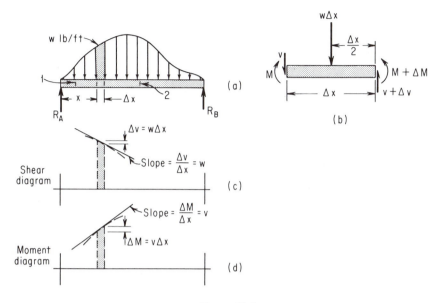

Figure 11-9

$$\sum M = 0$$

$$M - V\Delta x - w\Delta x\left(\frac{\Delta x}{2}\right) + (M + \Delta M) = 0$$

$$\Delta M = V\Delta x \qquad\qquad (b)$$

Since Δx is small, the term $w(\Delta x)^2/2$ approaches zero in value and is dropped from the computations. The physical interpretation of Eqs. (a) and (b) is the key to the method about to be described. The first equation states that the incremental change ΔV is equivalent to an area of a rectangle of height w and width Δx: the amount of external load on the elemental length of beam. *The difference in the value of the shear between points 1 and 2 in the beam is the sum of the incremental values ΔV or simply the area of the loading diagram between points 1 and 2.*

$$V_2 - V_1 = \sum \Delta V = \sum w\Delta x = \text{area of loading diagram between}$$
$$\text{points 1 and 2} \qquad\qquad (11\text{-}1)$$

Rearrangement of terms in Eq. (a) contributes a second important fact: *the slope of the shear diagram at a given point is equal to the magnitude of the load at that point.*

$$\frac{\Delta V}{\Delta x} = w \qquad\qquad (11\text{-}2)$$

The same reasoning, applied to Eq. (b), gives

$$\Delta M = V \Delta x$$

$$M_2 - M_1 = \sum \Delta M = \sum V \Delta x = \text{area of shear diagram}$$

between points 1 and 2   (11-3)

The difference in the value of the moment between points 1 and 2 in the beam is equal to the area enclosed by the shear diagram between these two points, and

$$\frac{\Delta M}{\Delta x} = V \qquad (11\text{-}4)$$

The slope of the moment diagram at a point is equivalent to the magnitude of the shear at that point. This statement is most significant, since it establishes a simple way of determining the maximum moment. Mathematically, maxima and minima occur in a curve when its slope is either zero or undefined—simply stated, *maxima or minima in the moment diagram occur at those points where the shear passes through zero.*

EXAMPLE 5. Use the relationship between loading, shear, and moment to sketch the shear and moment diagram for the beam shown in Fig. 11-10.

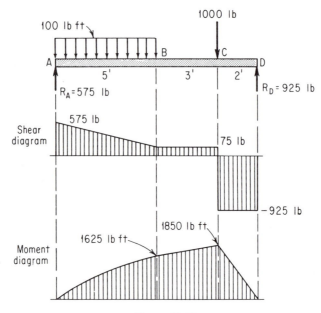

Figure 11-10

Solution: The reactions at *A* and *D* are found in the usual way.

$$\sum M_A = 0$$

$$10R_D = 100(5)2.5 + 1000(8)$$

$$R_D = 925 \text{ lb}$$

$$\sum M_D = 0$$
$$10R_A = 100(5)(7.5) + 1000(2)$$
$$R_A = 575 \text{ lb}$$

Check:

$$\sum F_y = 0$$
$$100(5) + 1000 - 925 - 575 = 0$$

The value of the shear at A is equal to the reaction $R_A = 575$ lb; it is an upward force and, therefore, positive. Next, the change in shear between points A and B is equal to the area of the loading diagram between these two points; thus

$$\Delta V = V_B - V_A = [\text{Area}_{AB}]_{\text{loading}}$$
$$= V_B - 575 = -100(5)$$
$$= V_B = 575 - 500 = +75 \text{ lb}$$

Note: The area representing the distributed load is negative since the load is downward.

A portion of the moment diagram can now be sketched. Since the slope of the moment curve is the magnitude of the shear, the moment diagram has a positive. but diminishing, slope between points A and B. At A the slope is $+575$ and at B, $+75$. The general shape of the curve is thus established.

Equation (11-3) is next used to find the value of the moment at B.

$$\Delta M = M_B - M_A = [\text{Area}_{AB}]_{\text{shear}}$$
$$M_B = \left(\frac{575 + 75}{2}\right)5 + 0 = 1625 \text{ lb ft}$$

Between points B and C, the shear is unchanged, since the area of loading is zero.

$$V_C - V_B = [\text{Area}_{BC}]_{\text{loading}} = 0$$
$$V_C = V_B = +75 \text{ lb}$$

Since $\Delta w / \Delta x = 0$ between these two points, the shear is represented by a horizontal line.

The moment continues to increase, however, and the difference in moment between points B and C is the area under the shear diagram between these two points:

$$M_C - M_B = [\text{Area}_{BC}]_{\text{shear}}$$
$$M_C = 1625 + 75(3)$$
$$M_C = 1625 + 225 = 1850 \text{ lb ft}$$

Since the shear is constant and positive between B and C, the moment diagram is a straight line having a positive slope.

The concentrated 1000-lb load at C causes the abrupt change in the shear from $+75$ lb to -925 lb; no additional load acts on the beam until the reaction R_D is reached, which brings the shear to zero at D.

The moment diagram between C and D is a straight line having a negative slope, as shown. If the computations are correctly carried out, the moment at D for this beam should be zero.

$$\Delta M = M_D - M_C = [\text{Area}_{CD}]_{\text{shear}}$$
$$M_D - 1850 = -925(2)$$
$$M_D = 0$$

The shear passes through the zero axis at C, the point of maximum moment.

$$M_{\max} = +1850 \text{ lb ft, 8 ft to right of } A$$

11-4 Moment Diagrams by Superposition

There are computations involving the deflections of beams which require a knowledge of the properties of the areas enclosed by moment diagrams rather than the principal values. To aid in this sort of computation, moment diagrams are sometimes drawn "in parts;" each load is imagined to produce a separate moment diagram. The *superposition*, or sum, of these diagrams is equivalent to the diagram drawn in the usual manner.

Consider the four fundamental cantilever loadings and their respective moment diagrams shown in Fig. 11-11. These diagrams are simple curves: rectangles, triangles, and parabolas. Next, consider a cantilever beam subjected to a combined loading of three concentrated forces, as shown in Fig. 11-12(a). By considering the loads individually, three separate moment diagrams could be constructed, as in Fig. 11-12(b). The composite moment diagram is actually the sum of these three parts, shown in Fig. 11-12(c). If areas and centroids of the moment diagrams are required, it is far simpler to work with the diagrams drawn in parts rather than in composite form.

Before we consider examples, it would be well to review some of the properties of simple areas: the rectangle, triangle, parabola, and cubic parabola. The areas and their centroids are listed in Table 11-1. The order in which

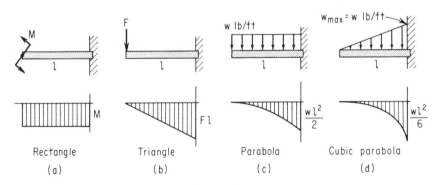

| Rectangle | Triangle | Parabola | Cubic parabola |
| (a) | (b) | (c) | (d) |

Figure 11-11

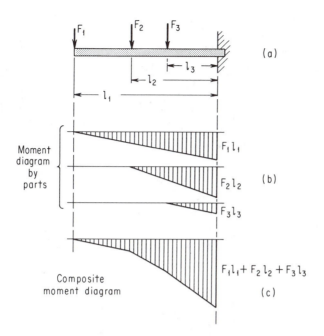

Figure 11-12

Table 11-1. Properties of Areas

		Area	*Centoid* $\bar{x}$
	Rectangle	bh	$b/2$
	Triangle	$\frac{1}{2}bh$	$b/3$
	Parabola	$\frac{1}{3}bh$	$b/4$
	Cubic Parabola	$\frac{1}{4}bh$	$b/5$

265

these are listed offers a rather interesting memory aid: both the areas and their centroids become progressively smaller as the degree of the curve increases.

EXAMPLE 6. A cantilever beam is subjected to a concentrated load of 100 lb and a distributed load of 50 lb per ft, as shown in Fig. 11-13(a). Draw the moment diagram by parts and determine the moment of the area of this diagram, first about end B, and then about end A.

Solution: The moment diagram consists of two areas, a triangle and a parabola. The moment at B caused by the concentrated load is

$$M_1 = 100(8) = 800 \text{ lb ft}$$

and the moment caused by the distributed load is

$$M_2 = 50(4)(\tfrac{4}{2}) = 400 \text{ lb ft}$$

The moments of these areas about a vertical line drawn at the extreme right side of the diagram would be

$$\text{Area}_{AB}\bar{X}_B = (A\bar{X})_1 + (A\bar{X})_2$$
$$= -[\tfrac{1}{2}(8)800]\tfrac{8}{3} - [\tfrac{1}{3}(4)400]\tfrac{4}{4}$$
$$= -9070 \text{ lb ft}^3$$

The sum of the moments of the areas about the extreme left end is

$$\text{Area}_{AB}\bar{X}_A = (A\bar{X})_1 + (A\bar{X})_2$$
$$= -[\tfrac{1}{2}(8)(800)(\tfrac{2}{3} \times 8)] - [\tfrac{1}{3}(4)(400)(4 + \tfrac{3}{4} \times 4)]$$
$$= -20,800 \text{ lb ft}^3$$

The concept of *moments by parts* can be extended to all varieties of beams

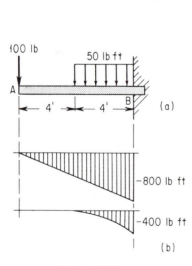

Figure 11-13

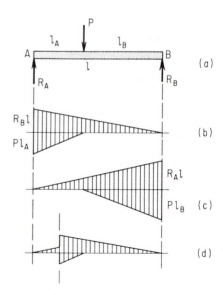

Figure 11-14

and modes of loading. Consider, for example, a simply supported beam, Fig. 11-14(a), acted upon by a concentrated load P. There are several ways of constructing a moment diagram by parts for this beam; if moments are drawn with respect to a reference line at the extreme left end, for instance, the diagram would appear as shown in Fig. 11-14(b). If a reference at the right end is selected, the diagram appears as shown in Fig. 11-14(c), and if an arbitrary line is drawn within the beam, the diagram appears as illustrated in Fig. 11-14(d). With care, a reference can usually be selected for any beam that will yield the simple areas described in Table 11-1.

EXAMPLE 7. Draw the moment diagram by parts for the loading arrangement shown in Fig. 11-15(a). Select a reference line that passes through the left support, and find the moment of the area of the moment diagram about this support.

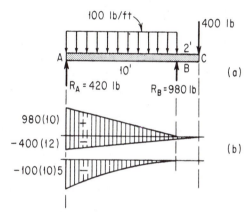

Figure 11-15

Solution: The reactions R_A and R_B are computed:

$$\sum M_A = 0$$
$$100(10)5 - 10R_B + 400(12) = 0$$
$$R_B = 980 \text{ lb}$$
$$\sum M_B = 0$$
$$10R_A - 100(10)5 + 400(2) = 0$$
$$R_A = 420 \text{ lb}$$

Check:

$$980 + 420 - 100(10) - 400 = 0$$

There are three parts to the moment diagram: the positive portion caused by the upward reaction R_B, and two negative portions, one caused by the concentrated load and the other by the distributed load.

Written symbolically, the desired quantity is $\text{Area}_{AC}\bar{X}_A$; thus

$$\text{Area}_{AC}\bar{X}_A = \tfrac{1}{2}(10)9800(\tfrac{10}{3}) - \tfrac{1}{2}(12)4800(\tfrac{12}{3}) - \tfrac{1}{3}(10)5000(\tfrac{10}{4})$$
$$= 6470 \text{ lb ft}^3$$

QUESTIONS, PROBLEMS,
AND ANSWERS

11-1. When the external forces and moments that support an object can be found by the equations of statics alone, the object is said to be _____
_____.

Ans. Statically determinate.

11-2. The beam shown in (a) is statically determinate; there is no question that the reactions are 150 lb each. Now, the loading is reversed as in (b); the supports become the forces and the forces become the supports. Are the reactions R_1, R_2, and R_3 100 lb each?

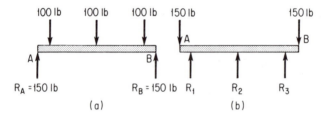

Problem 11-2

Ans. Try to find the three reactions by the equations of statics alone; you will find that it cannot be done. The best that you will be able to do is say that $R_1 = R_3$ and that $R_1 + R_2 + R_3 = 300$ lb; there are an infinite number of answers. Another way of looking at it is to ask yourself "Are there more supports than necessary to maintain stability"; the answer is yes, R_2 could be discarded.

11-3 Write the equations for the shear and moment as functions of x measured
to from the extreme left end A of the beams shown in the respective figures,
11-12. and then graph the shear and moment diagrams. Determine the maximum shear and maximum moment for each case. Neglect the weight of the beam in each instance.

Ans. (4-3) $V = -P$; $M = -Px$.
  (4-4) $V = 0$; $M = 200$ lb ft.
  (4-5) $0 < x < 2$: $V = M = 0$; $2 < x < 10$: $V = -100(x - 2)$,
    $M = -50(x - 2)^2$.
  (4-6) $V_{max} = -5$ kips; $M_{max} = -15$ kip ft.
  (4-7) $V_{max} = 150$ lb; $M_{max} = 400$ lb ft.
  (4-8) $V = 500 - 100x$; $M = 500x - 50x^2$.

(4-9) $M_{max} = 4$ ton ft at $x = 4$ ft.
(4-10) $V_{max} = -640$ lb; $M_{max} = 1710$ lb ft.
(4-11) $0 < x < 4$: $V = -120x$, $M = -60x^2$; $4 < x < 16$:
   $V = -120x + 1200$, $M = 1200(x - 4) - 60x^2$.
(4-12) $M_{max} = -800$ lb ft at $x = 8$ ft.

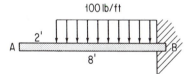

Problem 11-3

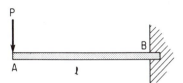

Problem 11-4

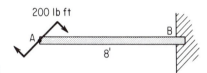

Problem 11-5

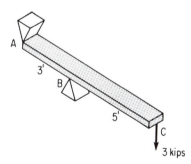

Problem 11-6

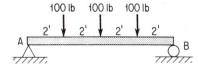

Problem 11-7

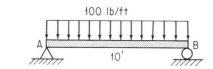

Problem 11-8

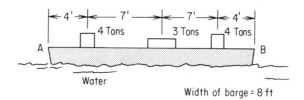

Water

Width of barge = 8 ft

Neglect weight of barge and
assume cargo weight to be
concentrated along lines drawn
from starboard to port.

Problem 11-9

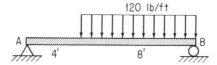

Problem 11-10

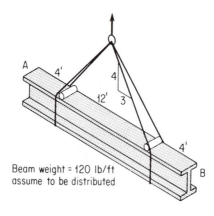

Beam weight = 120 lb/ft
assume to be distributed

Problem 11-11

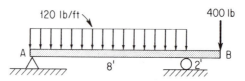

Problem 11-12

11-13 Without writing equations, sketch the shear and moment diagrams for the
to beams shown in the respective figures. Specify numerical values for all change
11-23. of loading positions and at all points of zero shear, and determine the maximum moment and its location.

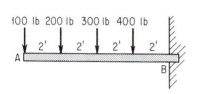

Problem 11-13

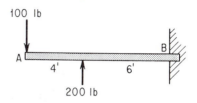

Problem 11-14

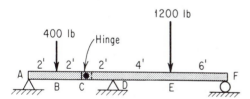

Problem 11-15

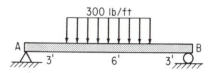

Problem 11-16

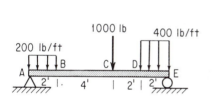

Problem 11-17

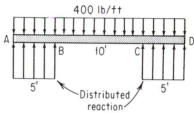

Problem 11-18

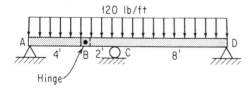

Problem 11-19

Problem 11-20

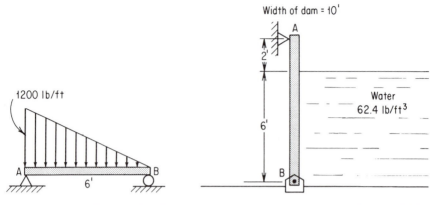

Problem 11-21   **Problem 11-22**

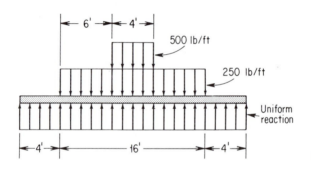

Problem 11-23

11-24. The wings of a jet airliner of weight W can be assumed to be a beam acted on by the downward weight forces of the engines and cabin. The *lift* or upward pressure that supports the plane can be assumed to be a uniformly distributed upward load as shown. Without writing equations, draw the shear and moment diagrams for the entire wing span.

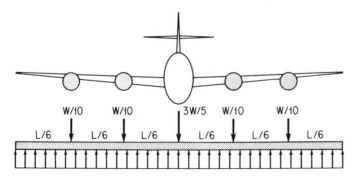

Problem 11-24

11-25. Draw the moment diagram for the beam of Prob. 11-7 by the method of superposition. Select a reference line through A and find the moment of the area of the diagram about this reference line.

Ans. 8000 lb ft^3.

11-26. Use the method of superposition to draw the moment diagram for the beam of Prob. 11-10. Select a reference line through the right reaction and find the moment of the area of the diagram about this line.

Ans. 90,112 lb ft^3.

11-27. Draw the moment diagram by the method of superposition for the beam of Prob. 11-12. Select a reference line through the left reaction and find the moment of the area of the diagram about this line.

Ans. −3520 lb ft^3.

11-28. Draw the moment diagram for the beam shown by the method of superposition. Select a reference line through the midspan M and find $\text{Area}_{MB}\bar{X}_B$

Ans. 81,000 lb ft^3.

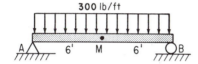

Problem 11-28

11-29. Find the moment of the area of the moment diagram, $\text{Area}_{AB}\bar{X}_B$, for the beam shown.

Ans. −28,300 lb ft^3.

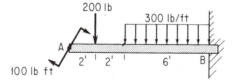

Problem 11-29

11-30. Find the moment of the area of the moment diagram, $\text{Area}_{AC}\bar{X}_C$, for the beam shown. *Hint:* Add and subtract a distributed load between B and C and then use a line through C as a reference.

Ans. 6890 lb ft^3.

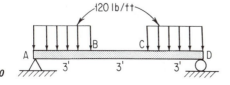

Problem 11-30

11-31 The figures represent the shear diagram for various beam loadings. For each
to   figure, draw the beam together with its loading arrangement and the moment
11-35. diagram.

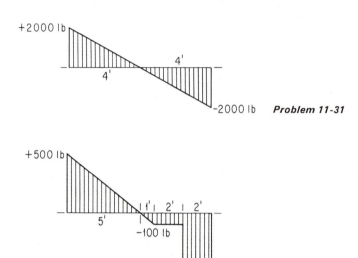

+2000 lb

4'

4'

-2000 lb   *Problem 11-31*

+500 lb

5'   1' | 2' | 2'
-100 lb

-500 lb   *Problem 11-32*

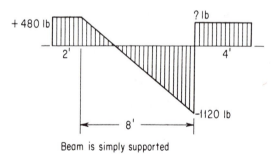

+ 480 lb

? lb

2'   4'

-1120 lb

←——— 8' ———→

Beam is simply supported   *Problem 11-33*

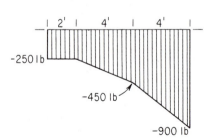

| 2' | 4' | 4' |

-250 lb

-450 lb

-900 lb   *Problem 11-34*

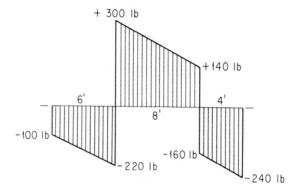

Problem 11-35

11-36 to 11-40. Moment diagrams by parts are illustrated in the respective figures for various beam loadings. Draw the loading arrangement for each.

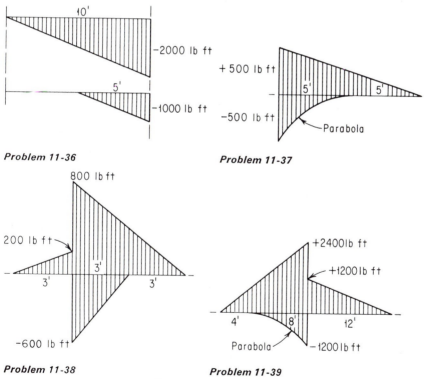

Problem 11-36

Problem 11-37

Problem 11-38

Problem 11-39

Problem 11-40

Chapter 12

STRESSES IN BEAMS

Some of the earliest studies in mechanics of materials were concerned with the strength and deflection of beams. Galileo, in 1638, was one of the first to propose a theory of stress distribution caused by bending, which, in part, was the accepted approach for many years to follow. Errors, however, in the thinking of the day were pointed out in 1773 by Coulomb, a French military engineer, and his concepts of stress distribution are now accepted as theoretically correct.

12-1 Tensile and Compressive Stress Caused by Bending

Consider a simple beam, Fig. 12-1(a), acted upon by end moments; its bending is exaggerated. The beam is assumed (a) *to be initially straight and of constant cross section*, (b) *to be elastic and have equal moduli of elasticity in tension and compression*, (c) *to be homogeneous*, and (d) *to obey Hooke's law.*

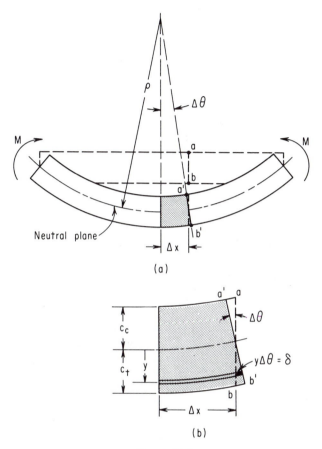

Figure 12-1

It is further assumed *that plane sections within the beam prior to bending remain plane*. To study the geometry of bending, a small segment of the beam is isolated as shown in Fig. 12-1(b). As the beam deforms, the top fibers contract and the bottom fibers elongate; the *neutral plane* of the beam is defined as *a plane whose length remains unchanged during the deformation*. The length of the segment of the neutral plane, Δx, can be expressed in terms of the radius of curvature ρ of the beam and the inclination $\Delta\theta$ of the transverse plane $a'b'$; thus

$$\Delta x = \rho\Delta\theta$$

At an arbitrary distance y from the neutral plane the deformation δ of a *fiber* within the beam is

$$\delta = y\Delta\theta$$

The strain in this fiber, its change in length per unit length, is

$$\epsilon = \frac{\delta}{\Delta x} = \frac{y\Delta\theta}{\rho\Delta\theta} = \frac{y}{\rho}$$

Stress in terms of strain and the modulus of elasticity is, therefore,

$$\sigma = \epsilon E = \frac{Ey}{\rho} \qquad (12\text{-}1)$$

The value of σ given by this equation is called the *bending*, or *flexural stress* and is directly proportional to the distance y from the neutral plane. The fibers elongate when they lie on the convex side of the neutral plane; the maximum tensile stress, therefore, occurs at the outermost fiber and has a magnitude of

$$\sigma_t = \frac{EC_t}{\rho} \qquad (a)$$

where c_t is the distance y from the neutral plane to the convex face of the beam. On the concave face of the beam, the fibers contract and are, therefore, in compression; in this case the maximum compressive stress is

$$\sigma_c = \frac{Ec_c}{\rho} \qquad (b)$$

Although these equations can be employed to find bending stresses, they are not convenient formulas to use, since the radius of curvature ρ is usually an unknown quantity. An equation must be obtained which will relate bending stress to the external moment and the geometric properties of the beam. This can be done by considering a segment of the beam whose internal reaction at a given transverse section is a moment M, as shown in Fig. 12-2(a).

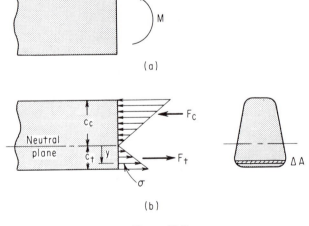

Figure 12-2

The stress distribution acting on the section, Fig. 12-2(b), varies directly with the distance from the neutral plane; in effect, the stress is a distributed horizontal load, partly negative and partly positive, whose resultant must be zero. Thus

$$\sum F_x = \sum \sigma \Delta A = 0 \tag{c}$$

where σ is an arbitrary stress acting on a cross-sectional element of area ΔA. Replacing the value of stress by its equivalent from Eq. (12-1) gives

$$\sum \sigma \Delta A = \sum \frac{Ey}{\rho} \Delta A = \frac{E}{\rho} \sum y \Delta A = \frac{E\bar{y}A}{\rho} = 0 \tag{d}$$

The only factor in Eq. (d) that can possibly be zero is $\bar{y}$; this places the neutral axis of the beam at the centroid of its cross-sectional area.

To satisfy equilibrium, the internal and external moments at the section must be equal.

$$M = \sum \sigma y \Delta A \tag{e}$$

Expressing the stress in terms of its equivalent from Eq. (12-1) gives

$$M = \sum \frac{Ey}{\rho} y \Delta A = \frac{E}{\rho} \sum y^2 \Delta A$$

The summation $\sum y^2 \Delta A$ is, by definition, the moment of inertia of the cross-sectional area with reference to the neutral plane; therefore,

$$M = \frac{EI}{\rho} \tag{f}$$

Equations (12-1) and (f), when combined, give the formula for bending stress in terms of M, y, and I.

$$\sigma = \frac{My}{I} \tag{g}$$

For maximum values of tensile or compressive stress, y is set equal to either c_t or c_c:

$$\sigma_t = \frac{Mc_t}{I}$$

$$\sigma_c = \frac{Mc_c}{I} \tag{12-2}$$

In symmetrical beams, c_t and c_c are equal, and the flexure formula is simply written as

$$\sigma = \frac{Mc}{I} \tag{12-3}$$

A common variation is obtained by writing Eq. (12-3) in terms of the *section modulus Z*, defined as the ratio I/c.

$$\sigma = \frac{M}{I/c} = \frac{M}{Z} \tag{12-4}$$

Table 12-1 Properties of Areas

Section	Moment of Inertia	Section Modulus
Square	$I_x = \dfrac{b^4}{12}$	$Z = \dfrac{b^3}{6}$
Rectangle	$I_x = \dfrac{bh^3}{12}$	$Z = \dfrac{bh^2}{6}$
Hollow Rectangle	$I_x = \dfrac{bh^3 - b_1 h_1^3}{12}$	$Z = \dfrac{bh^3 - b_1 h_1^3}{6h}$
Equal Rectangles	$I_x = \dfrac{b(h^3 - h_1^3)}{12}$	$Z = \dfrac{b(h^3 - h_1^3)}{6h}$
Circle	$I_x = \dfrac{\pi d^4}{64}$	$Z = \dfrac{\pi d^3}{32}$
Hollow Circle	$I_x = \dfrac{\pi(d^4 - d_1^4)}{64}$	$Z = \dfrac{\pi(d^4 - d_1^4)}{32d}$

Table 12-1 lists the geometric properties, including the section moduli, of some common beam sections; the properties of standard structural shapes are tabulated in the appendix.

EXAMPLE 1. A simply supported timber beam 20 ft long carries a concentrated load P, as shown in Fig. 12-3(a). The beam has a 4-in. by 12-in. rectangular cross-section. Determine the maximum value of P if the fiber stress is not to exceed 1200 psi, and (a) the 12-in. side is horizontal, (b) the 12-in. side is vertical.

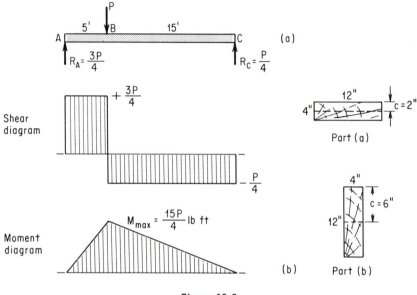

Figure 12-3

Solution: The reactions $R_A = 3P/4$ and $R_C = P/4$ are first computed, and the shear and moment diagrams are drawn as shown in Fig. 12-3(b). The maximum moment, which occurs at B, is equal to the area of the shear diagram between points A and B.

$$M_{max} = 5\left(\frac{3P}{4}\right) = \frac{15P}{4} \text{ lb ft} = 45\,P \text{ lb in.}$$

Part (a). The moment of inertia about the neutral plane is

$$I_{NA} = \tfrac{1}{12} bh^3 = \tfrac{1}{12}(12)(4)^3 = 64 \text{ in.}^4$$

By Eq. (12-3),

$$\sigma = \frac{Mc}{I}$$

$$1200 = \frac{45P(2)}{64}$$

$$P = 853 \text{ lb}$$

Part (b). With the 12-in. side vertical, the moment of inertia increases appreciably; thus

$$I_{NA} = \tfrac{1}{12} bh^3 = \tfrac{1}{12}(4)(12)^3 = 576 \text{ in.}^4$$

By Eq. (12-3).

$$\sigma = \frac{Mc}{I}$$

$$1200 = \frac{45P(6)}{576}$$

$$P = 2560 \text{ lb}$$

Thus, three times the load can be supported when the beam is oriented with its long dimension vertical. This greater strength is readily apparent by comparing the section moduli.

$$Z_a = \frac{I}{c} = \frac{1}{12}\frac{bh^2}{(h/2)} = \frac{bh^2}{6} = \frac{12(4)^2}{6} = 32 \text{ in}^3.$$

With the short dimension vertical:

$$Z_b = \frac{bh^2}{6} = \frac{4(12)^2}{6} = 96 \text{ in.}^3$$

EXAMPLE 2. A steel band-saw blade $\frac{1}{2}$ in. wide and $\frac{1}{32}$ in. thick is driven by two 4-ft-diameter pulleys, as shown in Fig. 12-4. Determine the maximum stress developed in the blade and the magnitude of the internal moment.

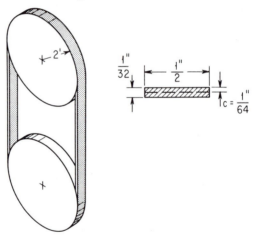

Figure 12-4

Solution: By Eq. (12-1),

$$\sigma = \frac{Ec}{\rho} = \frac{30 \times 10^6}{2(12)}\left(\frac{1}{64}\right) = 19,500 \text{ psi}$$

The moment of intertia about the neutral plane is

$$I_{NA} = \tfrac{1}{12}bh^3 = \tfrac{1}{12}(\tfrac{1}{2})(\tfrac{1}{32})^3 = 1.27 \times 10^{-6} \text{ in.}^4$$

By Eq. (12-3),

$$M = \frac{\sigma I}{c} = \frac{19,500(1.27 \times 10^{-6})}{\frac{1}{64}} = 1.58 \text{ lb in.}$$

Since the radius of curvature is constant, the internal moment is constant in the curved section of the blade.

EXAMPLE 3. The 10-ft tee-beam illustrated in Fig. 12-5 supports a distributed load of 1000 lb per ft. Determine the maximum tensile and compressive bending stresses in the beam.

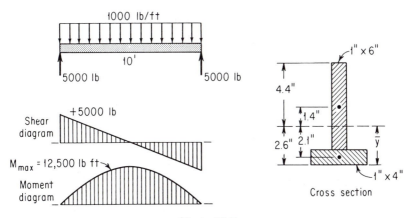

Figure 12-5

Solution: Before we attempt to find the stress, the centroidal axis must be located and the moment of inertia with respect to this axis computed. The centroid is computed by taking the first moment of the area with reference to the base.

$$A\bar{y} = A_1 y_1 + A_2 y_2$$

$$\bar{y} = \frac{4(1)(\frac{1}{2}) + 6(1)4}{4(1) + 6(1)} = 2.6 \text{ in. above the base}$$

The moment of inertia is computed by assuming the section to consist of two rectangles. The *parallel-axis theorem* must be employed to transpose the inertia of each rectangle to the neutral, or centroidal, axis.

$$\begin{aligned}
I_{NA} &= I_1 + I_2 \\
&= (\bar{I} + Ad^2)_1 + (\bar{I} + Ad^2)_2 \\
&= [\tfrac{1}{12}(4)(1)^3 + 4(1)(2.1)^2] + [\tfrac{1}{12}(1)6^3 + 6(1)(1.4)^2] \\
&= 0.33 + 17.64 + 18.00 + 11.76 \\
&= 47.7 \text{ in.}^4
\end{aligned}$$

The bottom fibers of the beam are in tension, and the top fibers are in compression; hence,

$$\sigma_t = \frac{Mc_t}{I} = \frac{12,500(12)(2.6)}{47.7} = 8180 \text{ psi}$$

and

$$\sigma_c = \frac{Mc_c}{I} = \frac{12,500(12)(4.4)}{47.7} = 13,800 \text{ psi}$$

12-2 Longitudinal Shear in Beams

A second important factor to be considered in the determination of the strength of beams is *horizontal*, or *longitudinal*, shear. Many materials, wood, for example, are primarily weak in shear, and for this reason the load that can be supported may depend upon the ability of the beam to resist horizontal shearing forces.

The shearing tendency can be best descibed in terms of a beam, Fig.

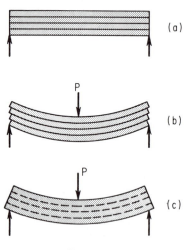

(a)

(b)

(c)

Figure 12-6

12-6(a), that is composed of a number of thin layers placed with no adhesion one on top of the other. With a force P applied, as shown in Fig. 12-6(b), the laminations slide relative to one another as the beam sags. If the beam is solid, Fig. 12-6(c), this slipping tendency would be prevented by internal forces acting parallel to the axis of the beam. The intensity of these forces, the shearing stresses, act on every horizontal plane, since every such plane is a potential sliding surface.

To arrive at a relationship for the shearing stress, consider the beam shown in Fig. 12-7(a). At section *a-d* the moment is M_1, and at section *b-c*, an incremental distance Δx to the right, the

moment is M_2. Next, consider an enlarged view of this section, Fig. 12-7(b), drawn to show the distribution of tensile and compressive bending stresses. In the element *abef*, the forces F_1 and F_2 are the resultants of the bending stress forces that act on the transverse planes *af* and *be*. The shear force $\tau b \Delta x$ that acts on plane *ef* is required to maintain equilibrium.

$$\tau b \Delta x = F_1 - F_2 \qquad \text{(a)}$$

Since

$$F_1 = \sum \sigma_{y1} \Delta A \quad \text{and} \quad F_2 = \sum \sigma_{y2} \Delta A$$

it follows that

$$\tau b \Delta x = \sum \sigma_{y1} \Delta A - \sum \sigma_{y2} \Delta A \qquad \text{(b)}$$

Next, substituting $\sigma_{y1} = M_1 y / I$ and $\sigma_{y2} = M_2 y / I$ gives

$$\tau b \Delta x = \frac{M_1}{I} \sum y \Delta A - \frac{M_2}{I} \sum y \Delta A$$

$$= \frac{(M_1 - M_2)}{I} \sum y \Delta A \qquad \text{(c)}$$

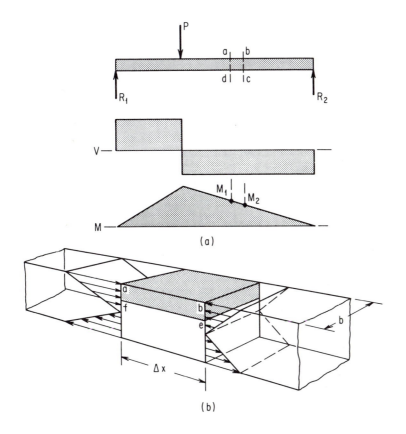

(a)

(b)

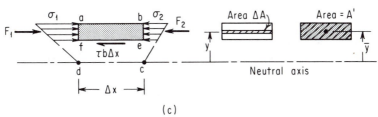

(c)

Figure 12-7

Since the difference $(M_1 - M_2)$ is small, Eq. (c) can be written

$$\tau b \Delta x = \frac{\Delta M}{I} \sum y \Delta A$$

Rearranging terms gives

$$\tau = \frac{\Delta M}{\Delta x} \sum \frac{y \Delta A}{Ib} \tag{d}$$

Noting that $\Delta M/\Delta x = V$ and $\sum y\Delta A = \bar{y}A'$, the shear-stress equation can be written in its final form.

$$\tau = \frac{V}{Ib}A'\bar{y} \qquad (12\text{-}5)$$

It is important to understand the meaning of $A'\bar{y}$; the term represents the first moment of the cross-sectional area lying above the plane at which shear is to be computed. The term b represents the width of the material comprising the beam at the shear plane under investigation.

In Eq. (12-5) V and I are constants for any given section; it follows, therefore, that the shearing stress is a maximum when the quantity $A'\bar{y}/b$ is a maximum. In a rectangular beam, Fig. 12-8(a), the maximum shear stress

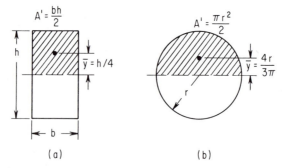

(a)   (b)   **Figure 12-8**

occurs at the neutral axis and is equal to

$$\tau_{\max} = \frac{V}{Ib}A\bar{y} = \frac{V}{b^2h^3/12}\left(\frac{bh}{2}\right)\left(\frac{h}{4}\right) = \frac{3V}{2A} \qquad (12\text{-}6)$$

where A is the area of the entire cross section. In a circular section, Fig. 12-8(b), the maximum shearing stress is again at the neutral axis and is equal to

$$\tau_{\max} = \frac{4V}{3A} \qquad (12\text{-}7)$$

When the beam is one of the standard shapes, an I-beam or a wide flange beam, it is generally safe to assume that the flanges contribute very little in the way of resistance to shear; the beam is considered simply as a rectangular web section. It is common practice in structural design to approximate the mximum shearing stress from the equation

$$\tau_{\max} = \frac{3V}{2td} \qquad (12\text{-}8)$$

where t and d are the web thickness and web depth, respectively.

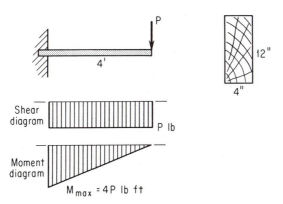

Figure 12-9 $M_{max} = 4P$ lb ft

EXAMPLE 4. The timber cantilever beam shown in Fig. 12-9 carries a load P at its free end. Find the safe value of P if the working stresses in tension or compression and in shear are $\sigma = 1200$ psi and $\tau = 100$ psi.

Solution: The moment of inertia with reference to the neutral axis is first computed:

$$I = \tfrac{1}{12}bh^3 = \tfrac{1}{12}(4)(12)^3 = 576 \text{ in.}^4$$

Next, a value of the load P is computed on the basis of the allowable flexural stress

$$\sigma = \frac{Mc}{I}$$

$$1200 = \frac{4(12)P(6)}{576}$$

$$P = 2400 \text{ lb}$$

A value of P is next determined on the basis of the permissible shearing stress; the vertical shear V in this instance is constant throughout the beam.

$$\tau = \frac{3}{2}\frac{V}{A}$$

$$100 = \frac{3}{2} \times \frac{P}{(4)12}$$

$$P = 3200 \text{ lb}$$

The lesser of the two values, $P = 2400$ lb, is the safe load.

EXAMPLE 5. Determine the maximum value of the shearing stress in the box beam shown in Fig. 12-10.

Solution: The reactions R_1 and R_2 are computed by the usual methods:

$$R_1 = \frac{600(12)2}{8} = 1800 \text{ lb}$$

$$R_2 = \frac{600(12)6}{8} = 5400 \text{ lb}$$

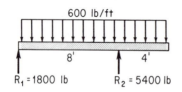

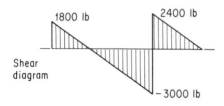

Figure 12-10

When plotted, the shear diagram indicates the magnitude of maximum vertical shear to be 3000 lb just to the left of R_2. The terms I, $A'\bar{y}$, and b are computed prior to substitution into Eq. (12-5); thus,

$$I = \tfrac{1}{12}(6)(8)^3 - \tfrac{1}{12}(4)(6)^3 = 184 \text{ in.}^4$$
$$A'\bar{y} = A_1 y_1 - A_2 y_2 = 6(4)2 - 4(3)(1.5) = 30 \text{ in.}^4$$
$$b = 1 + 1 = 2 \text{ in.}$$

Therefore,

$$\tau_{max} = \frac{V}{Ib} A'\bar{y} = \frac{3000(30)}{184(2)} = 245 \text{ psi}$$

12-3 Stresses in Built-up Beams

It is common construction practice to fabricate beams by joining various lightweight elements together to form a single strong section. The sections illustrated in Fig. 12-11 are typical examples of various built-up beams. The first, Fig. 12-11(a), is a box beam fabricated by nailing or lagging planks together as shown. Another type of built-up beam, Fig. 12-11(b), consists

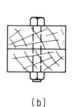

(a)　　　　　　(b)　　　　　(c)　　　Figure 12-11

simply of rectangular timbers bolted together to form a solid beam. A third type, the plate and angle girder shown in Fig. 12-11(c), consists of angle sections bolted or riveted to a web plate.

Of primary concern in built-up sections is the *spacing*, or *pitch*, between bolts or rivets that are used to secure the assembly. Consider the beam of Fig. 12-12(a), which is composed of three timbers bolted together to form a

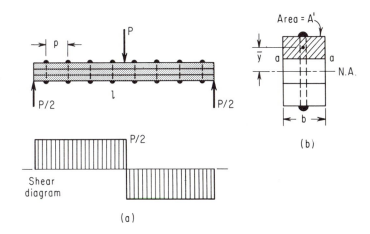

Figure 12-12

composite section. If the beam were truly solid, the shearing stress across *a-a* would be given by

$$\tau = \frac{V}{Ib} A'\bar{y}$$

where $A'\bar{y}$ represents the first moment of the cross-sectional area of the top plank with reference to the neutral axis. Since the beam is not solid, a typical bolt in the assembly must support a force in shear equal to the product of the shearing stress and the area bp, where p represents the distance or *pitch* between adjacent bolts.

$$F = \tau bp = p\frac{VA'\bar{y}}{I} \qquad (12\text{-}9)$$

In built-up beams the usual practice is to maintain a constant pitch, even though the vertical shear varies from point to point, and to base this pitch on the greatest value of vertical shear.

It is interesting to note that Eq. (12-9) can be used to determine pitch for all types of built-up sections; in every case the term $A'\bar{y}$ represents the first moment of the area that tends to slide relative to the beam itself. The example that follows will illustrate this concept.

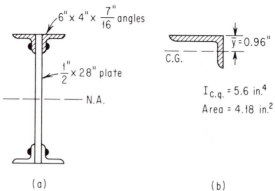

(a)   (b)   **Figure 12-13**

EXAMPLE 6. The built-up beam of Fig. 12-13(a) consists of four 6-in. by 4-in. by $\frac{7}{16}$-in. angles connected to a web plate by equally spaced $\frac{3}{4}$-in. rivets. The web plate is $\frac{1}{2}$ in. thick and has a depth of 28 in. Determine the rivet spacing if the allowable shearing stress in the rivets is 8000 psi. The beam carries a uniformly distributed load of 4000 lb per ft on a span of 20 ft.

Solution: The maximum vertical shear occurs at the supports and is equal to

$$V = \frac{wl}{2} = \frac{4000(20)}{2} = 40,000 \text{ lb}$$

If it were not for the rivets, the angles at the top (or bottom) would move relative to the web. The force necessary to prevent this motion is represented by Eq. (12-9), where $A'\bar{y}$ is the first moment of the shaded area about the neutral axis, and I is the moment of inertia of the entire cross section.

$$A'\bar{y} = 2(4.18)(14 - 0.96) = 109 \text{ in.}^3$$
$$I = I_{web} + 4I_{angle}$$
$$= \tfrac{1}{12}bh^3 + 4(\bar{I} + Ad^2)$$
$$= \tfrac{1}{12}(\tfrac{1}{2})(28)^3 + 4[5.6 + 4.18(14 - 0.96)^2]$$
$$= 3780 \text{ in.}^4$$

By Eq. (12-9),

$$F = p\frac{VA'\bar{y}}{I} = p\frac{(40,000)(109)}{3780}$$

$$F = 1150\,p \tag{a}$$

Each rivet is in *double shear* and is capable of supporting a load of

$$F = \tau A = (2)8000\frac{\pi}{4}\left(\frac{3}{4}\right)^2 = 7070 \text{ lb} \tag{b}$$

Equating Eqs. (a) and (b) gives the required pitch; thus,

$$1150p = 7070$$
$$p = 6.15 \text{ in.}$$

12-4 Beams of Two or More Materials

Certain advantages are often derived by combining two or more different materials to form a single beam; appearance, weight reduction, and strength are a few of the factors that often lead to the design of composite beams.

To simplify the computations, it is convenient to *transform* the composite section into a single material; to imagine, in other words, that the beam is composed of one homogeneous substance. The transformed beam and the composite beam must be of *equivalent sections;* they must deform equally under the action of the transverse loads. Consider the two beams, one wood and the other steel, shown in Fig. 12-14. Both sections have the same depth

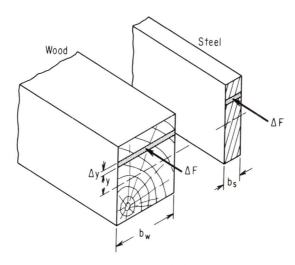

Figure 12-14

but vary in width. To be equivalent, the strains at corresponding distances from the neutral axis of both beams must be equal.

$$\epsilon_w = \epsilon_s \tag{a}$$

Since stress and strain are proportional, and since it is assumed that the stress does not exceed the proportional limit, it follows that

$$\frac{\sigma_w}{E_w} = \frac{\sigma_s}{E_s} \tag{12-10}$$

The beams have equal strengths, and both can sustain a similar force ΔF applied at equivalent distances from their neutral axes. This force, in terms of stress and area, is

$$\Delta F = \sigma_s b_s \Delta y = \sigma_w b_w \Delta y \tag{b}$$

Eqs. (12-10) and (b) can be combined to give

$$\sigma_s b_s = \frac{E_w}{E_s}\sigma_s b_w$$

$$b_s = \frac{E_w}{E_s}b_w \qquad\qquad (c)$$

If the moduli of wood and steel are 1.5×10^6 psi and 30×10^6 psi, respectively, the widths of the beams differ by a factor of 20.

$$b_s = \frac{1.5(10)^6}{30(10)^6}b_w = \frac{b_w}{20}$$

To be more general, the equivalent width of material 1 in terms of a second material 2 in a composite beam is

$$b_1 = \frac{E_2}{E_1}b_2 \qquad\qquad (12\text{-}11)$$

The various beam sections of Fig. 12-15 illustrate the types of trans-

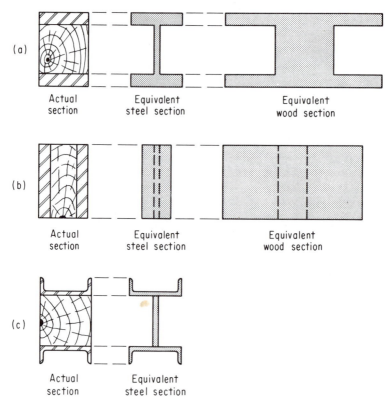

(a)

Actual section · Equivalent steel section · Equivalent wood section

(b)

Actual section · Equivalent steel section · Equivalent wood section

(c)

Actual section · Equivalent steel section

Figure 12-15

formations possible. Wood and steel are used as examples, although any two or more materials can be combined to form a composite beam. In Fig. 12-15(a) a wood block is laminated with steel plates at the top and bottom. For analysis, this section can be transformed into an all-steel section or an all-wood section, as shown. In the second example, Fig. 12-15(b), a wood core is faced with steel plates; as before, the depth of the equivalent section remains the same; as an imaginary solid steel beam, the section is much narrower than as an imaginary solid wooden beam. This is realistic, since much less steel would be required to match the strength of the wood. It would be mathematically difficult to transform the steel channel, Fig. 12-15(c), into an equivalent wood section; for this case the wood is exchanged for an equivalent amount of steel.

Once the transformed section is obtained, the analysis of bending proceeds as though the beam were either steel or wood. The stresses computed, however, are those sustained by the imaginary section. One value of stress will be the actual value, whereas the other must be corrected through use of the proportion given in Eq. (12-11):

$$\frac{\sigma_w}{E_w} = \frac{\sigma_s}{E_s}$$

EXAMPLE 7. A composite beam, Fig. 12-16(a), consists of a timber core reinforced by steel plates. The beam supports a distributed load of 500 lb per ft on a

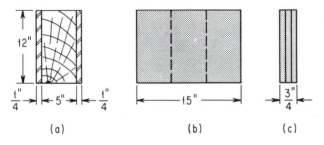

$$12''$$

$$\frac{1}{4}'' \quad 5'' \quad \frac{1}{4}''$$

$$15''$$

$$\frac{3}{4}''$$

(a)   (b)   (c)

Figure 12-16

12-ft span. Determine the bending stress in both materials. Steel and wood have moduli of elasticity of 30×10^6 psi and 1.5×10^6 psi, respectively.

Solution: The problem can be solved by transforming the composite section either to solid timber or to solid steel. Both methods will be applied to illustrate the approach.

Transformation to timber. An equivalent width of each steel plate in terms of the wood is given by Eq. (12-11):

$$b_w = \frac{E_s b_s}{E_w} = \frac{30(10)^6}{1.5(10)^6} \times \frac{1}{4} = 5 \text{ in.}$$

The composite section is 15 in. wide, as shown in Fig. 12-16(b), and has a moment of inertia of

$$I_{NA} = \tfrac{1}{12}bh^3 = \tfrac{1}{12}(15)(12)^3 = 2160 \text{ in.}^4$$

The maximum bending stress is

$$\sigma = \frac{Mc}{I}$$

where the moment M is

$$M = \frac{3000(6)}{2} \times 12 = 108,000 \text{ lb in.}$$

Therefore,

$$\sigma = \frac{108,000(6)}{2160} = 300 \text{ psi}$$

Numerical values are substituted into Eq. (12-10) to find the true stress in the steel.

$$\sigma_s = \frac{E_s}{E_w}\sigma_w = \frac{30(10)^6}{1.5(10)^6}(300) = 6000 \text{ psi}$$

Transformation to steel. An equivalent width of steel to replace the wood is

$$b_s = \frac{E_w}{E_s}b_w = \frac{1.5(10)^6}{30(10)^6}(5) = \frac{1}{4} \text{ in.}$$

The transformed steel section is, therefore, $\tfrac{3}{4}$ in. wide and has a moment of inertia of

$$I_{NA} = \tfrac{1}{12}bh^3 = \tfrac{1}{12}(\tfrac{3}{4})(12)^3 = 108 \text{ in.}^4$$

The flexure equation gives the true bending stress in the steel, whereas the stress in the wood must be corrected.

$$\sigma_s = \frac{Mc}{I} = \frac{108,000(6)}{108} = 6000 \text{ psi}$$

$$\sigma_w = \frac{E_w}{E_s}\sigma_s = \frac{1.5(10)^6}{30(10)^6}(6000) = 300 \text{ psi}$$

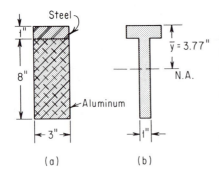

(a) (b)

Figure 12-17

EXAMPLE 8. Steel and aluminum are combined to form the composite beam shown in Fig. 12-17(a). Determine the greatest uniformly distributed load which can be supported on a span of 16 ft. Working stresses for the steel and aluminum are $\sigma_s = 15,000$ psi and $\sigma_a = 6000$ psi, $E_s = 30 \times 10^6$ psi; $E_a = 10 \times 10^6$ psi.

Solution: The midspan moment, by the usual methods, is

$$M = \tfrac{1}{2}(8w)8(12) = 384\,w \text{ lb in.}$$

A choice as to whether to convert the steel to aluminum or the aluminum to steel is arbitrary; however, if the latter transformation is employed, computations will involve smaller numbers.

The width of steel equivalent to the aluminum is

$$b_s = \frac{E_a}{E_s}b_a = \frac{10(10)^6}{30(10)^6}(3) = 1 \text{ in.}$$

First moments taken with respect to the top of the beam will locate the neutral axis.

$$\bar{y} = \frac{A_1 y_1 + A_2 y_2}{A_1 + A_2}$$

$$= \frac{2(1)(0.5) + 9(1)(4.5)}{3(1) + 8(1)} = 3.77 \text{ in.}$$

The parallel axis theorem is employed to find the moment of inertia of the transformed section with respect to the neutral axis.

$$I = (\tfrac{1}{12}bh^3 + Ad^2) + (\tfrac{1}{12}bh^3 + Ad^2)$$

$$I_{NA} = [\tfrac{1}{12}(3)(1)^3 + 3(1)(3.77 - 0.5)^2] + [\tfrac{1}{12}(1)(8)^3 + 8(1)(5.23 - 4)^2]$$

$$= 0.25 + 32.08 + 42.67 + 12.10 = 87.1 \text{ in.}^4$$

The maximum compressive bending stress occurs at the top of the beam, and since both the original and transformed sections are of the same material in this region, the allowable stress is found by direct substitution.

$$\sigma_s = \frac{Mc}{I}$$

$$15,000 = \frac{384w(3.77)}{87.1}$$

$$w = \frac{15,000(87.1)}{384(3.77)} = 900 \text{ lb per ft}$$

The bottom portion of the original section is aluminum, and its corrected allowable stress in terms of the transformed beam is

$$\sigma_s = \frac{E_s}{E_a}\sigma_a = \frac{30(10)^6}{10(10)^6}(6000) = 18,000 \text{ psi}$$

A permissible load in terms of the fiber stress at the bottom becomes

$$\sigma_s = \frac{Mc}{I}$$

$$18,000 = \frac{384w(5.23)}{87.1}$$

$$w = \frac{18,000(87.1)}{384(5.23)} = 780 \text{ lb per ft}$$

Aluminum governs the design, and the safe load is 780 lb per ft.

12-5 Reinforced Concrete Beams

Transformation concepts, developed in the previous section, can be applied in a rather interesting manner to reinforced concrete beams. Theory must be adjusted, however, to account for the fact that concrete has virtually no strength in tension. The steel reinforcing bars shown in Fig. 12-18(a)

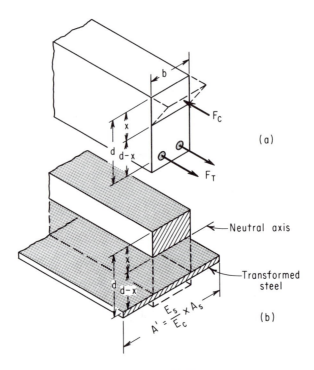

Figure 12-18

must assume a tensile load equal and opposite to the compressive load in the concrete. Furthermore, the moment of the transformed area of the steel bars about the neutral axis must just equal the moment of the area of the *active* concrete that lies above the neutral axis; this computation locates the neutral axis.

Using standard notation, d is the distance from the top of the beam to the center line of the steel bars, and x the height of the concrete above the neutral axis. The width of the beam is designated as b and the transformed steel area as A', where

$$A' = \frac{E_{\text{steel}}}{E_{\text{concrete}}} \times A_{\text{steel}} \qquad (12\text{-}12)$$

To simplify computations, the height of the transformed steel area is assumed to be negligible, and the single dimension $(d - x)$ locates the area below the neutral axis. Equating the first moments of the two areas will locate the neutral axis, and the moment of inertia can be computed. Stresses can then be determined by the methods outlined in the previous section. The examples that follow illustrate the complete approach.

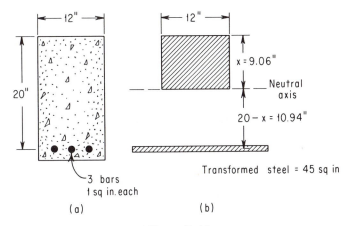

Figure 12-19

EXAMPLE 9. A reinforced-concrete beam, Fig. 12-19, is subjected to a bending moment of 500,000 lb in. Determine the maximum bending stress in the concrete and in the steel. Assume the ratio of the moduli of elasticity of steel to concrete to be 15.

Solution: The transformed area of the steel is

$$A' = \frac{E_s}{E_c} A_s = 15(3) = 45 \text{ sq in.}$$

The moments of the areas representing the *active* concrete and the steel must equate.

$$12(x)\left(\frac{x}{2}\right) = 45(20 - x)$$

$$x^2 = 7.5x - 150 = 0$$

Solving for x by the quadratic formula (only the positive root is of interest) gives

$$x = \frac{-7.5 \pm \sqrt{(7.5)^2 + 4(1)150}}{2(1)} = 9.06 \text{ in.}$$

Thus,

$$x = 9.06 \quad \text{and} \quad 20 - x = 10.94$$

The moment of inertia, with respect to the neutral axis, and the bending stresses are then computed.

$$I_{NA} = \tfrac{1}{3}bh^3 + Ad^2$$
$$= \tfrac{1}{3}(12)(9.06)^3 + 45(10.94)^2 = 8360 \text{ in.}^4$$

$$(\sigma_c)_{\max} = \frac{Mc}{I} = \frac{500,000(9.06)}{8360} = 542 \text{ psi}$$

$$(\sigma_s)_{\max} = 15\frac{Mc}{I} = \frac{15(500,000)10.94}{8360} = 9810 \text{ psi}$$

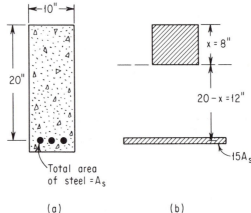

(a)   (b)   *Figure 12-20*

EXAMPLE 10. The ideal design of a reinforced-concrete beam requires both the steel and the concrete to be stressed to their allowable limits; this is called *balanced reinforcement*. Determine the number of $\frac{5}{8}$-in.-diameter steel reinforcing bars required for balanced reinforcement and the safe moment for the beam section shown in Fig. 12-20(a). The allowable stresses are $\sigma_c = 800$ psi and $\sigma_s = 18,000$ psi. Use $E_s/E_c = 15$.

Solution: Substitution of the values of the allowable stresses in the flexure formula provides a relationship for the distance x.

For the concrete:
$$\sigma_c = \frac{Mc}{I}$$

$$800 = \frac{M}{I}x \qquad \text{(a)}$$

For the transformed steel:
$$\sigma_s = \frac{15Mc}{I}$$

$$18,000 = \frac{15M}{I}(20 - x) \qquad \text{(b)}$$

Dividing Eq. (a) by Eq. (b) eliminates the common ratio M/I.

$$\frac{800}{18,000} = \frac{x}{15(20 - x)}$$
$$8(15)(20 - x) = 180x$$
$$20 - x = \tfrac{3}{2}x$$
$$x = 8 \text{ in.}$$

The required area of steel can now be found by equating first moments.

$$12(15)A_s = 10(8)(\tfrac{8}{2})$$
$$A_s = 1.78 \text{ sq in.}$$

The moment of inertia, based on the transformed area, and the safe bending moment are then computed.

$$I_{NA} = \tfrac{1}{3}bh^3 + Ad^2 = \tfrac{1}{3}(10(8)^3 + (15 \times 1.78)(12)^2 = 5550 \text{ in.}^4$$

$$M = \frac{\sigma I}{c} = \frac{800}{8}(5550) = 555,000 \text{ lb in. or } 555 \text{ kip in.}$$

The required number of $\tfrac{5}{8}$-in.-diameter bars is

$$N = \frac{1.78}{\pi(\frac{5}{16})^2} = 5.8 \,; \text{use 6 bars}$$

QUESTIONS, PROBLEMS, AND ANSWERS

12-1. A basic assumption in the derivation of stress equations is that Hooke's law prevails. In simple words, this means that stress and strain are

_____ .

Ans. Proportional.

12-2. How is radius of curvature of a stressed beam related to the modulus of elasticity?

Ans. By direct proportion.

12-3. The stress in a beam is _____ proportional to the moment and _____ proportional to the section modulus.

Ans. Directly; inversely.

12-4. A 0.1-in.-diameter steel wire is coiled around a 4-ft-diameter mandrel. What is the moment and the bending stress in the wire?

Ans. 6.13 lb in.; 62,500 psi.

12-5. A flat steel spring is subjected to end moment as shown. Determine the minimum radius of curvature to which the spring may be bent and the internal moment at this radius if the flexural stress may not exceed 25,000 psi.

Ans. 75 in.; 65.1 lb in.

Problem 12-5

12-6. Strain-gage measurements made on a beam that lacks symmetry about the neutral axis indicate the maximum compressive stress (at the top) to be 20,000 psi and the maximum tensile stress (at the bottom) to be 5000 psi. If the beam is 8 in. deep, how far from the top is the neutral plane?

Ans. 6.4 in.

12-7. Determine, with respect to the dashed line, the section modulus of each area shown.

Ans. (a) 6.28 in.^{3};   (b) 144 in.^{3}.

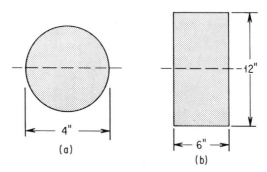

(a)

(b)   *Problem 12-7*

12-8. A rectangular bar is supported as a beam as shown. What is the maximum permissible value of P if the bending stress may not exceed 24,000 psi?

Ans. 4000 lb.

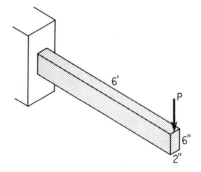

Problem 12-8

12-9. Solve Prob. 12-8, assuming the beam to be a hollow tube with a wall thickness of $\frac{1}{8}$ in. The outside dimensions are the same as those of the solid bar.

Ans. 920 lb.

12-10. A beam is supported as shown; the maximum stress must not exceed 20,000 psi. Graph the value of P as a function of wall thickness t. Use increments of $\frac{1}{4}$ in.

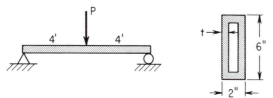

Problem 12-10

12-11. The maximum fiber stress at a certain section in a rectangular beam 4 in. wide by 6 in. deep is 3000 psi. What is the internal moment in the beam at this section?

Ans. 6000 lb, ft.

12-12. Select the most economical (lightest weight) wide-flanged beam that can be safely subjected to a moment of 50,000 lb ft. The maximum bending stress in the beam is not to exceed 15,000 psi.

Ans. 14 WF 30.

12-13. A wooden beam having a cross section of 4 in. wide by 12 in. deep carries two loads as illustrated. Find the magnitudes of *P* if the flexural stress is not to exceed 6000 psi and (a) the beam bends about an axis parallel to the 4-in. face; (b) the beam bends about an axis parallel to the 12-in. face.

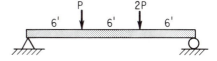

Problem 12-13

Ans. (a) 4800 lb; (b) 1600 lb.

12-14. A rectangular wooden beam supports the uniformly distributed load as shown. Determine the flexural stress in the beam at all points where the shear is zero.

Ans. 900 psi., 3 ft to right of *A*; 1600 psi., 8 ft to right of *A*.

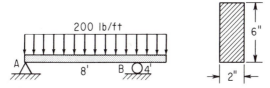

Problem 12-14

12-15. The beam shown is constructed by welding cover plates to two channels. What maximum uniformly distributed load can this beam support on a 20-ft span if the maximum fiber stress must not exceed 20,000 psi?

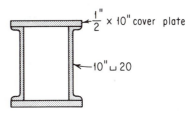

Ans. 2620 lb/ft.

Problem 12-15

12-16. Select the most economical beams, one an I-beam and the other a wide-flanged beam, to be used as a simply supported girder with a span of 20 ft. The beam carries a uniformly distributed load of 800 lb per ft. The flexural stress in either beam should not exceed 15,000 psi.

Ans. 12 I 31.8; 12 WF 27.

12-17. A wide-flanged beam is more economical than an I-beam. Why even consider I-beam construction?

Ans. An I-beam is narrow and space may be a factor.

12-18. The tee-beam carries two concentrated loads, as shown. What are the maximum tensile and compressive stresses in the beam?

Ans. 5040 psi tension; 9030 psi compression.

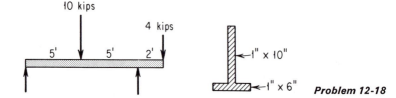

Problem 12-18

12-19. What is the greatest permissible distributed load w that the beam shown can support if the allowable flexural tensile and compressive stresses are 20,000 psi and 10,000 psi, respectively?

Ans. 4260 lb/ft.

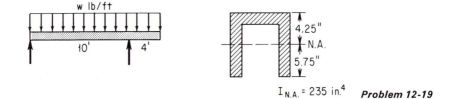

$I_{N.A.} = 235 \text{ in.}^4$ *Problem 12-19*

12-20. Select the most economical wide-flanged steel beam 21 ft long to carry the following total load: a uniform load of 500 lb per ft throughout the entire span, a concentrated load of 18 kips at a point 6 ft from the left end, and an additional uniform load of 600 lb per ft on the right-hand third of the beam. Base the selection on an allowable bending stress of 20,000 psi and neglect the weight of the beam.

Ans. 16 WF 40.

12-21. A log of diameter D is available from which a mill may cut a beam. Write an equation for the section modulus of the beam in terms of the diameter

D and the width *b*. Describe how you find the dimensions of the strongest beam that can be cut from the log.

Ans. Find a value of *b* that will give a maximum section modulus.

$$z = b(d^2 - b^2)/6$$

12-22. Two steel angles, each 6 in. by 4 in. by $\frac{1}{2}$ in., are combined and used with their 4-in. legs at the top to form a tee section for a cantilever beam 10 ft long. If the beam supports a uniform load of 500 lb per ft over its entire length, what are the maximum tensile and compressive stresses in the beam?

Ans. Tensile: 17,200 psi., compressive: 34,600 psi.

12-23. A 200-lb man stands in the middle of a 100-lb plank that is floating in water. The plank is 10 ft long, 3 in. thick, and 12 in. wide. What is the maximum bending stress in the plank?

Ans. 167 psi.

12-24. A rectangular timber beam carries a uniformly distributed load as shown. Determine the maximum shearing stress acting within the beam.

Ans. 104 psi.

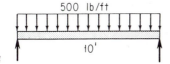

Problem 12-24

12-25. A rectangular section 4 in. wide and 12 in. deep is subjected to a vertical shear force of 30 kips. Determine the shearing stress on a horizontal plane (a) 2 in. below the top of the beam, and (b) 6 in. below the top of the beam.

Ans. (a) 521 psi; (b) 938 psi.

12-26. A simply supported beam with a 10-ft span carries a uniformly distributed load of *w* lb per ft. The cross section of the beam is a rectangle 3 in. wide by 9 in. deep. what is the maximum shearing stress in the beam if the maximum bending stress is 1200 psi?

Ans. 90 psi.

12-27. Find the maximum shearing stress in the beam shown. *Hint:* The maximum shearing stress occurs at the neutral plane.

Ans. 273 psi.

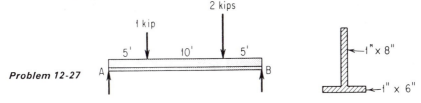

Problem 12-27

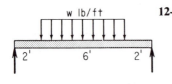

12-28. The cross-section of the timber beam shown is a rectangle 6 in. wide by 12 in. deep. Determine the maximum value of w if the allowable stresses in bending and shear are 1200 psi and 72 psi, respectively.

Problem 12-28

Ans. 1150 lb/ft.

12-29. The cross-section of an aluminum wing channel is illustrated. Determine the safe distributed load w lb per ft that can act on this section supported as a simple beam with a 4-ft span. The permissible stresses in bending and shear are 2500 psi and 1000 psi, respectively.

Ans. 37.2 lb/ft.

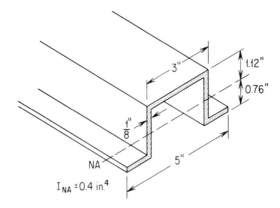

Problem 12-29

12-30. Two 8-in. by 8-in. timbers are bolted together to form a single solid beam as shown. The bolts each have a cross-sectional area of 0.625 in.^{2}, and the beam supports a uniformly distributed load of 800 lb per ft on a span of 20 ft. What is the required pitch of the bolts if their shearing stress must not exceed 8800 psi?

Ans. 5.5 in.

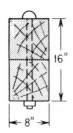

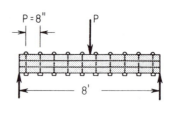

Problem 12-30   *Problem 12-31*

12-31. Three 4-in. by 4-in. full-dimension timbers are fastened as shown by carriage bolts having cross-sectional areas of 0.4 in.^{2} each. What concentrated load *P* can the beam support if the following restrictions are placed on the design? The maximum shearing stress in the bolts is not to exceed 8800 psi, and the maximum flexural stress in the wood is not to exceed 2000 psi.

Ans. 7920 lb.

12-32. Plywood and full-dimension 2-in. by 4-in. blocking are lagged together to form the beam section shown. What is the required pitch of the screws if the beam supports a uniform load of 100 lb per ft on a 16-ft span? The screws have cross-sectional areas of 0.01 in. and a permissible working stress in shear of 8000 psi.

Ans. 5.47 in.

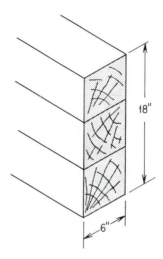

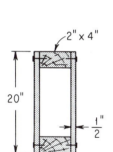

Problem 12-32   *Problem 12-33*

12-33. A timber beam is made of three square members glued together, as shown. The beam is simply supported and has a span of 10 ft. What is the safe load *P* that can be placed at midspan if the permissible shearing stresses in the wood and in the glue are 320 psi and 200 psi, respectively?

Ans. 32,400 lb.

12-34. A built-up steel girder has the cross-section shown. The moment of inertia of the cross-section with reference to the neutral axis is $I_{NA} = 4470$ in.^{4}. What is the proper rivet spacing if the girder is to support a maximum shear of 90 kips? The rivets have diameters of $\frac{3}{4}$ in. (area = 0.442 in.^{2}), and a permissible shearing stress of 8800 psi.

Ans. 3.92 in.

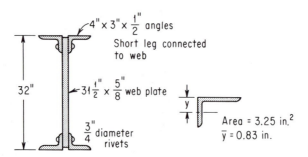

4" x 3" x $\frac{1}{2}$" angles
Short leg connected to web

32"

$3\frac{1}{2}$" x $\frac{5}{8}$" web plate

$\frac{3}{4}$" diameter rivets

Area = 3.25 in.²
$\bar{y}$ = 0.83 in.

Problem 12-34

12-35. A 15-ft timber beam 6 in. wide by 10 in. deep is reinforced with steel plates, 6 in. wide by 1 in. thick, top and bottom. What is the maximum fiber stress in each material if the beam is simply supported and carries a uniformly distributed load of 500 lb per ft? $E_w = 1.5 \times 10^6$ psi.

Ans. Steel: 2600 psi; wood: 108 psi.

12-36. A 10-ft-timber cantilever beam 10 in. wide by 18 in. deep is reinforced with steel plates, 18 in. wide by $\frac{1}{4}$ in. thick on either side. What is the maximum fiber stress in each material if the beam supports a concentrated load of 10,000 lb at its free end? $E_w = 1.5 \times 10^6$ psi.

Ans. Steel: 23,100 psi; Wood: 1160 psi.

12-37. A laminated-timber beam has the cross-section shown comprised of alternate layers of wood with the grain running at right angles to one another. If the moduli parallel to the grain and across the grain are 2×10^6 psi and 0.5×10^6 psi, respectively, what is the safe distributed load w lb per ft that the beam can carry on a 20-ft span? The permissible bending stresses in the strong and weak directions are 1200 psi and 300 psi, respectively.

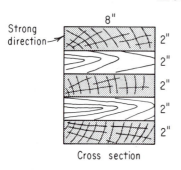

Strong direction

8"

2"
2"
2"
2"
2"

Cross section

Problem 12-37

Ans. 22.5 lb/ft.

12-38. Two 10-in., 30-lb channels are combined with timber to form the composite section shown. What is the safe bending moment for this section if the maximum permissible bending stresses are $\sigma_w = 1200$ psi and $\sigma_s = 18,000$ psi. $E_w = 1.5 \times 10^6$ psi; $E_s = 30 \times 10^6$ psi.

Ans. 71.8 kip ft.

12-39. A small cantilevered leaf spring is made of two pieces of brass and one piece of steel, each measuring 1 in. wide by 0.1 in. thick, brazed to form a solid

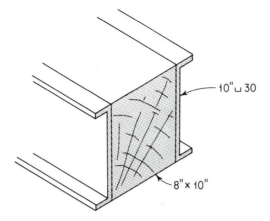

Problem 12-38

section. What is the stress in each material if a concentrated force acts as shown? $E_b = 15 \times 10^6$ psi; $E_s = 30 \times 10^6$ psi.

Ans. Brass: 3850 psi; steel: 2560 psi.

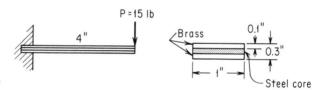

Problem 12-39

12-40. A rectangular concrete beam is reinforced with four steel bars, each having an area of $\frac{1}{2}$ in.^{2}, as shown. Locate the neutral axis of the section. Assume the ratio $E_s/E_c = 15$.

Ans. 8.88 in. below the top.

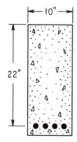

Problem 12-40

12-41. The reinforced concrete beam illustrated is subjected to a moment of 500,000 lb in. What is the maximum stress in the concrete and in the steel? The area of each of the four steel rods is 1 in.^{2}. Assume the ratio $E_s/E_c = 15$.

Ans. Concrete: 500 psi; steel: 7500 psi.

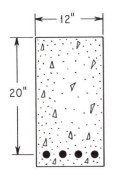

Problem 12-41

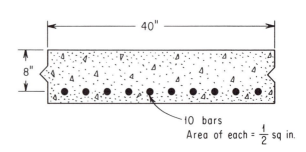

10 bars
Area of each = $\frac{1}{2}$ sq in.

Problem 12-42

12-42. A reinforced concrete floor slab has the dimensions shown. What is the allowable moment in this slab if the permissible stresses in the steel and concrete are 15,000 psi and 900 psi, respectively? Assume the ratio $E_s/E_c = 10$.

Ans. 390 kip in.

12-43. A 20-ft, simply supported reinforced concrete beam has the cross section shown. The beam supports a uniformly distributed load of w lb per ft (including its own weight) over the entire span. Determine w if both the concrete and the steel are stressed to their maximum values of 800 psi and 16,000 psi, respectively. Assume that $E_s/E_c = 10$. *Hint:* First find the required area of steel.

Ans. 1046 lb/ft.

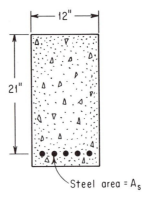

Steel area = A_s **Problem 12-43**

Chapter 13

DEFLECTION OF BEAMS

In design, deformation often shares an equal importance with strength. A girder, for example, may have sufficient strength to withstand a particular floor load, but it may be too "soft" or too "bouncy" underfoot. Its deflection under load may also far surpass the elasticity of architectural materials, such as mortar and plaster, that might be fastened to it. It is necessary, therefore, to understand how beam deflections, as well as stresses, are computed.

There are many ways of approaching the problem of beam deflections; two of these, the *moment-area* method and the *superposition* method, will be considered in this chapter. Either singly or in combination, these two concepts will handle all of the more common beam-deflection problems.

13-1 Moment-Area Method

The reciprocal of radius of curvature ρ of a beam in terms of bending moment, modulus of elasticity, and moment of inertia is defined as follows:

$$\frac{1}{\rho} = \frac{M}{EI} \qquad (13\text{-}1)$$

It is this equation that forms the basis of the moment-area method. Consider a portion AB of a beam, Fig. 13-1, distorted by some manner of loading

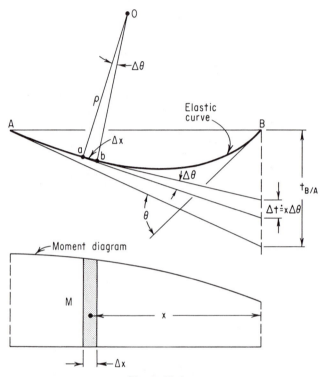

Figure 13-1

(not shown). An incremental length Δx of the *elastic curve* can be defined by the radius ρ and the angle $\Delta\theta$ as

$$\Delta x = \rho\Delta\theta \qquad (13\text{-}2)$$

Equations (13-1) and (13-2) are combined to give

$$\Delta\theta = \frac{M\Delta x}{EI}$$

Tangents drawn to the *elastic curve* at a and b are separated by the same angle as are the normals oa and ob. By imagining the beam to be composed of an infinite number of segments Δx, the slope at any point on the elastic curve relative to a second point would be the sum $\sum \Delta\theta$. Thus

$$\theta_{A/B} = \sum \Delta\theta = \frac{1}{EI} \sum M \Delta x \qquad (13\text{-}3)$$

where the term $\sum M\Delta x$ is the area of the moment diagram between corresponding limits A and B on the elastic curve. Theorem I, a formal statement of Eq. (13-3), is as follows:

Theorem I. The angle between tangenets drawn at A *and* B *on the elastic curve is equal to the area of the corresponding portion of the bending moment diagram, divided by* EI.

Consider, next, the deflection or *vertical deviation* of point B on the elastic curve relative to a tangent drawn at A. Returning to the small element ab, the distance Δt is very nearly equal to $x\Delta\theta$. This approximation is valid, since $\Delta\theta$, and for that matter θ, is very small. The vertical deviation of point B relative to a tangent drawn at A is the sum of all vertical deviations Δt between A and B.

$$t_{B/A} = \sum_A^B \Delta t = \sum_A^B x\Delta\theta = \sum_A^B x\left(\frac{M\Delta x}{EI}\right)$$

$$= \frac{1}{EI}\sum_A^B x(M\Delta x).$$

The summation $\sum_A^B x(M\Delta x)$ is physically equal to *the moment of the area of the moment diagram between* A *and* B *with respect to a vertical line drawn from* B. It is essential to understand the significance of the subscripts in terms representing tangential deviations. Thus, $t_{A/B}$ and $t_{B/A}$ are not necessarily equal. The first represents the vertical distance between point A and a tangent drawn to the elastic curve at point B, whereas the second is the vertical distance between point B and a tangent drawn at point A. This notation is illustrated in Fig. 13-2. A memory aid: *Moments of the areas of the moment diagram are always taken relative to the deviation line.* Hence,

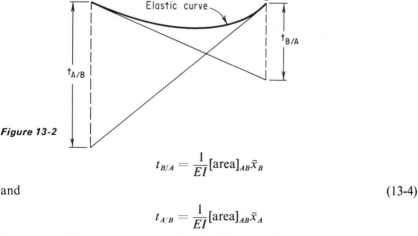

Figure 13-2

$$t_{B/A} = \frac{1}{EI}[\text{area}]_{AB}\bar{x}_B$$

and (13-4)

$$t_{A/B} = \frac{1}{EI}[\text{area}]_{AB}\bar{x}_A$$

Equations (13-4) can be summarized in Theorem II:

Theorem II. The vertical deviation of point B *on the elastic curve away from a tangent drawn at* A *is equal to the moment of the area of the bending moment diagram with respect to* B, *divided by* EI.

To simplify the calculations, moment diagrams are best drawn by the *method of parts* described in Chapter 11. For convenience, a table listing the properties of areas is presented again.

Table 13-1

		Area	*Centroid* $\bar{x}$
	Rectangle	bh	$b/2$
	Triangle	$\frac{1}{2}bh$	$b/3$
	Parabola	$\frac{1}{3}bh$	$b/4$
	Cubic	$\frac{1}{4}bh$	$b/5$

13-2. Statically Determinate Cantilever Beams

Deflections of fixed-end beams are easily computed by the moment-area method. Consider the beam shown in Fig. 13-3. Since the tangent drawn on the elastic curve at B is horizontal, the deflection δ_A at point A and the tangential deviation $t_{A/B}$ are exactly equal.

$$\delta_A = t_{A/B} = \frac{1}{EI}[\text{area}]_{AB}\bar{x}_A$$

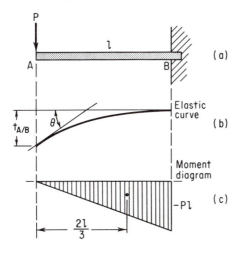

Figure 13-3

$$= \frac{1}{EI}\left[\frac{1}{2}(l)(-Pl)\left(\frac{2}{3}l\right)\right]$$

$$= -\frac{Pl^3}{3EI}$$

The negative sign indicates the deflection to be below the unloaded, or free, position of point A.

The slope of the beam at the free end, by Theorem I, is

$$\theta = \frac{1}{EI}[\text{area}]_{AB} = \frac{1}{EI}\left[\frac{1}{2}(l)(-Pl)\right]$$

$$= -\frac{Pl^2}{2EI} \text{ rad}$$

The minus sign merely indicates the slope of a tangent at A to be negative when measured relative to a tangent at B.

EXAMPLE 1. The timber cantilever beam ($E = 1.5 \times 10^6$ psi) shown in Fig. 13-4 supports two loads, as indicated. Determine the maximum deflection of the beam.

Solution: A sketch is drawn to represent the probable elastic curve. From this sketch it can be seen that the maximum deflection occurs at the free end and is, therefore, equal to the tangential deviation $t_{A/B}$. The moment diagram, drawn by *parts*, and Theorem II are used to compute the required deflection.

$$\delta_{\max} = t_{A/B} = \frac{1}{EI}[\text{area}]_{AB}\bar{x}_A$$

$$= \frac{1}{EI}\left[\frac{1}{2}(6)(-600)(10) + \frac{1}{2}(12)(-1200)(8)\right]$$

$$= -\frac{75{,}600}{EI}\text{lb ft}^3 = -\frac{75{,}600(1728)}{EI}\text{lb in.}^3$$

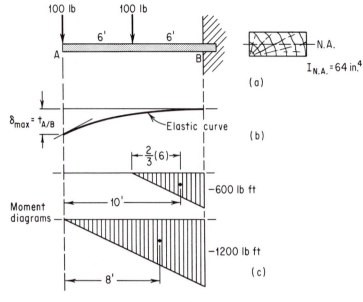

Figure 13-4

Numerical values are substituted to obtain the desired answer:

$$\delta_{max} = \frac{-75,600(1728)}{1.5(10)^6(64)} = -1.36 \text{ in.}$$

EXAMPLE 2. Find the distributed weight w lb per ft, Fig. 13-5, that will limit the maximum deflection to 1 in. $E = 1.5 \times 10^6$ psi; $I = 50$ in.^{4}.

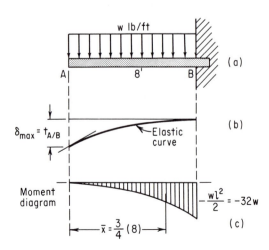

Figure 13-5

Solution: The moment diagram, which is a parabola, has an area and centroid of

$$A = \tfrac{1}{3}bh, \quad \bar{x} = \tfrac{3}{4}b$$

The maximum deflection and the deviation $t_{A/B}$ are equal; hence,

$$\delta_{\max} = t_{A/B} = \frac{1}{EI}[\text{area}]_{AB}\bar{x}_A$$

When symbolic terms are used, $\delta_{\max}$ is equal to

$$\delta_{\max} = \frac{1}{EI}\left[\frac{1}{3}(l)\left(-\frac{wl^2}{2}\right)\left(\frac{3}{4}l\right)\right]$$

$$= -\frac{wl^4}{8EI}\ \text{lb ft}^3$$

Substitution of numerical values gives

$$-1 = -\frac{w(8)^4(1728)}{8(1.5)(10^6)50}$$

$$w = 84.8\ \text{lb per ft}$$

EXAMPLE 3.   Find the maximum value of $EI\delta$ for the cantilever beam shown in Fig. 13-6(a).

Solution: To simplify computations, the moment diagram is drawn in parts, Fig. 13-6(b); the left end of the beam is used as a reference. The diagram, then, consists of a triangle, a rectangle, and a parabola. To accomplish this step, the reactions are first computed.

$$\sum M = 0$$
$$M = 50(5)(4.5) = 1125\ \text{lb ft}$$
$$\sum F_y = 0$$
$$V = 50(5) = 250\ \text{lb}$$

The heights of the triangular, rectangular, and parabolic areas are, respectively,

$$M_1 = 250(7) = 1750\ \text{lb ft}$$
$$M_2 = -1125\ \text{lb ft}$$
$$M_3 = -50(5)(\tfrac{5}{2}) = -625\ \text{lb ft}$$

By Theorem II,

$$\delta_{\max} = t_{A/B} = \frac{1}{EI}[\text{area}]_{AB}\bar{x}_A$$

$$EI\delta_{\max} = [\tfrac{1}{2}(7)(1750)(\tfrac{7}{3}) + 7(-1125)(\tfrac{7}{2}) + \tfrac{1}{3}(5)(-625)(\tfrac{5}{4})]$$
$$= -14{,}600\ \text{lb ft}^3$$

Alternate Solution: The moment diagram can be drawn in total form, as shown in Fig. 13-6(c). To simplify computations, the diagram is considered to be the sum

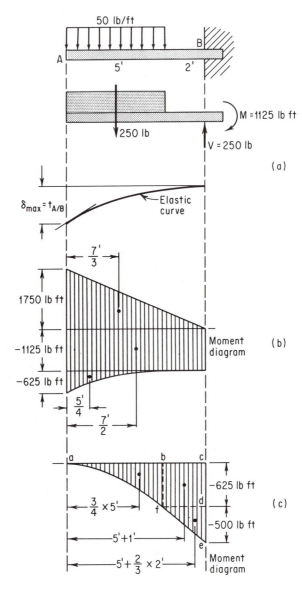

Figure 13-6

of three areas: parabola *abf*, rectangle *bcdf*, and triangle *def*. Ordinate *bf* represents the moment of the distributed load about a point 5 ft from the extreme left.

$$bf = -50(5)\tfrac{5}{2} = -625 \text{ lb ft}$$

Ordinate *de* is simply the numerical difference:

$$de = ce - cd$$
$$= -1125 + 625 = -500 \text{ lb ft}$$

By Theorem II,

$$\delta_{max} = \frac{1}{EI}[\text{area}]_{AB}\bar{x}_A$$

$$EI\delta_{max} = [\tfrac{1}{3}(5)(-625)(\tfrac{3}{4} \times 5) + 2(-625)(5 + 1) + \tfrac{1}{2}(2)(-500)(5 + \tfrac{2}{3} \times 2)]$$
$$= -14,600 \text{ lb ft}^3$$

13-3 Simply Supported Beams

Determination of deflections in simply supported beams by the moment-area method is not quite as direct as with cantilever beams. The existence of a horizontal tangent at the support, in the latter, simplified computations, since the tangential deviation and the maximum deflection were equal.

A series of steps are necessary, for example, to find the deflection δ_C on the simply supported beam of Fig. 13-7(a). First, a diagram of the elastic curve and its associated geometry must be carefully drawn, Fig. 13-7(b). In this example δ_C is the desired deflection, $t_{C/A}$ is the vertical deviation of point C relative to a tangent at A, and $t_{B/A}$ is the vertical deviation of point B relative

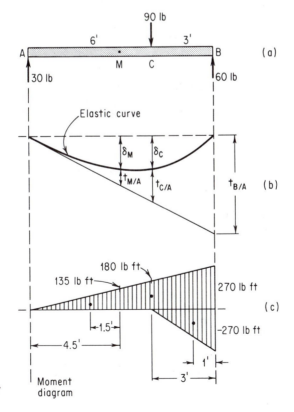

Figure 13-7

also to a tangent at A. By similar triangles,

$$\frac{\delta_C + t_{C/A}}{6} = \frac{t_{B/A}}{9}$$

Solving for the deflection gives

$$\delta_C = \tfrac{2}{3}t_{B/A} - t_{C/A}$$

By Theorem II:

$$EI\delta_C = \tfrac{2}{3}[\text{area}]_{AB}\bar{x}_B - [\text{area}]_{AC}\bar{x}_C \qquad \text{(a)}$$

Thought given to the construction of the moment diagram can greatly simplify computations; in this example the right end of the beam is selected as a reference line, and the diagram appears as shown in Fig. 13-7(c).

Numerical data are substituted into Eq. (a):

$$EI\delta_C = \tfrac{2}{3}[\tfrac{1}{2}(9)(270)(\tfrac{9}{3}) + \tfrac{1}{2}(3)(-270)(\tfrac{3}{3})] - [\tfrac{1}{2}(6)(180)(\tfrac{6}{3})]$$
$$= 1080 \text{ lb ft}^3 \text{ directed as shown}$$

The deflection at midspan is computed in a similar manner.

$$\frac{\delta_M + t_{M/A}}{4.5} = \frac{t_{B/A}}{9}$$

$$\delta_M = \tfrac{1}{2}t_{B/A} - t_{M/A}$$
$$EI\delta_M = \tfrac{1}{2}[\text{area}]_{AB}\bar{x}_B - [\text{area}]_{AM}\bar{x}_M$$

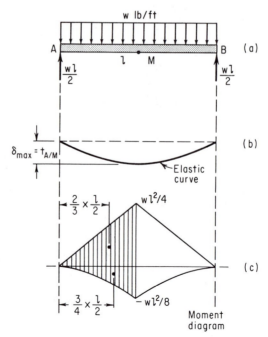

Figure 13-8

$$= \tfrac{1}{2}[\tfrac{1}{2}(9)(270)(\tfrac{9}{3}) + \tfrac{1}{2}(3)(-270)(\tfrac{3}{3})] - [\tfrac{1}{2}(4.5)(135)(\tfrac{4.5}{3})]$$
$$= 1164 \text{ lb ft}^3 \text{ directed as shown}$$

It will be shown in later discussion that the exact value of $EI\delta_{max}$ is 1176 lb ft^3. The answer obtained for midspan deflection is, therefore, approximately equal to the maximum deflection.

EXAMPLE 4. A uniformly distributed load w lb per ft acts across the entire span of a simply supported beam, as shown in Fig. 13-8. Determine the maximum value of $EI\delta$.

Solution: The moment diagram is constructed by the method of parts, as shown in Fig. 13-8(c). Since the loading is symmetrical, the maximum deflection occurs at the midspan M, a point at which the elastic curve has a zero slope. Tangential deviation of A relative to M, therefore, equals the maximum deflection.

Theorem II is applied to the area of the bending moment diagram between points A and M.

$$\delta_{max} = t_{A/M} = \frac{1}{EI}[\text{area}]_{MA}\bar{x}_A$$

Numerical values are substituted, and δ_{max} is computed.

$$EI\delta_{max} = \left[\left(\frac{1}{2}\right)\left(\frac{l}{2}\right)\left(\frac{wl^2}{4}\right)\left(\frac{2}{3} \times \frac{l}{2}\right) - \frac{1}{3}\left(\frac{l}{2}\right)\left(\frac{wl^2}{8}\right)\left(\frac{3}{4} \times \frac{l}{2}\right)\right]$$
$$= wl^4(\tfrac{1}{48} - \tfrac{1}{128})$$
$$= \tfrac{5}{384} wl^4 \text{ directed as shown}$$

13-4 Overhanging Beams

The treatment of beams which extend beyond their supports can be best explained by considering the example that follows.

The beam of Fig. 13-9(a) supports a concentrated load of 120 lb midway between supports and a second load of 60 lb extending beyond the right support, as shown. It is conceivable that point C could be either below or above the original axis of the beam. These two possibilities are illustrated in Figs. 13-9(b) and 13-9(c). If C is assumed to be below the axis, geometry indicates δ_C to be related by similar triangles to tangential deviations $t_{C/A}$ and $t_{B/A}$ in the following manner:

$$\frac{\delta_C + t_{C/A}}{14} = \frac{t_{B/A}}{12}$$

$$\delta_C = \tfrac{7}{6}t_{B/A} - t_{C/A} \tag{a}$$

If the second manner of bending is assumed, the geometry indicates that

$$\frac{t_{C/A} - \delta_C}{14} = \frac{t_{B/A}}{12}$$

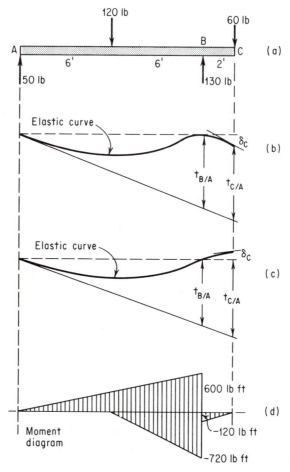

Figure 13-9

$$\delta_C = -\tfrac{7}{6}t_{B/A} + t_{C/A} \tag{b}$$

Numerically, Eqs. (a) and (b) are equal; they differ, however, in algebraic sign. One or the other of the two possibilities must be assumed in writing the geometric relationship; an incorrect guess will be acknowledged by a minus sign.

The moment diagram is constructed in parts with reference to reaction B, as shown in Fig. 13-9(d). Here again, there are many ways of drawing the moment diagram, and the method selected should be one which will keep the computations simple and direct.

It is assumed that δ_C will be below the axis, and numerical data are substituted into Eq. (a).

$$\delta_C = \tfrac{7}{6}t_{B/A} - t_{C/A}$$

$$\delta_c = \left(\frac{7}{6}\right)\frac{1}{EI}[\text{area}]_{AB}\bar{x}_B - \frac{1}{EI}[\text{area}]_{AC}\bar{x}_C$$

$$EI\delta_c = \tfrac{7}{6}[\tfrac{1}{2}(12)(600)(\tfrac{12}{3}) + \tfrac{1}{2}(6)(-720)(\tfrac{6}{3})] - [\tfrac{1}{2}(12)(600)\tfrac{12}{3} + 2)$$
$$+ \tfrac{1}{2}(6)(-720)(\tfrac{6}{3} + 2) + \tfrac{1}{2}(2)(-120)(\tfrac{2}{3} \times 2)]$$
$$= -1040 \text{ lb ft}^3$$

Since the answer is negative, point C is above the horizontal axis of the beam, opposite to the assumption.

13-5 Propped Beams

The propped beam is the first encounter in this text with statically indeterminate beam loading. It is unfortunate that the phrase *statically indeterminate* carries a feeling of something mysterious and incalculable. In truth, however, it simply means that, although the equations of statics apply, they are not sufficient; there are more unknowns than there are equations of equilibrium. The missing "link" in the computations can be supplied by the moment-area concept.

The propped beam, Fig. 13-10(a), is an example of an indeterminate structure. Three unknown reactions R, M, and V are shown in the free-body diagram, Fig. 13-10(b). If points A and B remain on the same horizontal line, the tangential deviation $t_{A/B}$ is zero. Thus, the reaction R can be obtained by applying Theorem II to the moment diagram of Fig. 13-10(c).

$$t_{A/B} = \frac{1}{EI}[\text{area}]_{AB}\bar{x}_A = 0$$

Hence,

$$[\text{area}]_{AB}\bar{x}_A = 0$$
$$\tfrac{1}{2}(12)(12R)(\tfrac{2}{3} \times 12) + \tfrac{1}{2}(9)(-2700)(3 + \tfrac{2}{3} \times 9) = 0$$
$$576R = 109{,}400$$
$$R = 190 \text{ lb}$$

Two unknown reactions remain, and these can be found through the equations of statics.

$$\sum M = 0$$
$$M = 12(190) - 2700 = -420 \text{ lb ft}$$
$$\sum F_y = 0$$
$$V = 300 - 190 = 110 \text{ lb}$$

To obtain maximum values of V and M, the shear and the moment diagrams can be drawn in their total forms, Fig. 13-10(d). Although the moment-area method can be used to find deflections in indeterminate beams, the computa-

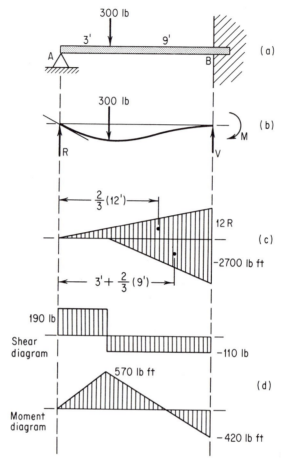

Figure 13-10

tion becomes rather tedious. For this reason, the superposition method, which is described in Sec. 13-7, is the more practical approach.

13-6 Restrained Beams

The restrained beam, Fig. 13-11(a), is *doubly indeterminate*. Four unknown reactions, consisting of two end moments and two vertical forces, as shown in Fig. 13-11(b), must be found before a moment diagram can be constructed. Here again, the moment-area method proves to be an efficient approach.

Consider the restrained beam of Fig. 13-12(a). Two moment-area equations can be written for this beam:

$$t_{AB} = \frac{1}{EI}[\text{area}]_{AB}\bar{x}_A = 0 \qquad (a)$$

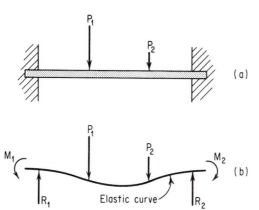

Figure 13-11

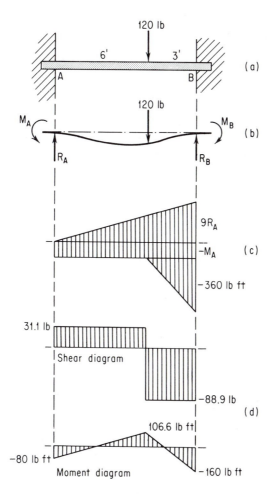

Figure 13-12

323

and

$$t_{B/A} = \frac{1}{EI}[\text{area}]_{AB}\bar{x}_B = 0 \qquad (b)$$

The walls act as perfect restraints; the slope of the beam at A and B is zero, and the two points are on a common horizontal tangent. For this reason Theorem I also applies and can be used to replace either Eq. (a) or Eq. (b).

$$\theta_{A/B} = \frac{I}{EI}[\text{area}]_{AB} = 0 \qquad (c)$$

For the sake of illustration, Eqs. (a) and (b) will be used in this example.

Either wall may be selected as a reference for the moment diagram, and in this example the right wall is used. Three areas comprise the diagram, as shown in Fig. 13-12(c); one is completely dimensioned, whereas the other two contain unknowns R_A and M_A.

Numerical data are substituted into Eq. (a):

$$t_{A/B} = 0$$

$$\frac{1}{EI}[\text{area}]_{AB}\bar{x}_A = 0$$

$$\tfrac{1}{2}(9)(9R_A)(\tfrac{2}{3} \times 9) - 9(M_A)(\tfrac{9}{2}) - \tfrac{1}{2}(3)(360)(6 + \tfrac{2}{3} \times 3) = 0$$

$$18R_A - 3M_A = 320 \qquad (a)$$

A second moment-area equation is written, and data are substituted in a similar manner.

$$t_{B/A} = 0$$

$$\frac{1}{EI}[\text{area}]_{AB}\bar{x}_A = 0$$

$$\tfrac{1}{2}(9)(9R_A)(\tfrac{9}{3}) - 9(M_A)(\tfrac{9}{2}) - \tfrac{1}{2}(3)(360)(\tfrac{3}{3}) = 0$$

$$9R_A - 3M_A = 40 \qquad (b)$$

The unknowns R_A and M_A can be computed by solving Eqs. (a) and (b) simultaneously.

$$18R_A - 3M_A = 320$$
$$\underline{-9R_A + 3M_A = -40}$$
$$9R_A \qquad = 280$$
$$R_A = \tfrac{280}{9} = 31.1 \text{ lb}$$

and

$$3M_A = 9R_A - 40 = 9(\tfrac{280}{9}) - 40$$

$$M_A = 80 \text{ lb ft directed as shown}$$

Two unknowns remain, R_B and M_B, and these are found through equations of static equilibrium

$$\sum F_y = 0$$
$$R_A + R_B = 120$$
$$R_B = 120 - 31.1 = 88.9 \text{ lb}$$
$$\sum M = 0$$
$$M_B = 9R_A - M_A - 360$$
$$= 9(31.1) - 80 - 360 = -160 \text{ lb ft}$$

The shear and moment diagrams can now be drawn, Fig. 13-12(d), and the critical values calculated.

When the loading is symmetrical, computations are greatly simplified, since forces R_A and R_B are equal, as are moments M_A and M_B. The reactions can easily be found, since they share the external load equally.

The example that follows illustrates a complete analysis of a symmetrically loaded, restrained beam.

EXAMPLE 4. Find the reactions and maximum deflection of the restrained beam shown in Fig. 13-13(a). Sketch the shear and moment diagrams and indicate critical values.

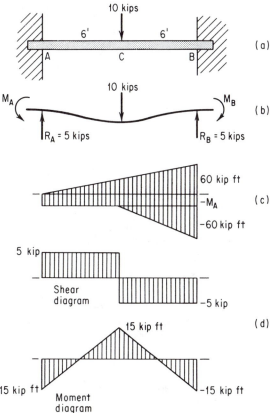

Figure 13-13

Solution: The free-body diagram is drawn as shown in Fig. 13-13(b). Since the loading is symmetrical, reactions R_A and R_B are each equal to one-half the external load.

$$R_A = R_B = \tfrac{10}{2} = 5 \text{ kips}$$

Next, Theorem I is applied to the moment diagram constructed as shown in Fig. 13-13(c). Tangents to the elastic curve at points A and B are horizontal; thus $\theta_{A/B} = 0$.

$$\theta_{A/B} = \frac{1}{EI}[\text{area}]_{AB} = 0$$

$$\tfrac{1}{2}(12)(60) - 12M_A - 1(6)(60) = 0$$

$$M_A = M_B = 15 \text{ kip ft directed as shown}$$

Shear and moment diagrams are constructed as shown in Fig. 13-13(d). The moment at midspan, computed in terms of the area of the shear diagram, is

$$\Delta M = \text{area under shear diagram}$$

$$M_C - M_A = 5(6) = 30$$

$$M_C = 30 + M_A = 30 - 15 = 15 \text{ kip ft}$$

Because of symmetry, a tangent drawn to the elastic curve at midspan is horizontal. Theorem II, therefore, can be applied directly to find $\delta_{\max}$.

$$\delta_{\max} = \frac{1}{EI}[\text{area}]_{AC}\bar{x}_C$$

$$\delta_{\max} = \frac{1}{EI}\left[\frac{1}{2}(6)(30)\left(\frac{6}{3}\right) + 6(-15)3\right]$$

$$= -\frac{90}{EI}\text{ ft}$$

13-7 Deflections by Superposition

Superposition, which simply means "to place one upon the other," is a powerful concept with many applications in engineering. It can be used to find deflections, moments, and stresses in both simple and complex beams.

Consider two identical beams, Fig. 13-14(a), made of the same material and with the same physical dimensions. One beam supports a concentrated load at midspan, and the other, a uniformly distributed load over its entire length. Each member deflects through a known distance, as indicated. By superposing the two loads, deflections and reactions become the algebraic sum of those of the individual beams. This is illustrated in Fig. 13-14(b).

Tabulated values of deflection, Table 13-2, are generally employed with the superposition method. Ingenuity often plays an important part in the use of these tables, as the examples which follow illustrate.

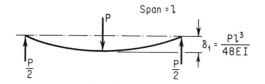

(a)

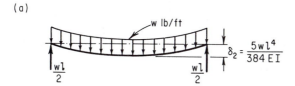

(b)

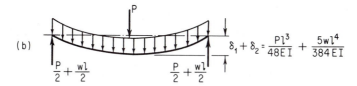

Figure 13-14

Table 13-2. Beam Deflection Equations

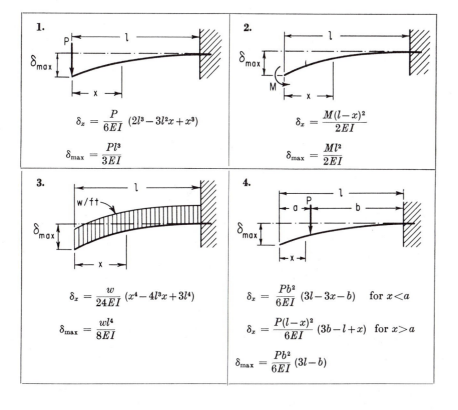

Table 13-2 (cont.)

5.

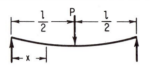

$$\delta_x = \frac{Px}{48EI}(3l^2 - 4x^2) \quad \text{for } x < \frac{l}{2}$$

$$\delta_{max} = \frac{Pl^3}{48EI} \quad \text{at } x = \frac{l}{2}$$

6.

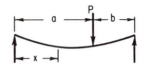

$$\delta_x = \frac{Pbx}{6lEI}(l^2 - x^2 - b^2) \quad \text{for } x < a$$

$$\delta_x = \frac{Pb}{6lEI}\left[\frac{l}{b}(x-a)^3 + (l^2 - b^2)x - x^3\right]$$
$$\text{for } x > a$$

$$\delta = \frac{Pb}{48EI}(3l^2 - 4b^2) \quad \text{at center if } a > b$$

$$\delta_{max} = \frac{Pb(l^2 - b^2)^{3/2}}{9\sqrt{3}\,lEI} \quad \text{at } x = \sqrt{\frac{l^2 - b^2}{3}}$$

7.

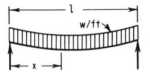

$$\delta_x = \frac{wx}{24EI}(l^3 - 2lx^2 + x^3)$$

$$\delta_{max} = \frac{5wl^4}{384EI} \quad \text{at center}$$

8.

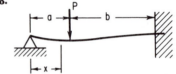

$$\delta_x = \frac{Pb^2x}{12EIl^3}(3al^2 - 2lx^2 - ax^2) \text{ for } x < a$$

$$\delta_x = \frac{Pa(l-x)^2}{12EIl^3}(3l^2x - a^2 x - 2a^2l)$$
$$\text{for } x > a$$

$$\delta = \frac{Pa^2b^3}{12EIl^3}(3l + a) \quad \text{at point of load}$$

9.

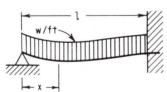

$$\delta_x = \frac{wx}{48EI}(l^3 - 3lx^2 + 2x^3)$$

$$\delta_{max} = \frac{wl^4}{185EI} \quad \text{at } x = 0 \cdot 422l$$

10.

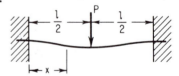

$$\delta_x = \frac{Px^2}{48EI}(3l - 4x)$$

$$\delta_{max} = \frac{Pl^3}{192EI} \quad \text{at center}$$

Table 13-2 (cont.)

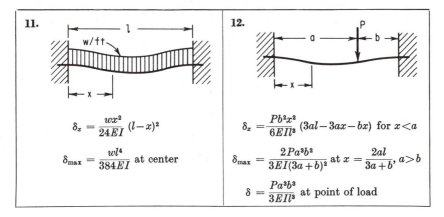

11.

$$\delta_x = \frac{wx^2}{24EI}(l-x)^2$$

$$\delta_{max} = \frac{wl^4}{384EI} \text{ at center}$$

12.

$$\delta_x = \frac{Pb^2x^2}{6EIl^3}(3al-3ax-bx) \text{ for } x<a$$

$$\delta_{max} = \frac{2Pa^3b^2}{3EI(3a+b)^2} \text{ at } x=\frac{2al}{3a+b}, a>b$$

$$\delta = \frac{Pa^3b^3}{3EIl^3} \text{ at point of load}$$

EXAMPLE 5. Use the superposition method to find the deflection at the free end of the cantilever beam shown in Fig. 13-15.

Solution: Two basic loadings, Case 3 and Case 4 of Table 13-2, are superposed. The deflection at the free end is the sum:

$$\delta_A = \frac{wl^4}{8EI} + \frac{Pb^2}{6EI}(3l-b)$$

Numerical data are substituted to find δ_A:

$$\delta_A = \frac{180(8)^4}{8EI} + \frac{1200(6)^2}{6EI}[3(8)-6]$$

$$= \frac{222,000}{EI} \text{ ft}$$

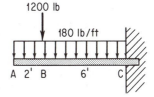

Figure 13-15

EXAMPLE 6. Find the midspan value of the deflection for the restrained beam shown in Fig. 13-16(a).

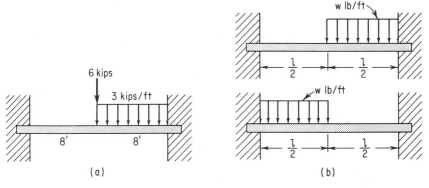

(a)

(b)

Figure 13-16

Solution: Case 10 of Table 13-2 represents the concentrated load at midspan; the deflection caused by the distributed load, which does not appear in the table, lies hidden in Case II. This can be explained as follows: consider two identical beams, Fig. 13-16(b), each supporting a distributed load over half a span. Superposed, the beams are equivalent to Case 11. The "half beams" deflect equally; therefore,

$$\delta + \delta = \frac{wl^4}{384EI}$$

and

$$\delta = \frac{wl^4}{768EI}$$

The midspan deflection of the original beam can now be calculated.

$$\delta = \frac{Pl^3}{192EI} + \frac{wl^4}{768EI}$$

$$= \frac{6(16)^3}{192EI} + \frac{3(16)^4}{768EI}$$

$$EI\delta = 384 \text{ kip ft}^3$$

13-8 Continuous Beams

The superposition method can be applied to analyze *continuous beams*— beams that have more supports than necessary to maintain equilibrium. While this is not the only method of attack,[1] it is one that can be easily used.

Consider the continuous beam of Fig. 13-17(a). The reactions are to be determined, and both a shear and a moment diagram are to be drawn.

The member can be viewed in the following way: imagine first, Fig. 13-17(b), that the distributed load is simply supported and is typical of Case 7 given in Table 13-2. The midspan deflection is $5wl^4/384EI$ and the end reactions are $wl/2$, each. Now imagine that a load P acts upward on a similar beam with sufficient force to cause the midspan deflection to equal the former value, Fig. 13-17(c). This second loading is typical of Case 5, where $\delta = Pl^3/48EI$.

Equating the deflections will give a value of P which, in reality, is the center reaction R_2.

$$\frac{Pl^3}{48EI} = \frac{5l^4}{384EI}$$

$$P = \tfrac{5}{8}wl = R_2$$

The end reactions, Fig. 13-17(d), are the sums of those for the two cases; of course, in the first instance the reaction is upward and in the second, downward.

---

[1]For a classical method of approach, the reader is referred to the *three-moment method* described in advanced texts on the subject of mechanics of materials.

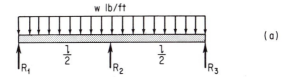

(a)

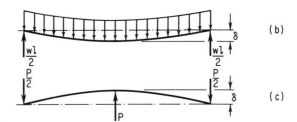

(b)

(c)

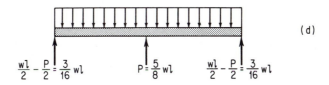

(d)

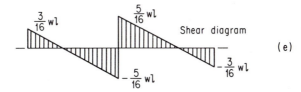

Shear diagram

(e)

Figure 13-17

$$R_1 = R_3 = \frac{wl}{2} - \frac{1}{2}\left(\frac{5}{8}\right)wl = \frac{3}{16}wl$$

The shear diagram, from which the moment diagram could be constructed, is shown in Fig. 13-17(e).

EXAMPLE 7. Find the moment over the center support for the continuous beam shown in Fig. 13-18(a).

Solution: The deflections δ_1 and δ_2 must add to give δ_3, since all three supports lie on the same horizontal line. Case 5 and Case 6 of Table 13-2 are used to find these deflections.

For the beam of Fig. 13-18(b):

$$\delta_1 = \frac{Pb}{6lEI}\left[\frac{l}{b}(x-a)^3 + (l^2 - b^2)x - x^3\right] \text{ for } x > a$$

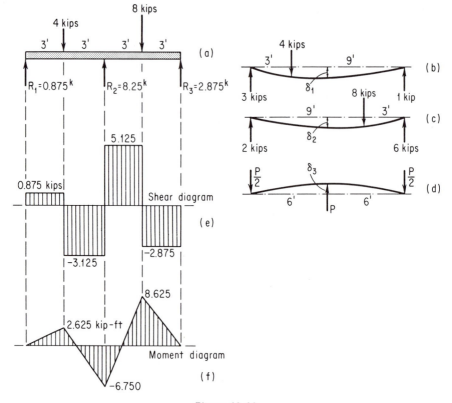

Figure 13-18

$$= \frac{4(9)}{6(12)EI}\left[\frac{12}{9}(6-3)^3 + (12^2 - 9^2)6 - 6^3\right] = \frac{99}{EI}$$

For the beam of Fig. 13-18(c):

$$\delta_2 = \frac{Pbx}{6lEI}(l^2 - x^2 - b^2) \quad \text{for } x < a$$

$$= \frac{8(3)6}{6(12)EI}(12^2 - 6^2 - 3^2) = \frac{198}{EI}$$

For the beam of Fig. 13-18(d):

$$\delta_3 = \frac{Pl^3}{48EI}$$

$$= \frac{P(12)^3}{48EI} = \frac{36P}{EI}$$

Thus

$$\delta_3 = \delta_1 + \delta_2$$

$$\frac{36P}{EI} = \frac{99}{EI} + \frac{198}{EI}$$

$$P = \frac{99 + 198}{36} = 8.25 \text{ kips}$$

With reference to the individual free-body diagrams, the reactions are found to be

$$R_1 = 3 + 2 - \frac{8.25}{2} = 0.875 \text{ kips}$$

$$R_2 = P = 8.25 \text{ kips}$$

$$R_3 = 1 + 6 - \frac{8.25}{2} = 2.875 \text{ kips}$$

The shear and moment diagrams are constructed in the usual manner and appear as shown in Figs. 13-18(e) and (f); the maximum moment is found to be under the 8-kip load.

13-9 Special Techniques

Interesting situations arise when loads act on beams which, in turn, are supported by deformable bodies. These bodies may be cables, springs, or columns. In general, equations can be written which relate the deformations of the various components, and these equations are then combined with those of statics to complete the solution. Examples that follow illustrate typical problems involving this type of loading.

EXAMPLE 8. The steel cantilever beam, Fig. 13-19, is partially supported at its free end by a coil spring. Find the force in the spring. The moment of inertia of the beam is 1.728 in.[4] and the constant of the spring is 1000 lb per in.

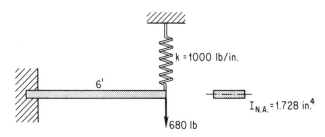

Figure 13-19

Solution: The beam and spring deflect equally as they share the load.

$$\delta_b = \delta_s$$

$$\left(\frac{Pl^3}{3EI}\right)_b = \left(\frac{P}{k}\right)_s$$

$$\frac{P_b(6 \times 12)^3}{3(30 \times 10^6)(1.728)} = \frac{P_s}{10^3}$$

$$2.4P_b = P_s \tag{a}$$

and

$$P_b + P_s = 680 \tag{b}$$

The forces P_s and P_b are obtained by solving Eqs. (a) and (b) simultaneously.

$$P_b + 2.4P_b = 680$$

$$3.4P_b = 680$$

$$P_b = 200 \text{ lb}$$

$$P_s = 2.4(200) = 480 \text{ lb}$$

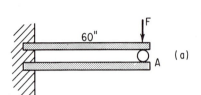

(a)

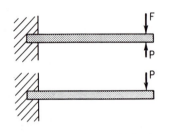

(b)

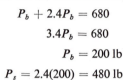

EXAMPLE 9. Two steel cantilever beams help support load F as shown in Fig. 13-20(a). Find F if the deflection at A is to be $\frac{1}{2}$ in. The moment of inertia of each member is 1.5 in.[4].

Figure 13-20

Solution: Two conditions at point A provide the necessary equations for this problem. First, a mutual force P acts on both beams, Fig. 13-20(b), and second, the deflections of both beams are equal.

With reference to Table 13-2, the deflection of the upper beam is

$$\delta_A = \frac{Fl^3}{3EI} - \frac{Pl^3}{3EI} \tag{a}$$

and for the lower beam

$$\delta_A = \frac{Pl^3}{3EI} \tag{b}$$

Equations (a) and (b) are combined to eliminate the mutual force P.

$$2\delta_A = \frac{Fl^3}{3EI}$$

$$F = \frac{6\delta_A EI}{l^3}$$

$$= \frac{6(0.5)(30 \times 10^6)1.5}{(60)^3}$$

$$= 625 \text{ lb}$$

QUESTIONS, PROBLEMS, AND ANSWERS

13-1. The cantilevered beam shown is acted on by the force P. The deflection d changes by what factor when:
(a) P is doubled?
(b) b is doubled?
(c) h is doubled?
(d) l is doubled?
(e) E is doubled?

Ans. (a) 2; (b) $\frac{1}{2}$; (c) $\frac{1}{8}$; (d) 8: (e) $\frac{1}{2}$.

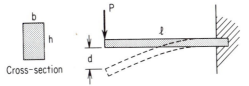

Problem 13-1

13-2. The steel beam shown has a moment of inertia of 172.8 in.^{4}. What is the magnitude of load P if the deflection at the free end is 0.40 in?

Ans. 11,100 lb.

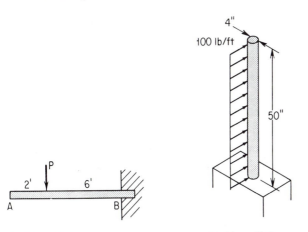

Problem 13-2   *Problem 13-3*

13-3. The cantilever beam shown is subjected to a distributed load over the entire span. Find the slope and deflection at the free end. $E = 1.5 \times 10^6$ psi.

Ans. 0.00922 rad., 3.46 in.

13-4. What is the depth d of the beam shown if the maximum deflection is to equal 1.2 in.? $E = 1.5 \times 10^6$ psi.

Ans. 6.78 in.

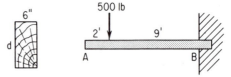

Problem 13-4

13-5. What is the safe load P for the beam shown if the deflection is not to exceed 1 in., the bending stress is not to exceed 1800 psi, and the permissible shearing stress is limited to 800 psi? $E = 1.5 \times 10^6$ psi.

Ans. 633 lb.

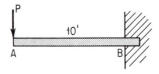

Problem 13-5

13-6. The three limiting design conditions, deflection, bending, and shear, illustrated in Prob. 13-5, usually cannot be satisfied simultaneously; as an engineer, what recourse do you have?

Ans. To search for the most economical section—the section that weighs the least or costs the least. Sometimes this is like looking for the lost chord.

13-7. What is the deflection at the free end of the cantilever beam shown? E is 10×10^6 psi and $I = 17.28$ in.[4]

Ans. 0.850 in. upward.

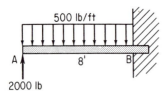

Problem 13-7

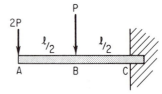

Problem 13-8

13-8. The maximum bending stress in the steel 10 WF 33 cantilever beam illustrated is 20,000 psi. What is the deflection at the free end if $l = 12$ ft.?

Ans. 0.873 in.

13-9. The wing of a jet-liner of total weight W can be assumed to be a cantilever beam as shown. Assume the moment of inertia to be constant (for a preliminary computation) and equal to I. Find the deflection of end A relative to the cabin.

Ans. 11 $Wl^3/432\ EI$ upward.

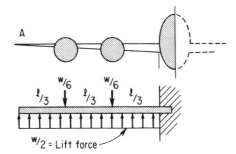

Problem 13-9

13-10. A simply supported beam of length l supports a concentrated load P at the midspan. Determine the maximum deflection by the moment-area method.

Ans. $-Pl^3/48EI$.

13-11. Use the moment-area method to find the midspan value of $EI\delta$ for the beam shown.

Ans. 3100 lb/ft^3.

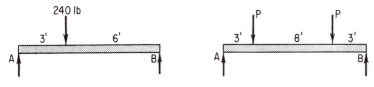

Problem 13-11   **Problem 13-12**

13-12. The permissible midspan deflection for the oak beam shown is 2 in. The beam has a cross section 6 in. wide by 12 in. deep and a modulus of elasticity of 1.5×10^6 psi. What is the safe value of P?

Ans. 21,700 lb.

13-13. A 1-ton flywheel is supported at the center of a steel shaft as shown. Find (a) the maximum deflection by the moment-area method, (b) the maximum bending stress in the shaft.

Ans. (a) 0.605 in.; (b) 22,650 psi.

13-14. Use the moment-area method to calculate the midspan value of $EI\delta$ for the simply supported beam shown.

Ans. 21,600 lb/ft^3.

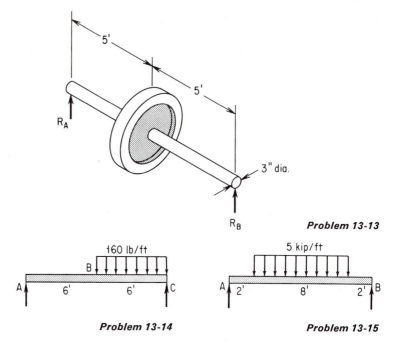

R_A

5'

5'

3" dia.

R_B **Problem 13-13**

160 lb/ft

B

A

6'

6'

C

5 kip/ft

A

2'

8'

2'

B

Problem 13-14 **Problem 13-15**

13-15. A 12 WF 65 steel beam supports a distributed load symmetrically arranged as shown. Determine the midspan value of deflection by the moment-area method.

Ans. 0.127 in.

13-16. A steel line-shaft is subjected to belt pulls as shown. Use the moment-area method to determine the deflection at midspan.

Ans. 0.111 in.

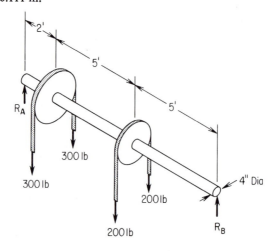

2'

5'

5'

R_A

300 lb

300 lb

200 lb

200 lb

4" Dia

R_B

Problem 13-16

13-17. Steel coils, each weighing 5000 lb, are lashed to a trailer bed as shown. What is the deflection of the bed at the center coil? Assume that the trailer bed consists of two 12-in., 35-lb-per-ft standard I beams, as shown in the section. Neglect the weights of the beams and the effects of the flooring, blocking, and guy wires. Use the moment-area method.

Ans. 0.351 in.

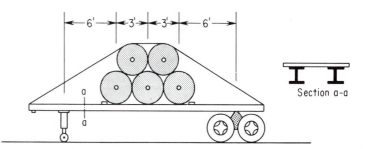

Problem 13-17

13-18. Compare the deflections and stresses in two identical beams; one has a total load of W lb distributed uniformly over its entire span, and the second has the load of W lb concentrated at the midspan.

Ans. $\delta_c = 1.6\delta_d$; $\sigma_c = 2\sigma_d$.

13-19. For the beams shown in the figures, find the value of $EI\delta$ at the free end by
to the moment-area method.
13-22.

Ans. (13-19) 5400 lb ft³ with C above B.
(13-20) 5760 lb ft³ with C below B.
(13-21) 38,400 lb ft³.
(13-22) $-Pl^3/27$.

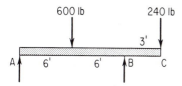

Problem 13-19

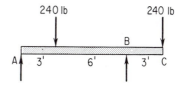

Problem 13-20

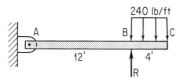

Problem 13-21

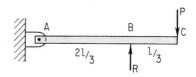

Problem 13-22

13-23. What value of P, in the figure shown, will make the deflection at C equal to zero?

Ans. 720 lb.

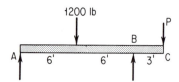

Problem 13-23

13-24 For the beams shown in the figures, find the reaction at the prop and then
to   draw the shear and moment diagrams. Indicate the maximum values of V
13-28. and M.

Ans. 13-24. 0.375 kips.
13-25. 600 lb.
13-26. 300 lb.
13-27. 900 lb.
13-28. 800 lb.

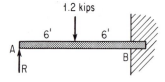

Problem 13-24

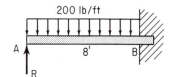

Problem 13-25

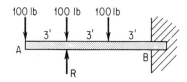

Problem 13-26

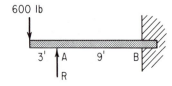

Problem 13-27

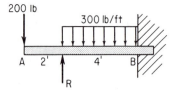

Problem 13-28

13-29. Two props are employed to help support the beam shown. Find the reactions R_1 and R_2.

Ans. $R_1 = 34.3$ lb; $R_2 = 199$ lb.

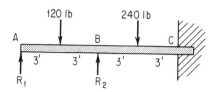

Problem 13-29 R_1 R_2

13-30 to 13-34. For the restrained beams shown in the figures, find, by the moment-area method, the forces and moments that act at the supports. Draw the shear and moment diagrams for each and indicate maximum values.

Ans. 13-30. $R_A = R_B = 1.5$ tons; $M_A = M_B = -4.5$ ton ft.

13-31. $R_A = 1420$ lb; $R_B = 3380$ lb; $M_A = 3200$ lb ft; $M_B = 4800$ lb ft.

13-32. $R_A = R_B = 3$ kips; $M_A = M_B = -6$ kip ft.

13-33. $R_A = R_B = 1440$ lb; $M_A = M_B = -2880$ lb ft.

13-34. $R_A = 19.8$ kips; $R_B = 6.19$ kips; $M_A = -34.1$ kip ft; $M_B = -18.4$ kip ft.

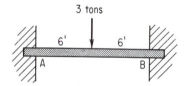

Problem 13-30

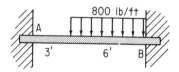

Problem 13-31

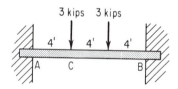

Problem 13-32

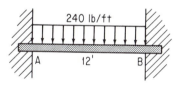

Problem 13-33

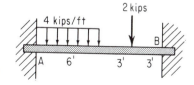

Problem 13-34

13-35. Use the superposition method to determine the deflection at the free end of the cantilever beam shown. See figure on p. 342.

Ans. 13 $Pl^3/24\ EI$ down.

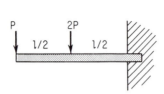

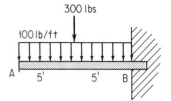

Problem 13-35   *Problem 13-36*

13-36. What is the maximum deflection of the cantilever beam shown? The beam is wood ($E = 1.5 \times 10^6$ psi) and has a square cross-section 8 in. on edge. Use the method of superposition.

Ans. 0.527 in.

13-37. Use the method of superposition to find the midspan deflection of the timber beam shown. The beam is 4 in. wide by 9 in. deep and has a modulus of elasticity of 1.5×10^6 psi.

Ans. 2.56 in.

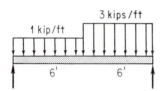

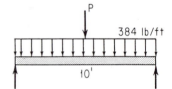

Problem 13-37   *Problem 13-38*

13-38. Use the superposition method to find the value of P, if the steel beam shown is to deflect 0.25 in. The moment of inertia of the beam is 17.28 in.[4].

Ans. 1200 lb.

13-39. Two loads of W lb act on the beam shown. One load is concentrated at the center of the span and the other is distributed uniformly over the entire length. Find the deflection, by the superposition method, in terms of E, I, l, and W.

Ans. $3\, Wl^3/384\, EI$.

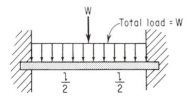

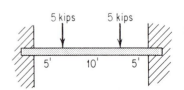

Problem 13-39   *Problem 13-40*

13-40. Find the least moment of inertia of the steel beam if the midspan deflection must not exceed $\frac{1}{400}$ of the span.

Ans. 20 in.^{4}.

13-41. A simple beam having a span of 12 ft supports a concentrated load P at the midspan. The beam has a rectangular cross-section, 6 in. wide and 12 in. deep, and a modulus $E = 2 \times 10^6$ psi. Determine the safe value of P under the following conditions:

Maximum allowable bending stress:	1000 psi
Maximum allowable longitudinal stress:	100 psi
Maximum allowable deflection:	span/400

Ans. 4000 lb.

13-42. A cantilever beam having an 8-ft span supports a uniformly distributed load of w lb per ft. The beam is a steel I-beam having a depth of 10 in. and a moment of inertia of 172.8 in. 4. Find the safe value of w under the following limitations:

Maximum bending stress:	20,000 psi
Maximum deflection:	span/500

Ans. 1130 lb/ft.

13-43. A timber cantilever beam ($E = 2 \times 10^6$ psi) having a depth of 12 in. supports a concentrated load P at its free end. Find the length of the beam if the following conditions are to be satisfied simultaneously:

Deflection:	span/400
Bending stress:	1000 psi

Ans. 7.5 ft.

13-44. Stress and deflection are criteria involved in design. Describe situations where (a) limiting stress is more critical than deflection, and (b) where deflection is more critical than stress.

13-45. A steel tie-bar having a cross-sectional area of 0.2 in.^{2} helps to support a timber cantilever beam, as shown on p. 344. The beam has a moment of inertia of 500 in.^{4} and a modulus of elasticity of 2×10^6 psi. What is the axial force in the tie-bar?

Ans. 7830 lb.

13-46. A steel post having a cross-sectional area of 0.5 in.^{2} is placed under a restrained steel beam, as shown on p. 344. The beam has a moment of inertia of 100 in.^{4}. What is the stress in the post if its temperature is raised by 100°F? The coefficient of linear expansion is 6.5×10^{-6} in./in./°F.

Ans. 15,200 psi.

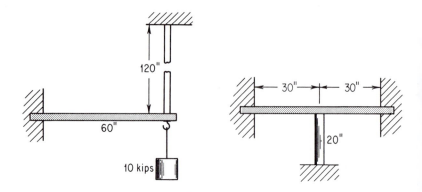

Problem 13-45   **Problem 13-46**

13-47. A brass tube having a cross-sectional area of $\frac{1}{2}$ in.^{2} is placed between two steel cantilever beams, as shown. The moments of inertia of the beams are 100 in.^{4} and 200 in.^{4}, respectively. What is the stress in the tube if its temperature is increased by $100°F$? $E_b = 12 \times 10^6$ psi; $E_s = 30 \times 10^6$ psi; $\alpha = 10 \times 10^{-6}$ in./in./$°F$.

Ans. 2180 psi.

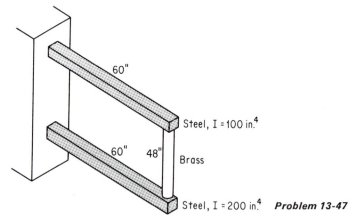

Problem 13-47

13-48. In the illustration, find the force F necessary to cause point A to move 0.2 in. downward. Each beam is aluminum and has a centroidal moment of inertia of 0.15 in.^{4}. $E_a = 10 \times 10^6$ psi.

Ans. 527 lb.

13-49. When a spring is placed under a steel cantilever beam, as shown, it limits the

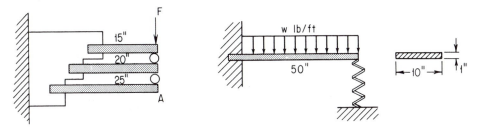

Problem 13-48   **Problem 13-49**

deflection to one-half the value obtained when the beam could deflect freely. The beam is 10 in. wide by 1 in. deep. What is the constant of the spring?

Ans. 600 lb/in.

Chapter 14

COMBINED
LOADING

In the development of basic concepts, loads have been applied to members in four ways: *along the axis of the member, in direct shear, in torsion, and in bending.* This chapter will consider the analysis and effects of combined loading.

14-1 General Considerations

More often than not, in an actual structure, the fundamental loading arrangements act in combination. Figure 14-1 illustrates how a single force can produce conditions equivalent to all four basic loads. On a section within the bar, the component F_z acts in two ways: it both pushes and bends, whereas the component F_y produces three distinct actions within the bar: direct shear, torsion, and bending.

If considered individually, the stresses in this example could be calculated by formulas developed in previous chapters. The problem of combined

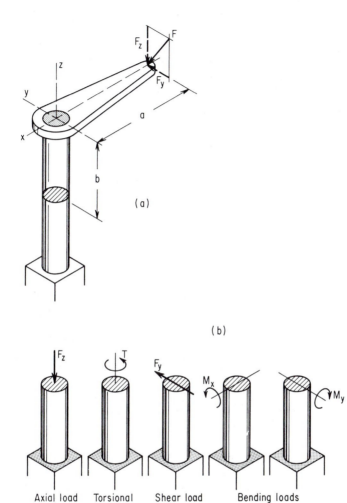

(a)

(b)

| Axial load F_z | Torsional load $T = F_y a$ | Shear load F_y | Bending loads $M_x = F_y b$, $M_y = F_z a$ | |

Figure 14-1

stress is approached by using these basic equations and the concepts of super-position.

14-2 Combined Axial and Bending Loads

Probably the most common case of combined loading occurs when bending moments and axial forces act together. Although the combination is usually not one of choice, there are instances when properties can be enhanced by superimposing compressive and bending loads; prestressed concrete, which will be discussed in detail, is an example.

All cases of combined axial loading and bending can be analyzed by superposition methods. Figure 14-2(a) shows a beam subjected to bending

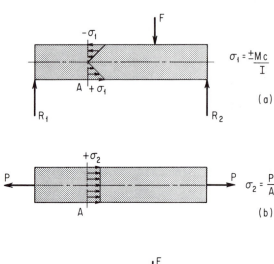

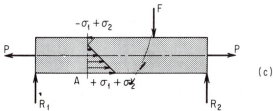

Figure 14-2

loads. For the sake of simplicity, the beam is assumed to have a rectangular cross-section. Stresses, then, vary directly with distance from the neutral axis, and a stress pattern exists within the member typical to that shown. If this same beam were subjected to an axial load, Fig. 14-2(b), the stresses acting on a transverse section would be uniformly distributed across the entire area. When the loads act together, Fig. 14-2(c), the resultant stress is the algebraic, or superposed, sum of the two previous patterns. In this illustration, stresses in the top fibers diminish, while those in the bottom fibers increase; in effect, the neutral axis is shifted upward. The combined stress at a point within the section is found by adding the axial stress and the flexural stress at that point:

$$\sigma = \pm\frac{P}{A} \pm \frac{My}{I} \qquad (14\text{-}1)$$

In terms of maximum values, the stress is given by

$$\sigma = \pm \frac{P}{A} \pm \frac{Mc}{I} \qquad (14\text{-}2)$$

It must be remembered that algebraic signs are as much a part of Eqs. (14-1) and (14-2) as are magnitudes and that tensile stress and compressive stress assume opposite signs.

EXAMPLE 1. A concentrated load of 600 lb is supported by a pin-connected truss, as shown in Fig. 14-3(a). Find the maximum values of the tensile and compressive stresses that act on a transverse section at D. The horizontal member is 2 in. wide and 6 in. deep.

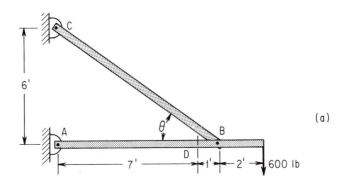

(a)

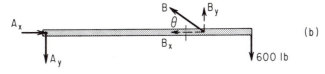

(b)

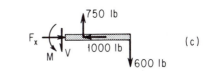

(c)

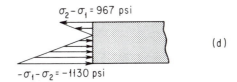

(d)

Figure 14-3

Solution: Figure 14-3(b) shows a free-body diagram of the horizontal member. The vertical component of the pin reaction at B is computed by summing moments about A.

$$\Sigma M_A = 0$$

$$8B_y = 600(10)$$

$$B_y = 750 \text{ lb}$$

The ratio B_y/B_x is equal to the tangent of the angle θ; hence,

$$\frac{B_y}{B_x} = \tan \theta = \frac{3}{4}$$

$$B_x = \tfrac{4}{3}B_y = \tfrac{4}{3}(750) = 1000 \text{ lb}$$

The internal reactions at D become apparent when the horizontal member is sectioned as shown in Fig. 14-3(c). The reactions F_x, V, and M are

$$F_x = 1000 \text{ lb}$$

$$V = 750 - 600 = 150 \text{ lb}$$

$$M = -600(3) + 750(1) = -1050 \text{ lb ft}$$

Since only maximum axial stresses are to be computed, the direct shear V does not enter into the computations. The direct stress and flexure stress can be computed as follows:

$$\sigma_1 = -\frac{P}{A} = -\frac{1000}{(2 \times 6)} = -83.3 \text{ psi}$$

$$\sigma_2 = \pm\frac{Mc}{I} = \pm\frac{(1050 \times 12)3}{2(6)^3/12} = \pm 1050 \text{ psi}$$

The maximum compressive stress occurs at the bottom fibers, where σ_1 and σ_2 add algebraically:

$$\sigma_{c\,max} = -\sigma_1 - \sigma_2 = -83.3 - 1050 = -1133.3 \text{ psi}$$

Fibers at the top of the section are in tension; the difference between direct stress and bending stress is equal to the tensile stress.

$$\sigma_{t\,max} = -\sigma_1 + \sigma_2 = -83.3 + 1050 = +966.7 \text{ psi}$$

Superposition of the direct and flexural stresses is shown in Fig. 14-3(d).

EXAMPLE 2. The vertical member of the jib, Fig. 14-4(a), is a 12 WF 40 beam. Find the axial stresses σ_t and σ_c in this member if $x = 5$ ft. Assume the column to be adequately braced to prevent buckling.

Solution: A force and a moment act at section *a-a*, as shown in the free-body diagram, Fig. 14-4(b). The axial compressive stress σ_1 is constant and equal to

$$\sigma_1 = -\frac{P}{A} = -\frac{10,000}{11.77} = -850 \text{ psi}$$

The bending stress σ_2 is a function of the moment M, and at $x = 5$ ft is

$$\sigma_2 = \pm\frac{M}{Z} = \pm\frac{10,000(5 \times 12)}{51.9} = \pm 11,600 \text{ psi}$$

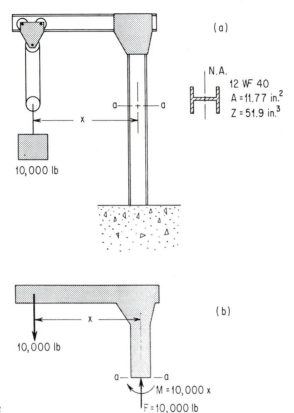

Figure 14-4

The bending stress σ_2 acts in compression at the inside face of the beam and in tension at the outside face; thus we have, at the inside face:

$$\sigma_c = -\sigma_1 - \sigma_2 = -850 - 11{,}600 = -12{,}450 \text{ psi}$$

at the outside face:

$$\sigma_t = -\sigma_1 + \sigma_2 = -850 + 11{,}600 = 10{,}750 \text{ psi}$$

14-3 Prestressed Concrete

In certain circumstances the effects of combined loading are a desirable feature rather than a condition one would wish to avoid. *Prestressed concrete* —or perhaps the more descriptive phrase, precompressed concrete—is an example.

Concrete is weak in tension, and for this reason steel reinforcing bars are used whenever concrete members must sustain tensile loads. For many cases, however, where the spans are great or the loads are heavy, the use of reinforced concrete becomes impractical. In such an instance, prestressed concrete is an economical solution.

The desire in prestressing is to eliminate tensile stress; this is done by stretching reinforcing steel so as to superimpose on the concrete compressive stresses equal or greater than the expected tensile stress. The strengthening effect is somewhat like the squeeze put on a horizontal row of boxes when they are to be moved from place to place; with sufficient pressure they may be lifted and carried, even though those in the center are unsupported.

The actual fabrication of prestressed beams consists of casting concrete in a form containing high-strength steel bars under tension. This tension is maintained, Fig. 14-5(a), until the concrete has hardened and cured. The

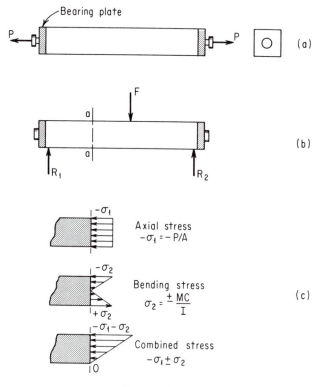

Figure 14-5

tensile load P is then released, Fig. 14-5(b), and the beam is *precompressed.* When the reinforcing bars are centrally located,[1] the compressive stresses

---

[1] Usual practice is to locate reinforcing bars closer to the tension side of the beam. This permits an even greater amount of precompression for a given amount of steel because of the eccentricity of the load. The reader is referred to the many fine texts on concrete design for a more detailed discussion of prestressed concrete.

and bending stresses add or subtract to give the superimposed stress pattern, Fig. 14-5(c).

EXAMPLE 3. A concrete beam, Fig. 14-6, having a width of 6 in. and a depth of 12 in. is precompressed by a force of 100,000 lb acting through centrally located reinforcing bars. Find the safe distributed load w lb per ft that can be supported by the beam.

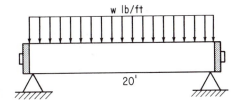

Figure 14-6

Solution: The precompressive stress is equal to the applied load divided by the cross-sectional area of the beam.

$$\sigma_1 = -\frac{P}{A} = -\frac{100,000}{6(12)} = -1390 \text{ psi}$$

The maximum bending stress occurs at midspan, where the moment has its greatest value, $wl^2/8$ lb ft.

$$\sigma_2 = \pm\frac{Mc}{I} = \pm\frac{M}{Z}$$

$$= \pm\left(\frac{wl^2}{8} \times \frac{1}{\frac{bh^2}{6}}\right) = \pm\left[\frac{w(20)^2(12)}{8}\right]\left[\frac{6}{6(12)^2}\right]$$

$$= \pm 4.17w \text{ psi}$$

The safe load is obtained by equating the tensile bending stress to the compressive prestress.

$$\sigma_2 = \sigma_1$$
$$4.17w = 1390$$
$$w = 333 \text{ lb per ft}$$

14-4 Bending in Two Directions

Still another from of combined stress occurs when members are acted upon by loads, Fig. 14-7(a), which induce bending in two directions, as illustrated in Fig. 14-7(b). At a distance a from the free end, bending moments M_x and M_y act to produce the stress distribution patterns shown in Fig. 14-7(c), where

$$\sigma_1 = \pm\frac{M_y}{Z_y} = \pm\frac{F_x a}{Z_y}$$

and   (14-3)

$$\sigma_2 = \pm\frac{M_x}{Z_x} = \pm\frac{F_y a}{Z_x}$$

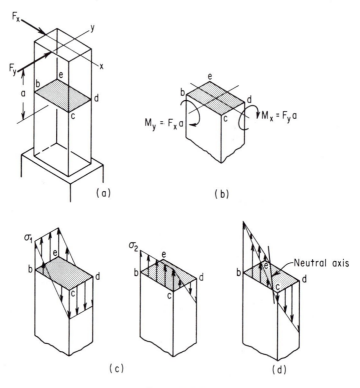

Figure 14-7

where Z_x and Z_y are the section moduli with respect to the x-axis and the y-axis. The *superposition* of bending stresses σ_1 and σ_2 is illustrated in Fig. 14-7(d). Stress magnitudes at each of the four corners are as follows:

$$\sigma_b = \sigma_1 + \sigma_2$$
$$\sigma_c = -\sigma_1 + \sigma_2$$
$$\sigma_d = -\sigma_1 - \sigma_2$$
$$\sigma_e = \sigma_1 - \sigma_2$$

In this illustration the maximum tensile stress occurs at b, and the maximum compressive stress at d.

There is always a line of zero stress associated with two-directional bending; in effect, this line is the neutral axis of the beam. For this example, the axis of zero stress is a line joining the points of zero stress on faces bc and ed, Fig. 14-7(d).

EXAMPLE 4. Find the maximum stress in the cantilever beam of Fig. 14-8. The beam has a width of 4 in. and a depth of 9 in.

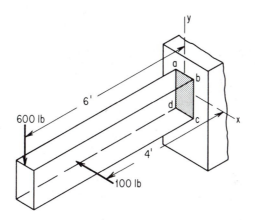

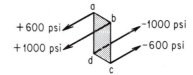

Figure 14-8

Solution: The section moduli are computed as follows:

$$Z_x = \frac{I}{c} = \frac{bh^3}{12\frac{h}{2}} = \frac{bh^2}{6} = \frac{4(9)^2}{6} = 54 \text{ in.}^3$$

$$Z_y = \frac{b^2h}{6} = \frac{(4)^29}{6} = 24 \text{ in.}^3$$

Stresses, produced by the individual bending loads, are

$$\sigma_1 = \pm\frac{M}{Z_x} = \pm\frac{600(6 \times 12)}{54}$$

$$= \pm 800 \text{ psi (tension on the top face and compression on the bottom face)}$$

$$\sigma_2 = \pm\frac{M}{Z_y} = \pm\frac{100(4 \times 12)}{24}$$

$$= \pm 200 \text{ psi (tension on the near side and compression on the far side)}$$

The maximum stresses occur along edges b and d at the supports.

$$\sigma_b = \sigma_1 + \sigma_2 = 800 + 200 = 1000 \text{ psi (tension)}$$

$$\sigma_d = -\sigma_1 - \sigma_2 = -800 - 200 = -1000 \text{ psi (compression)}$$

Minimum values of stress occur at corners a and c and are, respectively,

$$\sigma_a = \sigma_1 - \sigma_2 = 800 - 200 = 600 \text{ psi (tension)}$$

$$\sigma_c = -\sigma_1 + \sigma_2 = -800 + 200 = -600 \text{ psi (compression)}$$

EXAMPLE 5. Two loads are applied to a 2-in.-diameter circular shaft simply supported as shown in Fig. 14-9(a). Find the maximum normal stress in the shaft.

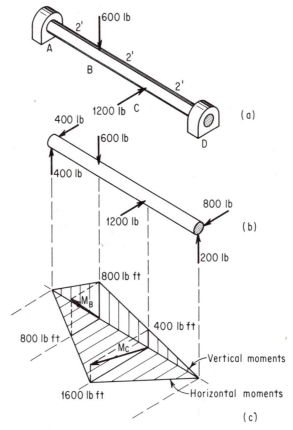

(a)

(b)

(c)   *Figure 14-9*

Solution: The bearing reactions are found in the usual way, and the free-body diagram is drawn as shown in Fig. 14-9(b). Vertical and horizontal bending-moment diagrams are next constructed, as illustrated in Fig. 14-9(c). Since moments are directed magnitudes, they can be added vectorially, just as force vectors are added. The maximum moment, which will occur either at B or C, is the vector sum M_B or M_C; thus

$$M_B = \sqrt{(800)^2 + (800)^2} = 1130 \text{ lb ft}$$

$$M_C = \sqrt{(400)^2 + (1600)^2} = 1650 \text{ lb ft}$$

The critical section, therefore, is at point C.

$$\sigma_{max} = \frac{Mc}{I}$$

where

$$c = 1 \text{ in.}$$

and

$$I = \frac{\pi D^4}{64} = \frac{\pi (2)^4}{64} = \frac{\pi}{4}$$

Thus

$$\sigma_{max} = \frac{(1650 \times 12)(1)}{\frac{\pi}{4}} = 25{,}200 \text{ psi}$$

14-5 Axial Loading Applied Eccentrically

Bending stresses and axial stresses are induced when loads are applied *eccentrically*, as shown in Fig. 14-10(a). The equivalent loading, Fig. 14-10(b),

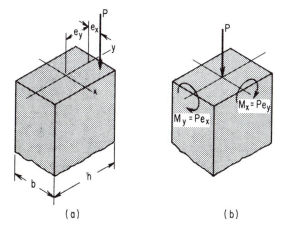

Figure 14-10   (a)   (b)

consists of bending moments M_x and M_y and a direct load of P. In a prismatic bar, like that shown in the figure, the stress at any corner would be

$$\sigma = \pm \frac{P}{A} \pm \frac{M_x}{Z_x} \pm \frac{M_y}{Z_y} \tag{14-4}$$

where Z_x and Z_y represent the section moduli with respect to the x-axis and the y-axis. The positive and negative signs indicate the stress to be either tension or compression.

Equation (14-4) is a particularly important relationship which governs the design of short columns constructed of materials weak in tension. A permissible amount of eccentricity can be found by equating the direct compressive stress to the tensile stresses induced by bending.

$$\frac{P}{A} = \frac{M_x}{Z_x} + \frac{M_y}{Z_y} = \frac{Pe_y}{bh^2/6} + \frac{Pe_x}{b^2h/6}$$

The section moduli of a prismatic bar, Fig. 14-11(a), in terms of area is

$$Z_x = \frac{bh^2}{6} = \frac{Ah}{6}$$

and

$$Z_y = = \frac{b^2h}{6} = \frac{Ab}{6}$$

Hence,

$$\frac{P}{A} = \frac{6Pe_y}{Ah} + \frac{6Pe_x}{Ab}$$

$$1 = \frac{6e_y}{h} + \frac{6e_x}{b}$$

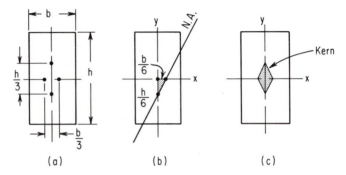

Figure 14-11

By setting $e_x = 0$, the permissible eccentricity e_y is

$$e_y = \frac{h}{6}$$

and similarly, if $e_y = 0$,

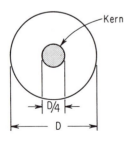

Figure 14-12

$$e_x = \frac{b}{6}$$

A neutral axis can be drawn by joining the points which represent the greatest permissible eccentricity in each coordinate direction, Fig. 14-11(b). If the compressive force acts anywhere to the right of this axis, tensile stresses will be created within the member. By symmetry, four neutral axes exist in the section; these lines form the boundaries of a *core* or *kern*, as shown in Fig. 14-11(c). A compressive force acting within this *core* will not produce tensile stresses within the member. Every geometric section has its characteristic *kern*. In a solid circular section, the *kern* is the central circular area shown in Fig. 14-12.

EXAMPLE 6. Determine the maximum and minimum stresses in the eccentrically loaded post of Fig. 14-13(a). The member is a 16 WF 50 section with the following geometric properties:

$$\text{Area:} \qquad 14.7 \text{ in.}^2$$
$$\text{Section modulus:} \qquad 80.7 \text{ in.}^3$$
$$\text{Depth:} \qquad 16.25 \text{ in.}$$

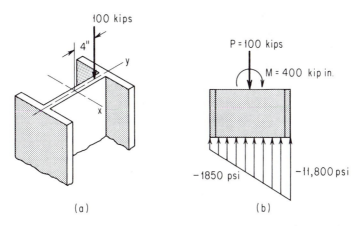

Figure 14-13

Solution: The eccentric load is equivalent to a direct load of 100 kips super-imposed on a bending load of 400 kip in., as shown in Fig. 14-13(b). On the face of the beam nearest the load, stresses add, and on the opposite face, they subtract.

$$\sigma_{\text{max}} = -\frac{P}{A} - \frac{M}{Z} = -\frac{100,000}{14.7} - \frac{400,000}{80.7}$$
$$= -11,800 \text{ psi}$$
$$\sigma_{\text{min}} = -\frac{P}{A} + \frac{M}{Z} = -\frac{100,000}{14.7} + \frac{400,000}{80.7}$$
$$= -1850 \text{ psi}$$

Both σ_{max} and σ_{min} are negative; the 4-in. eccentricity, therefore, is within the *kern* of the section. While it is not required, the limit of eccentricity in this example can be easily found by equating the direct stress to the bending stress.

$$\frac{P}{A} = \frac{M}{Z}$$

$$\frac{100}{14.7} = \frac{100e_y}{80.7}$$

$$e_y = \frac{80.7}{14.7} = 5.49 \text{ in.}$$

Hence, if e_y exceeds 5.49 in., tensile stresses will be developed in the post.

EXAMPLE 7. Two square members, one solid and the other hollow, are being considered for the eccentrically loaded strut of Fig. 14-14. The members have equal weights, and, therefore, equal cross-sectional areas and the eccentricity in each is the same. Compare the load-carrying capacity of the two sections.

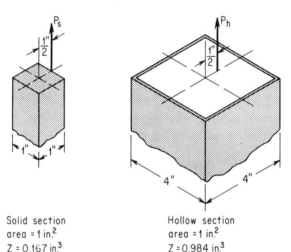

Solid section
area = 1 in.²
Z = 0.167 in.³

Hollow section
area = 1 in.²
Z = 0.984 in.³

Figure 14-14

Solution: For a given limiting stress, the allowable loads in each strut can be computed as follows:

Solid strut:
$$\sigma = \frac{P_s}{A} + \frac{P_s e}{Z}$$

$$= \frac{P_s}{1 \times 1} + \frac{P_s(0.5)}{0.167} = 4\,P_s \qquad (a)$$

Hollow strut:
$$\sigma = \frac{P_h}{A} + \frac{P_h e}{Z}$$

$$= \frac{P_h}{1 \times 1} + \frac{P_h(0.5)}{0.984} = 1.51\,P_h \qquad (b)$$

The ratio P_h/P_s can be found by equating (a) and (b):

$$1.51P_h = 4P_s$$

$$\frac{P_h}{P_s} = \frac{4}{1.51} = 2.65$$

For the same weight of material, the hollow strut will carry almost three times the load. This is why hollow members are favored in the design of eccentrically loaded struts.

14-6 Biaxial Stress

A familiar example of a member subjected to *biaxial stress*, or *normal stress in two directions*, is the pressure vessel described in Chapter 8. The walls of the vessel, Fig. 14-15, are stressed in both the longitudinal and circumferential directions.

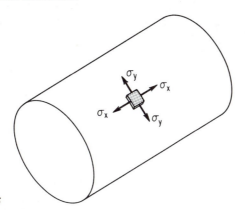

Figure 14-15

The effects of biaxial stress can be best analyzed by considering a member of uniform thickness, Fig. 14-16(a), acted upon by the *orthogonal forces* F_x and F_y. *The state of stress within the member* can be computed by isolating a small triangular element, as shown in Fig. 14-16(b). Let the diagonal plane of this element have an area A; the vertical and horizontal planes will then have areas $A \cos \theta$ and $A \sin \theta$, respectively. The element is subjected to normal and shear forces on the diagonal plane, Fig. 14-16(c), and these forces must statically balance the x- and y-components of force imposed by the loading. This is the same as saying that the vector addition of the forces must form a closed polygon, as shown in Fig. 14-16(d). The purpose of this analysis is, of course, to find the normal and shear forces that accompany biaxial stress, and this can be accomplished by applying trigonometry to the polygon of Fig. 14-16(d); thus

$$\sigma A = (\sigma_x A \cos \theta) \cos \theta + (\sigma_y A \sin \theta) \sin \theta$$

$$\sigma = \sigma_x \cos^2 \theta + \sigma_y \sin^2 \theta \tag{a}$$

and

$$\tau A = (\sigma_x A \cos \theta) \sin \theta - (\sigma_y A \sin \theta) \cos \theta$$

$$\tau = (\sigma_x - \sigma_y) \sin \theta \cos \theta \tag{b}$$

In Eqs. (a) and (b), *tensile stresses are assumed to be positive and compressive stresses negative: a shear stress that tends to impose clockwise rotation on the element is arbitrarily considered to be positive shear.*

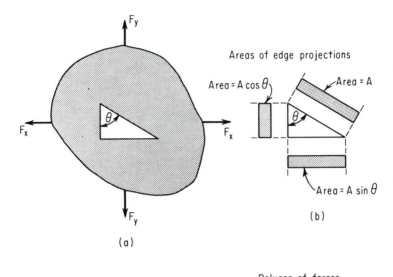

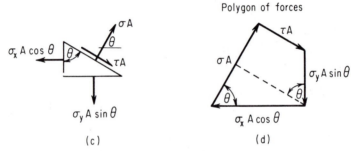

Figure 14-16

By making the following substitutions, Eqs. (a) and (b) can be expressed in terms of the double angle 2θ.

$$\cos^2 \theta = \tfrac{1}{2}(1 + \cos 2\theta)$$
$$\sin^2 \theta = \tfrac{1}{2}(1 - \cos 2\theta)$$
$$\sin \theta \cos \theta = \tfrac{1}{2} \sin 2\theta$$

Equations (a) and (b), written in terms of the double angle, are

$$\sigma = \frac{\sigma_x + \sigma_y}{2} + \frac{\sigma_x - \sigma_y}{2} \cos 2\theta \qquad (14\text{-}5)$$

$$\tau = \frac{\sigma_x - \sigma_y}{2} \sin 2\theta \qquad (14\text{-}6)$$

The maximum and minimum values of stress, obtained through these equations, are called *principal stresses:* the principal normal stresses occur when

$$\cos 2\theta = \pm 1$$
$$2\theta = 0°, 180°$$
$$\theta = 0°, 90°$$

In other words, there are no planes within the element subjected to simple biaxial loading that will have normal stresses greater than σ_x or σ_y. This can be demonstrated by first allowing $\cos 2\theta$ to equal $(+1)$ and then (-1):

$\cos 2\theta = 1$:

$$\sigma = \frac{\sigma_x + \sigma_y}{2} + \frac{\sigma_x - \sigma_y}{2}(1) = \sigma_x$$

$\cos 2\theta = -1$:

$$\sigma = \frac{\sigma_x + \sigma_y}{2} + \frac{\sigma_x - \sigma_y}{2}(-1) = \sigma_y$$

The shear stress is maximum or minimum at $\sin 2\theta = \pm 1$; hence,

$$\sin 2\theta = \pm 1$$
$$2\theta = 90°, 270°$$
$$\theta = 45°, 135°$$

Thus, *for biaxial stress the maximum (or minimum) shear stress occurs on planes inclined at 45 deg to the principal normal planes.*

$$\tau_{\max} = \frac{\sigma_x - \sigma_y}{2} \quad \text{at} \quad \theta = 45°$$

$$\tau_{\min} = -\frac{\sigma_x - \sigma_y}{2} \quad \text{at} \quad \theta = 135°$$

When the biaxial stresses are equal (in other words, when $\sigma_x = \sigma_y$), all planes within the member will be free of shearing stresses and will have a normal stress equal to the biaxial stress. This is observed when σ_x and σ_y are set equal to one another in Eqs. (14-5) and (14-6).

$$\sigma_x = \sigma_y$$

$$\sigma = \frac{\sigma_x + \sigma_x}{2} + \frac{\sigma_x - \sigma_x}{2} \cos 2\theta = \sigma_x$$

$$\tau = \frac{\sigma_x - \sigma_x}{2} \sin 2\theta = 0$$

Although substitution of numerical values into Eqs. (14-5) and (14-6) is not difficult, the equations are very rarely used as such. The reason is that a much simpler approach exists in their graphical interpretation.

A set of coordinate axes are drawn as shown in Fig. 14-17(a). These axes represent normal and shearing stresses, and must not be confused with an x-, y-coordinate system. The coordinates σ_x and σ_y are located on the hori-

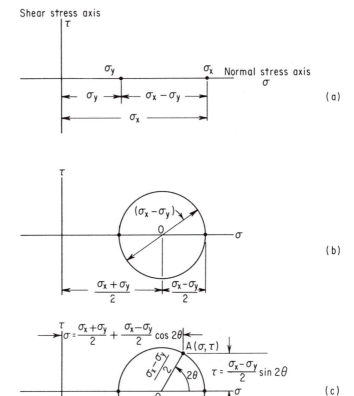

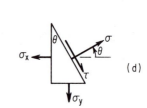

Figure 14-17

zontal axis, the axis that represents the coordinates of *all* normal stresses. A circle is then drawn having a diameter equal to the difference $(\sigma_x - \sigma_y)$. This is shown in Fig. 14-17(b). The radius of this circle is, therefore, $(\sigma_x - \sigma_y)/2$, and its center lies at a distance $(\sigma_x + \sigma_y)/2$ from the shear axis. All possible inclined planes within a biaxially stressed element are represented by this circle, as are the values of normal and shear stress that act on these planes. This is evident in Fig. 14-17(c), where point A represents the coordinates of stress σ and τ that act on a plane within an element inclined at an angle θ, as shown in Fig. 14-17(d). The angle is measured relative to the x-axis of the element and a normal to the plane under investigation. In the circle, the angle is doubled and is measured relative to a line drawn between the center of the circle O and σ_x, and the radius line OA. If the angle is measured counterclockwise in the element, it must be measured counterclockwise in the circle.

The circle of stress was devised by Otto Mohr, a German engineer, in 1882, and is called *Mohr's circle.*

In its symbolic form, Mohr's circle might appear to be the more cumbersome approach to the problem of combined biaxial stress. The examples that follow, however, will illustrate the simplicity of the graphical method.

EXAMPLE 8. A rectangular plate, Fig. 14-18(a), is subjected to the forces shown Determine the normal and shearing stresses that act on a plane within the member, inclined at 30 deg, as shown.

Solution: The biaxial stresses σ_x and σ_y are first computed:

$$\sigma_x = \frac{P}{A} = \frac{1800}{3(\frac{1}{4})} = 2400 \text{ psi}$$

$$\sigma_y = \frac{P}{A} = \frac{1000}{5(\frac{1}{4})} = 800 \text{ psi}$$

An element is isolated, Fig. 14-18(b), to represent the state of stress under consideration, and Mohr's circle is constructed, as shown in Fig. 14-18(c). The intersection of the radius line with the circumference of the circle represents all possible states of stress on all possible planes within the element. At $\theta = 30$ deg ($2\theta = 60$ deg), the state of stress, Fig. 14-18(d), is represented by the coordinates σ and τ of point A, where

$$\sigma = 1600 + 800 \cos 60°$$
$$= 1600 + 800(0.5) = 2000 \text{ psi}$$
$$\tau = 800 \sin 60°$$
$$= 800(0.866) = 693 \text{ psi}$$

Point B on the circle gives the coordinates of the stress which act on a face at right angles to that described by point A. In the element these faces are 90 deg apart; in Mohr's circle they are 180 deg apart, since all angles are doubled. The shear-stress coordinate at B is negative and is, therefore, drawn as a counterclockwise couple on the element.

To understand better the importance of algebraic signs, the previous problem is repeated with the 1800-lb force acting in compression.

EXAMPLE 9. A rectangular plate, Fig. 14-19(a), is subjected to the forces illustrated. Determine the normal and shearing stresses that act on a plane within the member, inclined at 30 deg, as shown.

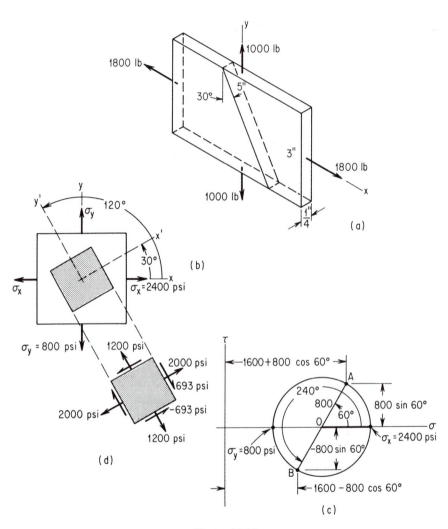

Figure 14-18

Solution: The plate is subjected to stresses that have the same magnitude as those in Example 8; they differ in sign, however, since σ_x acts in compression. The state of stress drawn on the original body is shown in Fig. 14-19(b). Mohr's circle, which represents an infinite variety of possible states of stress, is drawn as shown in Fig. 14-19(c). Since σ_x is negative, it appears to the left of the shear axis, and, for that matter, to the left of σ_y. This is important to note, because, for the case of biaxial stress, angles in Mohr's circle are measured relative to the radius line $O\sigma_x$.

The coordinates of point A represent the state of stress on the face of the element inclined at 30 deg, and shown in Fig. 14-19(d); thus,

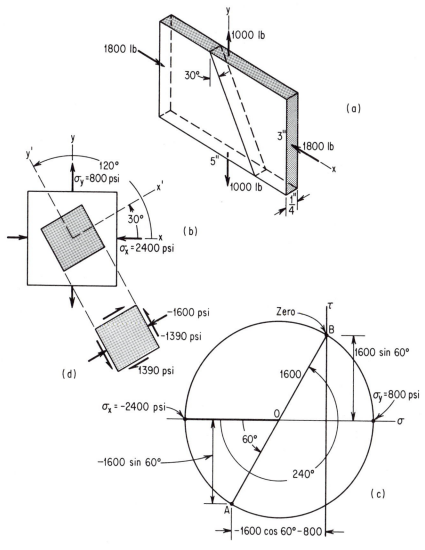

Figure 14-19

$$\sigma = -1600 \cos 60° - 800 = -1600 \text{ psi}$$
$$\tau = -1600 \sin 60° = -1390 \text{ psi}$$

Point B on the circumference of the circle represents the state of stress on the adjacent face of the element.

$$\sigma = 1600 \cos 60° - 800 = 0$$
$$\tau = 1600 \sin 60° = 1390 \text{ psi}$$

14-7 Pure Shear

Although pure shear is not considered as a combined load in itself, it does produce normal and shearing stresses within members on which it acts. By way of illustration, consider an element of area on the surface of a torsion bar, Fig. 14-20(a). The state of stress on this element, Fig. 14-20(b), consists

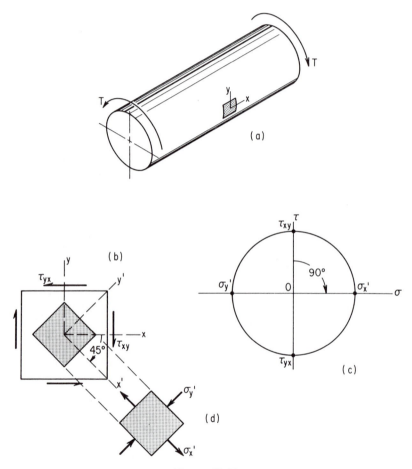

Figure 14-20

of shearing stresses τ_{xy} and τ_{yx}. The subscripts used imply that τ_{xy} *act in the y-direction on a face normal to the x-axis.* Since the element must be in static equilibrium, shearing stresses always act in pairs to form *stress-couples* of equal magnitudes. In other words, all four faces of the element are in shear, and the stresses are such that

$$\sum F_x = 0, \quad \sum F_y = 0, \text{ and } \sum M = 0$$

Mohr's circle for pure shear is constructed, Fig. 14-20(c), by locating points τ_{xy} and τ_{yx} on the shear axis and then drawing a circle having a diameter of $2\tau_{xy}$. In pure shear, Mohr's circle is always symmetrical about the normal and shear axis.

The radius of the circle, the line drawn from O to τ_{xy}, represents the x-axis on the element under investigation. As before, all angles measured in the circle are double those of the element and must be similarly directed. Thus, the principal stresses in this example are at $2\theta = 90$ deg and $2\theta = 270$ deg measured clockwise in the circle; these occur on the faces of the element oriented at $\theta = 45$ deg and $\theta = 135$ deg measured clockwise from the positive x-axis. This is illustrated in Fig. 14-20(d), where the x'- and the y'-axis are the principal axes, or the axes perpendicular to faces acted on by normal stress alone.

EXAMPLE 10. A state of stress on an element is shown in Fig. 14-21(a). (a) Determine the stress components that act on faces of the element rotated 15 deg clockwise. (b) Find the values of the principal stresses.

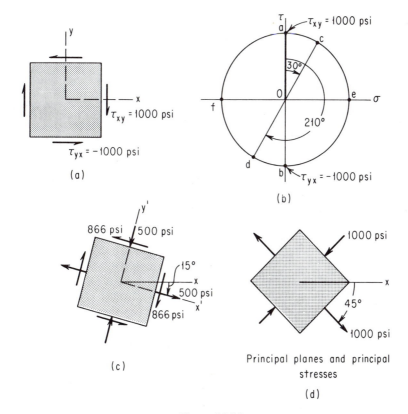

(a)

(b)

(c)

(d)

Principal planes and principal stresses

Figure 14-21

Solution: Part (a). Mohr's circle is drawn as shown in Fig. 14-21(b). Radius line *Oa* represents the state of stress on a face of the element normal to the *x*-axis, and *Ob*, the state of stress on a face normal to the *y*-axis. By rotating the radius line, first 30 deg and then 210 deg, stress components that act on faces of the element normal to the *x'*- and the *y'*-axis can be found.

$$\sigma_{x'} = 1000 \sin 30° = 500 \text{ psi}$$

$$\tau_{x'} = 1000 \cos 30° = 866 \text{ psi}$$

$$\sigma_{y'} = -1000 \sin 30° = -500 \text{ psi}$$

$$\tau_{y'} = -1000 \cos 30° = -866 \text{ psi}$$

Part (b). Points *e* and *f* on Mohr's circle are the coordinates of the principal stresses. These stresses, which are shown in Fig. 14-21(c), act on planes inclined at

$$2\theta = 90° \quad \text{(clockwise from } Oa\text{)}$$

$$\theta = 45°$$

and

$$2\theta = 270° \quad \text{(clockwise from } Oa\text{)}$$

$$\theta = 135°$$

Hence:

$$\sigma_{x'} = 1000 \text{ psi}$$

$$\sigma_{y'} = -1000 \text{ psi}$$

$$\tau_{x'y'} = \tau_{y'x'} = 0$$

EXAMPLE 11. Determine the horsepower rating of a 4-in.-diameter shaft rotating at a speed of 315 rpm. The normal stress in the shaft is not to exceed 12,000 psi. and the shear stress 8000 psi.

Solution: In the case of pure shear, maximum normal stresses, those that act on the principal planes, have magnitudes equal to the maximum shearing stress, a fact illustrated by the previous example. In this problem, therefore, the limiting shear stress governs the design, since τ_{max} is less than σ_{max}.

$$\tau = \frac{Tc}{J}$$

$$T = \frac{\tau J}{c} = \frac{8000\pi(4)^4}{32(2)} = 32{,}000\pi \text{ lb in.}$$

$$\text{hp} = \frac{Tn}{63{,}000} = \frac{32{,}000\pi(315)}{63{,}000} = 500 \text{ hp}$$

14-8 Generalized Plane Stress

The simplicity of Mohr's circle as a method of determining the stress at a point can be fully appreciated when it is applied to *generalized plane stress*: the case of biaxial and shearing stresses acting simultaneously.

To draw Mohr's circle for generalized plane stress, Fig. 14-22(a), locate the coordinates (σ_x, τ_{xy}) and (σ_y, τ_{yx}) on a set of normal and shear axes, as shown in Fig. 14-22(b). The straight line that joins these points is the diameter of the circle. The radius line Oa, Fig. 14-22(c), represents the x-axis of the ele-

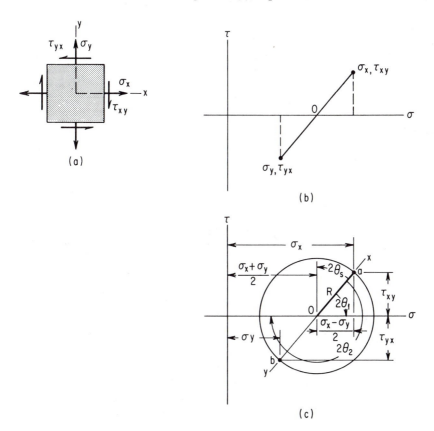

Figure 14-22

ment, and line Ob the y-axis. As before, all angles are doubled in the circle, and the x- and the y-axes appear to be 180 deg apart. The radius R has the value

$$R = \sqrt{\left(\frac{\sigma_x - \sigma_y}{2}\right)^2 + (\tau_{xy})^2}$$

and the principal stresses, therefore, are

$$\sigma = \frac{\sigma_x + \sigma_y}{2} \pm \sqrt{\left(\frac{\sigma_x - \sigma_y}{2}\right)^2 + (\tau_{xy})^2} \qquad (14\text{-}7)$$

The principal planes on which these stresses act are defined in terms of the double angles $2\theta_1$ and $2\theta_2$, where

$$\tan 2\theta_1 = \frac{2\tau_{xy}}{\sigma_x - \sigma_y}$$

$$\theta_1 = \frac{1}{2} \text{ arc } \tan\left(\frac{2\tau_{xy}}{\sigma_x - \sigma_y}\right)$$

and

$$\theta_2 = \theta_1 + 90°$$

As before, maximum and minimum shearing stresses are equal to the radius of Mohr's circle:

$$\tau_{\substack{max \\ min}} = \pm\sqrt{\left(\frac{\sigma_x - \sigma_y}{2}\right)^2 + (\tau_{xy})^2}$$

and the planes on which these act are defined by the angles θ_s and $(\theta_s + 90°)$.

EXAMPLE 12. A plane element in a body is subjected to the following stresses: $\sigma_x = 9$ ksi, $\sigma_y = -5$ ksi, $\tau_{xy} = 4$ ksi. Use Mohr's circle to determine: (a) the principal stresses and the principal planes, (b) the principal shearing stresses and the principal shearing planes.

Solution: The element and its accompanying state of stress is shown in Fig. 14-23(a). Mohr's circle is drawn as shown in Fig. 14-23(b). The radius of the circle is computed as follows:

$$R = \sqrt{7^2 + 4^2} = 8.06 \text{ ksi}$$

To determine the principal stresses, the radius line R must be rotated clockwise through an angle $2\theta_{p1}$ and then through an angle $(2\theta_{p1} + 180 \text{ deg})$. Stresses evaluated at those two positions are

$$\sigma_{max} = 2 + 8.06 = 10.06 \text{ ksi}$$

$$\sigma_{min} = 2 - 8.06 = -6.06 \text{ ksi}$$

The direction of the principal planes are next computed:

$$\tan 2\theta_{p1} = \tfrac{4}{7} = 0.5714$$

$$2\theta_{p1} = 29.7°$$

$$\theta_{p1} = 14.85° \text{ (clockwise from the positive } x\text{-axis)}$$

and

$$\theta_{p2} = \theta_{p1} + 90° = 104.85° \text{ (clockwise from the positive } x\text{-axis)}$$

Fig. 14-23(c) shows the principal stresses and the planes on which they act.

Points c and d on Mohr's circle represent the coordinates of the principal shearing stresses; thus

Point c: $\sigma = 2$ ksi, $\tau = 8.06$ ksi

Point d: $\sigma = 2$ ksi, $\tau = -8.06$ ksi

The planes of principal shear are located at

$$2\theta_{s1} = 90° - 2\theta_{p1} = 90° - 29.7° = 60.3°$$

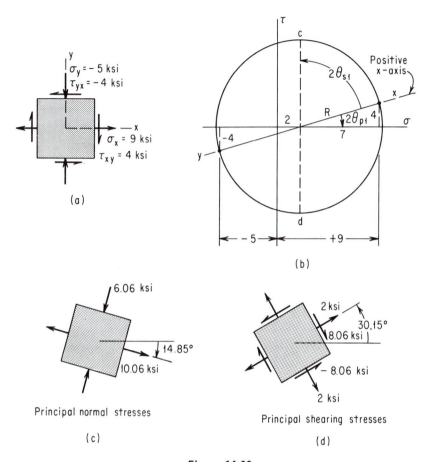

Figure 14-23

$$\theta_{s1} = 30.15° \quad \text{(counterclockwise from the positive } x\text{-axis)}$$

and

$$\theta_{s2} = \theta_{s1} + 90° = 30.15° + 90°$$
$$= 120.15° \quad \text{(counterclockwise from the positive } x\text{-axis)}$$

It is important to note that the angle to any plane is always measured relative to the positive x-axis as defined by the original state of stress. Thus, if the planes of principal shear are directed counterclockwise in the circle, they must be directed counterclockwise on the element. This is illustrated in Fig. 14-23(d).

EXAMPLE 13. A propeller mounted on a 4-in.-diameter solid steel shaft produces a torque of 40,000 lb in. and a thrust of 20,000 lb on the shaft. Determine the maximum normal stress and the maximum shearing stress developed in the member.

Solution: The combined loading produces shearing stresses and compressive stress within the member.

$$\tau = \frac{Tc}{J} = \frac{40,000(2)}{\pi(4)^4/32} = 3180 \text{ psi}$$

$$\sigma = \frac{P}{A} = \frac{20,000}{\pi(2)^2} = 1590 \text{ psi}$$

The state of stress at A in Fig. 14-24(a) can be represented as shown in Fig. 14-24(b). Mohr's circle for the loading, Fig. 14-24(c), is constructed; note that since σ_x is negative and τ_{xy} positive, the x-axis is up and to the left.

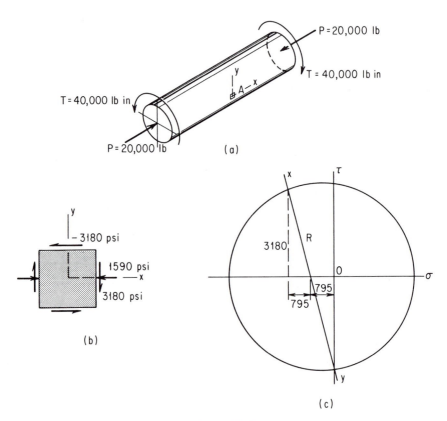

Figure 14-24

The maximum normal stress is compressive and equal to

$$\sigma_{max} = -795 - R = -795 - \sqrt{(3180)^2 + (795)^2}$$
$$= -795 - 3278 = -4073 \text{ psi}$$

The maximum shearing stress in the shaft is equal to the radius R.

$$\tau_{max} = 3278 \text{ psi}$$

QUESTIONS, PROBLEMS,
AND ANSWERS

14-1. With the aid of sketches, describe the principle of *superposition* as applied to the combination of axial and bending stresses.

14-2. A timber beam supports loads as shown. What are the maximum tensile and compressive stresses acting in the beam?

Ans. Tension: 400 psi; compression: 200 psi.

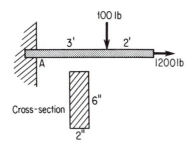

Problem 14-2

14-3. Graph the stress distribution on a section normal to the axis at *A* of the beam of Prob. 14-2.

14-4. What moment *M* in the figure is necessary to cause the stress at the top fibers to be zero? What is the magnitude of the stress in the bottom fibers under these conditions?

Ans. 1250 lb ft.

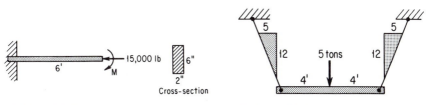

Problem 14-4   Problem 14-5

14-5. The 8 WF 17 beam shown is supported by cables. What are the maximum tensile and compressive stresses in the beam at a section just to the right of midspan?

Ans. Tension: 17,400 psi; compression: 16,600 psi.

14-6. Compare the load-carrying capacity of the two 1-in. square bars shown. The upper bar is slightly curved, and the lower bar is perfectly straight. *Hint:* Assume maximum stress in each member to be the same.

Ans. Straight bar is 13 times as strong.

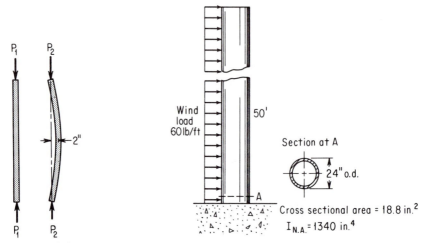

Problem 14-6   **Problem 14-7**

14-7. A circular steel smokestack, weighing 100 lb per lin ft, is securely supported at its base, as illustrated. Determine the maximum compressive stress in the stack if a wind load acts as shown.

Ans. 6710 psi.

14-8. Determine the maximum fiber stress on section *A-A* of the machine frame illustrated for a load of $P = 5000$ lb.

Ans. 18,750 psi.

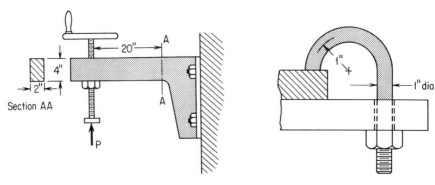

Problem 14-8   **Problem 14-9**

14-9. Determine the compressive force that can be exerted by the clamp shown if the allowable axial stress must not exceed 10,000 psi.

Ans. 462 lb.

14-10. Ten cubical blocks are held together by a 100-lb force to form a beam, as shown. What is the greatest force *P* that could act at midspan? Neglect the weight of the blocks. *Hint:* The bottom fibers cannot be in tension.

Ans. 6.67 lb.

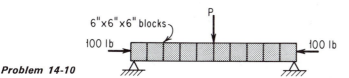

Problem 14-10

14-11. A 20-ft. beam, suspended by cables as shown, carries four loads. What is the maximum normal stress that acts at section *A*? The weight of the beam is 100 lb/ft, $A = 10$ in.^{2}, $Z = 25$ in.^{3}.

Ans. 688 psi.

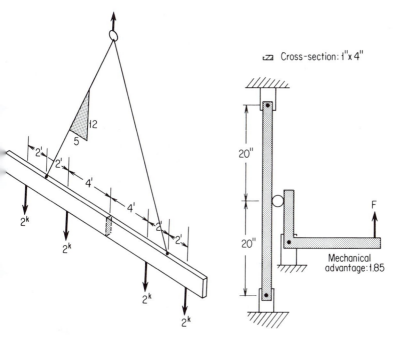

Problem 14-11   **Problem 14-12**

14-12. A steel bar is supported as shown. Find the load *F* that can act if the temperature of the bar increases by 50°F. The normal stress in the bar is not to exceed 20,000 psi.

Ans. 1480 lb.

14-13. A simply supported concrete beam 40 ft long is to support a uniformly distributed load of 1000 lb per ft, which includes its own weight. The beam is 30 in. deep and 12 in. wide, and is prestressed by high-strength steel bars located at the neutral axis. Find the necessary precompressive force and determine the required number of high-strength steel bars required to produce the precompression load. The bars have a cross-sectional area of 0.06 in.² and a working stress of 140,000 psi.

Ans. 480 kips; 58 bars.

14-14. A slab bridge of 80-ft span is to be made of prestressed concrete. The bridge is 10 ft wide and 2 ft deep, and is to carry a load of 1000 lb per sq ft, which includes its own weight. What is the required number of high-strength (140,000 psi) steel bars necessary in the structure? The bars have a cross-sectional area of 1.00 in.² and are located so that the neutral axis remains at the centroid of the beam.

Ans. 171 rods.

14-15. A concrete beam having the cross-section shown is placed in a state of precompression by 20 high-strength steel bars, each having a diameter of $\frac{1}{4}$ in. and each initially stressed to 150,000 psi. On curing, the concrete shrinks, and the stress in the steel is reduced by 10 per cent. What is the permissible bending moment that may be applied to the beam?

Ans. 36.8 kip ft.

14-16. Two loads act at right angles to one another at the free end of a cantilever beam, as illustrated. The beam has a square cross-section. Show that the maximum bending stress in the beam is equal to that obtained by applying the resultant *R* of these two forces along a diagonal. *Hint:* The moment of inertia of a square about the diagonal is $I = d^4/12$.

Ans. $12\,Pl/d^3$.

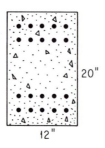

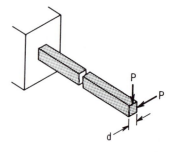

Problem 14-15   **Problem 14-16**

14-17. Compute the magnitudes of the maximum and minimum normal stresses that act in the beam illustrated. The beam is a standard 10 WF 33 girder.

Ans. 22,410 psi; 14,190 psi.

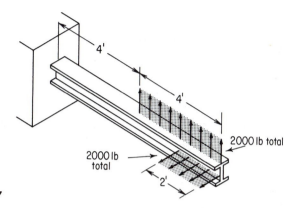

Problem 14-17

14-18. Find the width *b* of the beam shown; the maximum normal stress is to equal 1200 psi. Assume that the force passes through the geometric center of the cross-section.

Ans. 19 in.

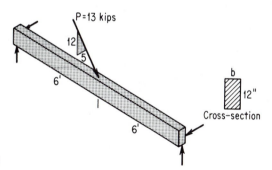

Problem 14-18

14-19. For the beam illustrated on p. 380 find the normal stress at each of the four edges, *a,b,c*, and *d*, at a section 4 ft from the free end.

Ans. $\sigma_a = 1400$ psi; $\sigma_b = 600$ psi; $\sigma_c = -1000$ psi; $\sigma_d = 200$ psi

14-20. Use the answers to Prob. 14-19 to construct a three-dimensional graph of the stress distribution at the section indicated. If it exists, locate the region of zero stress.

Ans. A line that intersects *ad* 10.5 in. down and *bc* 4.5 in. down.

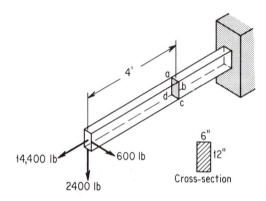

14,400 lb   600 lb

2400 lb   Cross-section

6"

12"

Problem 14-19

14-21. A three-dimensional stress distribution is illustrated. Locate the line of zero stress.

Ans. A line that intersects *ad* 12 in. down from *a* and *dc* 9.6 in. to the right of *d*.

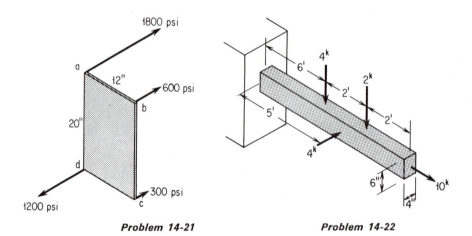

Problem 14-21   *Problem 14-22*

14-22. The beam shown is subjected to axial loads and bending loads. Determine maximum tensile and compressive stresses in the member.

Ans. +35.417 ksi.; 34.583 ksi.

14-23. Stresses on a section caused by three different loading arrangements are shown. Superimpose these stresses on a single sketch and locate the line of zero stress.

Ans. A line that intersects *ad* at 7 in. down from *a* and *dc* at 8.125 in. to the right of *d*.

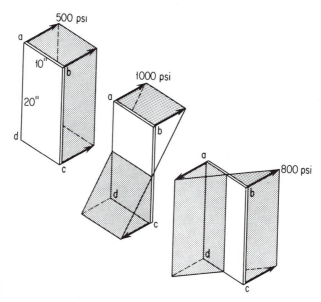

Problem 14-23

14-24. Determine the maximum bending stress in the circular shaft shown. Assume the belt "pulls" act at right angles to one another and that the section modulus of the shaft is 3 in.^{3}.

Ans. 9720 psi.

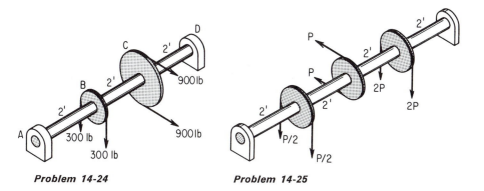

Problem 14-24   **Problem 14-25**

14-25. Three loads act on a 2-in.-diameter steel line shaft as shown. What is the maximum value of P if the bending stress is not to exceed 20 ksi?

Ans. 204 lb.

14-26. An eccentric load acts on a beam face as shown on p. 382. Find the stresses that act on the horizontal faces.

Ans. $|\sigma_{min}| = 62$ psi tension;
$\quad$ $|\sigma_{max}| = 313$ psi compression.

14-27. Find the permissible load P, in Prob. 14-26, if the following limitations are placed on the design:

Maximum compressive stress: 1000 psi
Maximum tensile stress: 300 psi

Ans. 12,800 lb; compression governs.

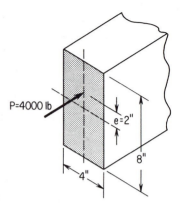

14-28. Find the permissible eccentricity e, in Prob. 14-26, if the load remains at 4000 lb and the stress limitations are:

Maximum compressive stress: 1000 psi
Maximum tensile stress: 300 psi

Ans. 4.53 in.; tension governs.

Problem 14-26

14-29. Determine the permissible load P that can act on the short column shown. The column is a 14 WF 68 section, and the maximum stress is not to exceed 20,000 psi.

Ans. 102 kips.

14-30. The short column illustrated supports an eccentric load as shown. What are the values of axial stress within the member at each of the four corners, A, B, C, and D?

Ans. A: $5P/A$; B: $-P/A$; C: $-7P/A$; D: P/A where $A = bh$.

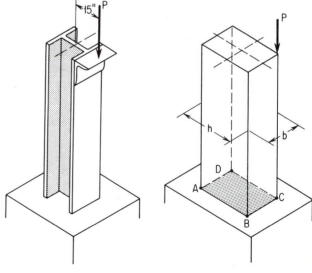

Problem 14-29   **Problem 14-30**

14-31. Compressive forces acting within the *kern* of a section (cannot, will sometimes, will always) produce tensile stresses.

Ans. Cannot.

14-32. The stress at the boundary of the *kern* is _____.

Ans. Zero.

14-33. In a circle, the kern is a circle; in a square, the kern is a square. It therefore follows that in a rectangle, the kern is rectangular. (True or false?)

Ans. False. See Fig. 14-11.

14-34. Prove that the *kern* (a) of a circular section of diameter D is a concentric circular area having a diameter of $D/4$; (b) prove that the kern of a square section is a concentric square of 1/9 the area of the original area.

14-35. What is the permissible eccentricity, in the figure, of a load P that acts on a post if the maximum tensile stress is not to exceed 20 per cent of the permissible compressive stress?

Ans. 4.5 in.

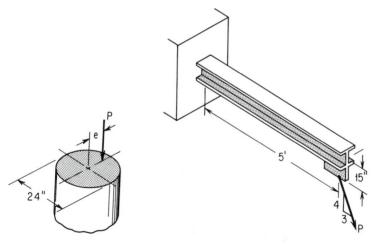

Problem 14-35   **Problem 14-36**

14-36. The cantilever beam supports an eccentric load P applied at an angle, as shown. The beam is a 14 WF 61 section. What is the safe value of P if the normal stress is not to exceed 20,000 psi?

Ans. 43,800 lb.

14-37 A state of biaxial stress is indicated in each figure. Determine, by means of
to   Mohr's circle, components of stress that act on the faces of an element
14-42. oriented as indicated.

> *Ans.* 14-37. $\sigma_{x'} = 500$ psi; $\tau_{x'} = 500$ psi; $\sigma_{y'} = 500$ psi; $\tau_{y'} = -500$ psi.
>
> 14-38. $\sigma_{x'} = 600$ psi; $\tau_{x'} = 600$ psi; $\sigma_{y'} = 600$ psi; $\tau_{y'} = -600$ psi.
>
> 14-39. Mohr's circle has a zero radius, hence $\sigma_\theta = 200$ psi and $\tau_\theta = 0$ for all angles
>
> 14-40. $\sigma_{x'} = 0$; $\tau_{x'} = 173.2$ psi; $\sigma_y = 200$ psi; $\tau_{y'} = -173.2$ psi.
>
> 14-41. $\sigma_{x'} = 67$ psi; $\tau_{x'} = 125$ psi; $\sigma_{y'} = 367$ psi; $\tau_{y'} = -125$ psi.
>
> 14-42. $\sigma_{x'} = -350$ psi; $\tau_{x'} = 86.6$ psi; $\sigma_{y'} = -250$ psi; $\tau_{y'} = -86.6$ psi.

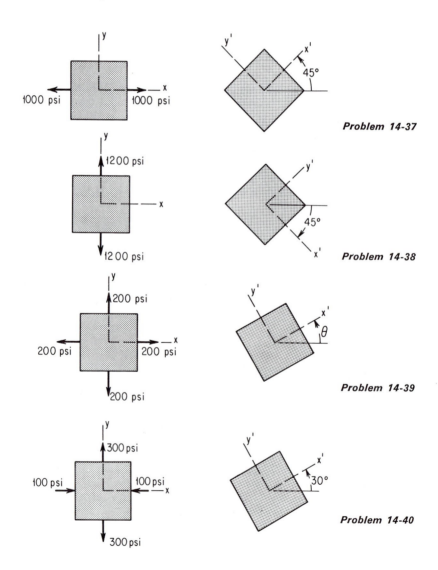

Problem 14-37

Problem 14-38

Problem 14-39

Problem 14-40

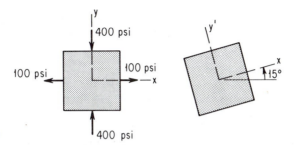

Problem 14-41

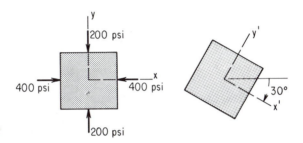

Problem 14-42

14-43. Is it always necessary to be concerned about the effects of biaxial stress? When can biaxial stress be ignored?

    Ans. A body may have normal stresses acting in two dimensions and yet fail in shear. Ignoring necessary computations is like playing Russian Roulette.

14-44. Two pieces of steel are welded along a diagonal as shown. Determine the safe value of *P*, using Mohr's circle, if the permissible stresses in the weld are $\sigma = 6000$ psi and $\tau = 1000$ psi.

    Ans. 18,500 lb; shear governs.

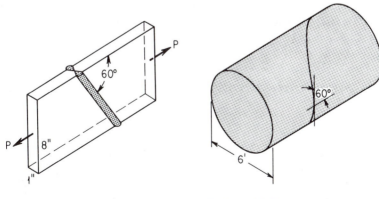

Problem 14-44   **Problem 14-45**

14-45. A closed pressure vessel is spirally welded as shown. What components of stress act in the weld if the pressure within the vessel is 25 psi? The vessel has a wall thickness of $\frac{3}{8}$ in.

Ans. $\sigma_{x'} = 2100$ psi; $\tau_{x'} = -520$ psi.

14-46. Three pieces of steel are welded to form a tee-beam as shown. What is the maximum shear stress developed at A? Each beam has a moment of inertia of $I_{NA} = 60$ in.^{4} and a depth of 4 in. *Hint:* Assume that the stress at A is *biaxial* and neglect the effect of stress concentration at the joints.

Ans. 3400 psi.

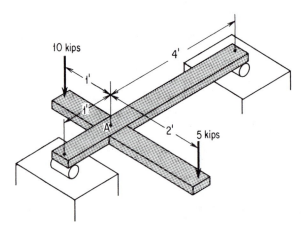

Problem 14-46

14-47. The tee-beam illustrated is used as a shock absorber in a towing rig. The steel beam is fabricated by welding steel plates together as shown. Determine the state of the stress at A in terms of P and the permissible value of P if the shearing stress at A is not to exceed 18,000 psi. Neglect the effects of stress concentration and assume that the weld is as strong as the parent metal.

Ans. 3000 lb.

14-48. What principal stresses are associated with a state of pure shear $\tau_{xy} = 10$ ksi, $\tau_{yx} = -10$ ksi?

Ans. $\sigma = \pm 10$ ksi.

14-49. A plane element on a body is subjected to the two states of stress superimposed upon one another as shown. Find the components of stress that act on a face of the element inclined at 15 deg with the x-axis, as indicated. *Hint:* Draw Mohr's circle for each case and add the results.

Ans. $\sigma = 5.93$ ksi; $\tau = 9.2$ ksi.

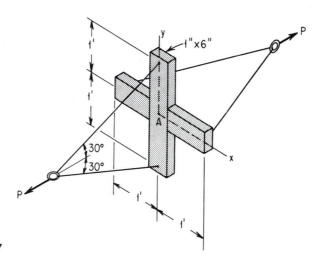

Problem 14-47

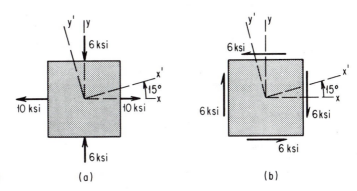

Problem 14-49   **(a)**   **(b)**

14-50. A plane element on a body is subjected to the state of stress shown in each
to   of the respective figures. Determine (a) the principal stresses and their
14-53. directions, (b) the maximum shearing stresses and the direction of the planes
            on which they occur.

      Ans. 14-50 (a) $\sigma_{max} = 8120$ psi at 37.8°CW from x-axis; $\sigma_{min} = -120$ psi
               at 127.8°CW from x-axis.
           (b) $\tau_{max} = 4120$ psi at 7.25°CCW from x-axis; $\tau_{min} = -4120$ psi
               at 97.25°CCW from x-axis.
       14-51 (a) $\sigma_{max} = 6.78$ ksi at 124.1°CW from x-axis; $\sigma_{min} = -14.78$
               ksi at 34.1° CW from x-axis.
           (b) $\tau_{max} = 10.78$ ksi at 79.1° CW from x-axis; $\tau_{min} = -10.78$
               ksi at 10.9° CCW from x-axis.

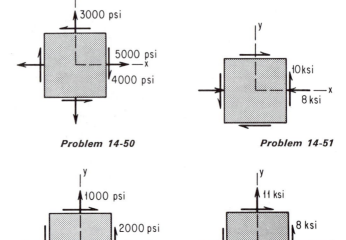

Problem 14-50   Problem 14-51

Problem 14-52   Problem 14-53

14-52 (a) σ_{max} = 3000 psi at 45°CCW from x-axis; σ_{min} = −1000 psi at 135°CCW from x-axis.

(b) Planes of maximum and minimum shear are those given in the problem.

14-53 (a) σ_{max} = 14.3 ksi at 112.5° CW from x-axis; σ_{min} = −8.3 ksi at 22.5° CW from x-axis.

(b) τ_{max} = 11.3 ksi at 67.5° CW from x-axis; τ_{min} = −11.3 ksi at 22.5° CCW from x-axis.

14-54. A solid circular shaft 4 in. in diameter is subjected to an axial compressive force of 100 kips and a twisting moment of 50 kip in. What is the maximum normal stress in the shaft?

Ans. 9.61 ksi.

14-55. A tensile force P lb and twisting moment of $P/4$ lb in. act together on a 2-in.-diameter circular shaft. What is P if the maximum tensile stress is not to exceed 20,000 psi?

Ans. 52.0 kips.

14-56. The 2-in.-diameter line shaft in the figure is subjected to both bending and torsion. What is the maximum shearing stress in the shaft? List the assumptions you make.

Ans. 61,000 psi.

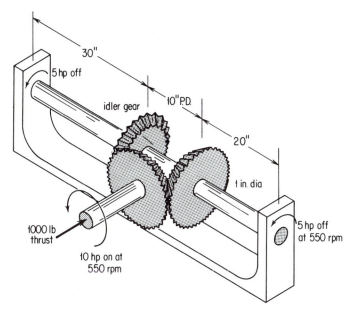

Problem 14-56

14-57. A fly-cutter is acted upon by the components of force shown. What is the maximum shearing stress in the shaft at A?

Ans. 16,030 psi.

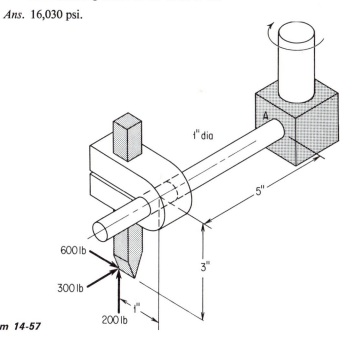

Problem 14-57

14-58. A thin-walled closed tube having a diameter of 2 in. and a wall thickness of 0.10 in. is subjected to an internal pressure of 200 psi and a twisting moment of 314 lb in. Determine the maximum normal and shearing stresses that act in the tube.

Ans. $\sigma_{max} = 2620$ psi; $\tau_{max} = 1120$ psi.

Chapter 15

WELDED, BOLTED, AND RIVETED CONNECTIONS

There are many well-established methods of connecting components to form structures and machines; these include fusion welding, bolting, riveting, brazing, soldering, and gluing. Size, shape, and service requirements of the structure or machine are among the many factors which influence the selection of connectors. By far, however, the most important methods are fusion welding, bolting, and riveting.

15-1 Welded Connections

Gas and arc welding techniques have improved to the point that welding is now the most important single method of joining metallic components. There is virtually no size limit to welding, and the process can be used with equal dependability in both shop and field.

Some of the more common types of welded joints are shown in the composite drawing of Fig. 15-1. With the exception of the seam weld, all require an addition of *weld metal* to the *parent metal*. The strength of a welded con-

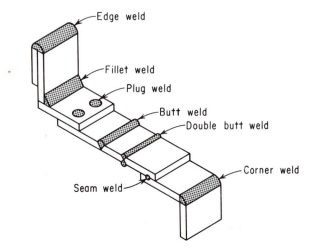

Edge weld
Fillet weld
Plug weld
Butt weld
Double butt weld
Corner weld
Seam weld

Figure 15-1

nection depends, therefore, on the properties of both the parent metal and the weld metal, and the geometry of the weld. Involved in the latter is the minimum cross-sectional area through which failure can occur, and specified values for the allowable stress. In a butt weld, Fig. 15-2(a), the minimum

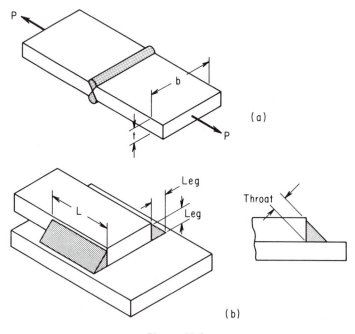

P

b

(a)

t

P

Leg

L

Leg

Throat

(b)

Figure 15-2

cross-sectional area is assumed to be the length of the joint times the thickness of the thinner plate; thus

$$P = \sigma_t bt \qquad (15\text{-}1)$$

where σ_t is the permissible tensile stress in the weld.

In a fillet weld, Fig. 15-2(b), the minimum cross-sectional area depends upon the width of the *throat* (roughly the length of the *fillet leg* times the sine of 45 deg), and the length of the weld.

$$\text{minimum area} = 0.707 \times (\text{fillet size}) \times L$$

Hence,

$$P = \tau \times (0.707) \times (\text{fillet size}) \times L \qquad (15\text{-}2)$$

where τ is the permissible shearing stress.

Working stresses for welded joints are governed largely by codes, established either by industries involved or by local building regulations. Typical values of working stress for steel weldments are

Shear stress:	14,000 psi
Tensile stress:	16,000 psi
Compressive stress:	18,500 psi

Since practically all fillet welds are subject to shear, the strength can be given in terms of fillet size. Thus, for an $\frac{1}{8}$-in. fillet weld:

$$P = (14,000)(0.707)(0.125) = 1250 \text{ lb/in.}$$

Based on this computation, the allowable shearing loads per inch of weld for various fillet sizes are as follows:

Size of fillet (in.)	Safe load (lb per in.)
$\frac{1}{8}$	1250
$\frac{3}{16}$	1875
$\frac{1}{4}$	2500
$\frac{5}{16}$	3125
$\frac{3}{8}$	3750
$\frac{1}{2}$	5000
$\frac{5}{8}$	6250
$\frac{3}{4}$	7500

These are easy numbers to remember, since each contains the decimal equivalent of the fillet size.

EXAMPLE 1. Two plates are joined by three sections of $\frac{3}{8}$ in. fillet weld, as shown in Fig. 15-3(a). What length L is required if the joint is to be 100 per cent efficient, i.e., if the weld is to be as strong as the parent metal? Assume an allowable tensile strength of the plate to be 20,000 psi.

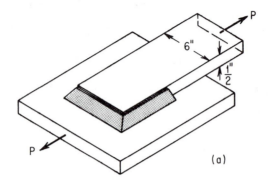

(a)

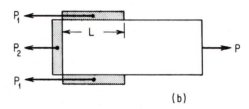

(b)   *Figure 15-3*

Solution: The load P is determined as follows

$$P = \sigma A = 20{,}000(6 \times \tfrac{1}{2}) = 60{,}000 \text{ lb}$$

In terms of the free-body diagram, Fig. 15-3(b), the load P must be supported by two longitudinal welds and one transverse weld.

$$2P_1 + P_2 = 60{,}000 \text{ lb}$$

Substitution of numerical values from the table of permissible loads gives

$$2(3750)L + 3750(6) = 60{,}000$$
$$L = 5.0 \text{ in.}$$

EXAMPLE 2. A 4-in. by 3-in. by $\tfrac{5}{8}$-in. standard steel angle is attached to the face of a gusset plate by $\tfrac{1}{2}$-in. fillet welds, as shown in Fig. 15-4(a). Determine the lengths L_1 and L_2 if the angle is to support a tensile load of 75 kips applied at its centroid.

Solution: Forces and moments must be balanced.

$$\sum F = 0$$
$$P_1 + P_2 = 75{,}000$$

and

$$\sum M = 0$$
$$2.63P_1 = 1.37P_2$$

The permissible load on a $\tfrac{1}{2}$ in. fillet is 5000 lb per in.; therefore,

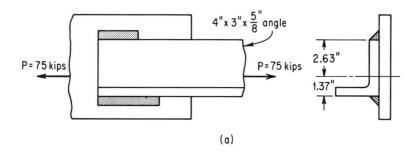

(a)

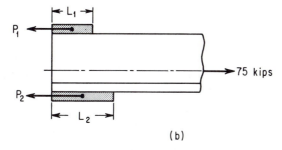

(b)

Figure 15-4

$$5000L_1 + 5000L_2 = 75,000$$
$$L_1 + L_2 = 15 \qquad\qquad (a)$$

and

$$2.63(5000)L_1 = 1.37(5000)L_2$$
$$L_2 = 1.92L_1 \qquad\qquad (b)$$

Equations (a) and (b) are solved simultaneously for L_1 and L_2.

$$L_1 + 1.92L_1 = 15$$
$$L_1 = \frac{15}{2.92} = 5.14 \text{ in.}$$
$$L_2 = 15 - 5.14 = 9.86 \text{ in.}$$

15-2 Riveted and Bolted Joints: Single Connector

It is customary to consider the strength of a riveted or bolted connection to depend on three quantities: the shearing strength of the rivets or bolts, the crushing or bearing strength of the connected plates, and the tensile strength of the connected plates. It is presumed that the joint will ultimately fail in one of the three ways shown in Fig. 15-5.

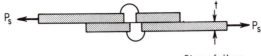

Shear failure

Bearing failure

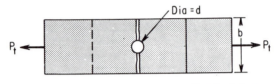

Tensile failure   *Figure 15-5*

The allowable load P_s based on permissible shearing stress is

$$P_s = \tau A_s \tag{15-3}$$

where τ and A_s are the working stress in shear and the area being sheared, respectively.

In bearing, failure is caused by the pressure force between the cylindrical surfaces; the bearing strength P_b, therefore, depends upon the ultimate compressive stress of the connector or the plate and the projected area td of the rivet or bolt hole:

$$P_b = \sigma_b A_b = \sigma_b td \tag{15-4}$$

where σ_b is the permissible bearing stress.

In tension, the strength of the joint depends upon the tensile strength of the plates and the minimum area that must sustain the force. If a joint has a single connector, the permissible force P_t is

$$P_t = \sigma_t A_t = \sigma_t (b - d)t \tag{15-5}$$

where

σ_t = the working stress in tension.
b = the width of the plate.
d = the diameter of the rivet or bolt hole.
t = the thickness of the plate.

Efficiency of a riveted or bolted joint is defined as the *ratio of the permissible load to the strength of the plate itself.*

As with welded joints, suitable working stresses for riveted and bolted connections are governed by either industrial practice or local building codes.

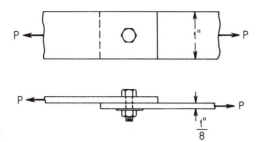

Figure 15-6

EXAMPLE 3. A single $\frac{3}{8}$-in.-diameter bolt is used to connect two plates, as shown in Fig. 15-6. Determine the strength and the efficiency of the joint if the following working stresses apply: $\tau = 15,000$ psi, $\sigma_b = 32,000$ psi, $\sigma_t = 18,000$ psi.

Solution: The safe loads, based on the shearing strength of the bolt, the strength of the plate in the vicinity of the hole, and the crushing strength of the plate, are each investigated.

Shear: $P_s = \tau A_s = 15,000\pi(\frac{3}{16})^2 = 1660 \text{ lb}$

Tension: $P_t = \sigma_t A_t = 18,000(\frac{1}{8})(1 - \frac{3}{8}) = 1410 \text{ lb}$

Bearing: $P_b = \sigma_b A_b = 32,000(\frac{1}{8})\frac{3}{8} = 1500 \text{ lb}$

The smallest value, 1410 lb, is the design load. Efficiency e is the ratio of the design load to the strength of the plate expressed in per cent; thus

$$e = \frac{1410}{18,000(\frac{1}{8})(1)} \times 100 = 62.7 \text{ per cent}$$

15-3 Riveted and Bolted Joints: Multiple Connectors

When multiple connectors are used in a fitting or joint, and the line of action of the load passes through the centroid of these connectors, as in Fig. 15-7(a), the rivets (or bolts) are assumed to deform equally. Each connector, therefore, carries an equal share of the load. The shear and bearing forces are distributed so that

$$\tau = \frac{P}{nA_s} \tag{15-6}$$

and

$$\sigma_b = \frac{P}{nA_b} \tag{15-7}$$

where τ and σ_b are the shear and bearing stresses, respectively, and n is the number of connectors. As before, A_s and A_b are the areas in shear and in bearing, respectively.

The tensile stress in the plate varies with the rivet or bolt geometry. The full load is carried by the plate across section 1-1, in Fig. 15-7, while two-thirds of the load is carried by the plate at section 2-2. This is the case

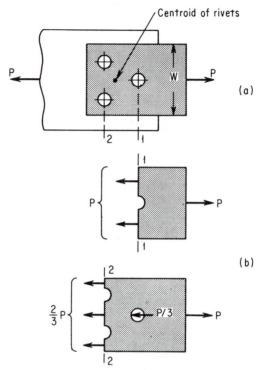

Figure 15-7

since one-third of the load is partially supported by the rivet in row 1–1, as illustrated in the free-body diagram, Fig. 15-7(b).

In the analysis of concentrically loaded joints fastened by more than one connector, shear and bearing are generally considered first. Free-body diagrams are then drawn and tension is investigated along each transverse row of rivets or bolts. The least value of the load P is the design load.

EXAMPLE 4. A *triple-riveted lap joint* is shown in Fig. 15-8. Determine the safe load P and the efficiency of the joint. The rivets are $\frac{7}{8}$ in. in diameter, and the following permissible stresses apply: $\tau = 15,000$ psi, $\sigma_b = 32,000$ psi, $\sigma_t = 20,000$ psi.

Solution: Since eight rivets are acting to support the load, each rivet carries a force of $P/8$.

Permissible load in shear:

$$P_s = \tau A_s = 15,000(8)\pi(\tfrac{7}{16})^2 = 72,100 \text{ lb}$$

Permissible load in bearing:

$$P_b = \sigma_b A_b = 32,000(8)(\tfrac{7}{8})(\tfrac{3}{8}) = 84,000 \text{ lb}$$

Permissible load in tension:

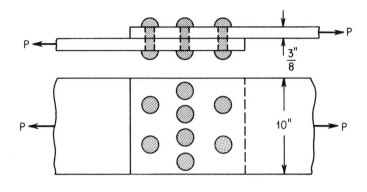

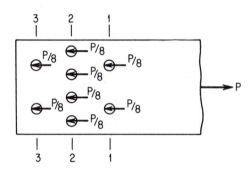

Tensile forces

Section 1-1: P
Section 2-2: 3/4 P
Section 3-3 1/4 P

Figure 15-8

At section 1–1

$$P_t = \sigma_t A_t = 20{,}000(10 - \tfrac{7}{8} \times 2)(\tfrac{3}{8}) = 61{,}900 \text{ lb}$$

At section 2–2

$$\tfrac{3}{4}P_t = \sigma_t A_t = 20{,}000(10 - \tfrac{7}{8} \times 4)(\tfrac{3}{8})$$

$$P_t = 65{,}000 \text{ lb}$$

At section 3–3

$$\frac{P_t}{4} = \sigma_t A_t = 20{,}000(10 - \tfrac{7}{8} \times 2)(\tfrac{3}{8})$$

$$P_t = 247{,}500 \text{ lb}$$

Thus, tension at section 1–1 governs the design, and the allowable load is 61,900 lb. The efficiency of this connection is the ratio of the safe load to the strength of the plate.

$$e = \frac{61{,}900}{20{,}000(10)(\tfrac{3}{8})} \times 100 = 82.5 \text{ per cent}$$

EXAMPLE 5. Figure 15-9 shows a standard AISC (American Institute of Steel Construction) connection for joining a 16 WF 88 beam to a column. The fastening consists of two 4-in. by $3\frac{1}{2}$-in. by $\frac{3}{8}$-in. angles each $11\frac{1}{2}$ in. long joined to the web of the beam by four $\frac{7}{8}$-in.-diameter rivets. The AISC code provides the following working stresses: 15 ksi in shear, 20 ksi in tension, 32 ksi in bearing for single shear, and 40 ksi in bearing for double shear. Find the load capacity of the connection.

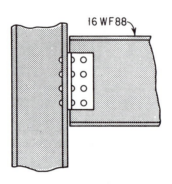 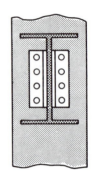

16 WF 88

Figure 15-9

Solution: Since a tensile failure cannot occur, only shear and bearing will be *investigated.*

Shear. Four rivets are in double shear; hence,

$$P_s = \tau A_s = 15(8)\pi(\tfrac{7}{16})^2 = 72.2 \text{ kips}$$

Bearing on web. The four rivets that bear on the web of the beam are in double shear; the allowable bearing stress, therefore, is 40 ksi. The web thickness, found in the Appendix, is 0.504 in.; hence,

$$P_b = \sigma_b A_b = 40(4)(0.504)\tfrac{7}{8} = 70.6 \text{ kips}$$

Bearing on angles. Eight bearing surfaces are involved, each supporting a rivet that is in single shear; thus

$$P_b = \sigma_b A_b = 32(8)(\tfrac{3}{8})(\tfrac{7}{8}) = 84 \text{ kips}$$

The capacity of the connection is, therefore, 70.6 kips.

In an actual design, the thickness of the plate to which the connection is made must also be investigated. If this plate is thin, the permissible load may be reduced.

15-4 Eccentrically Loaded Connections

When the line of action of the applied load does not pass through the centroid of the connectors, *torsion* as well as a *direct shear* acts on the fastening. This can be shown by considering the eccentrically loaded rivet joint of Fig. 15-10(a). The external load P acts at a distance e measured from the centroid of the rivet array. This is *statically equivalent* to a direct load P and a moment Pe, as shown in Fig. 15-10(b). The resisting force of the rivets acts

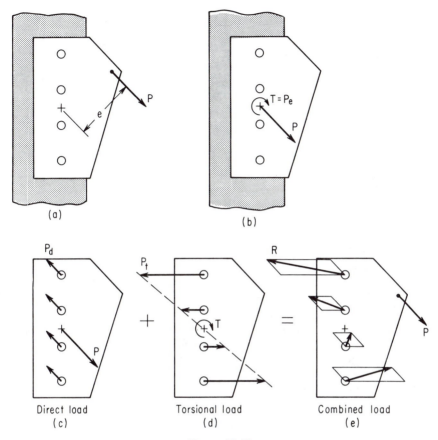

(a)

(b)

Direct load
(c)

Torsional load
(d)

Combined load
(e)

Figure 15-10

to oppose the direct force P by sharing the load equally, Fig. 15-10(c). In resisting the torque, however, the rivets act as the pins of a coupling, and the reactive forces on each are proportional to their respective centroidal distances. The sum of the moments of these reactive forces must balance the couple Pe. This is shown in Fig. 15-10(d). The load carried by any one rivet is the *vector sum of the direct force and the coupling force*, as shown in Fig. 15-10(e).

Eccentrically applied loads are not a desirable design feature, since the effective strength of the joint is appreciably reduced. The coupling effect can be minimized, however, by keeping the eccentricity small and by careful consideration of rivet geometry.

EXAMPLE 6. Compare the load-carrying capacity of the two rivet arrangements shown in Fig. 15-11. Half-inch-diameter rivets are used in each, and the working stress in shear is $\tau = 15,000$ psi.

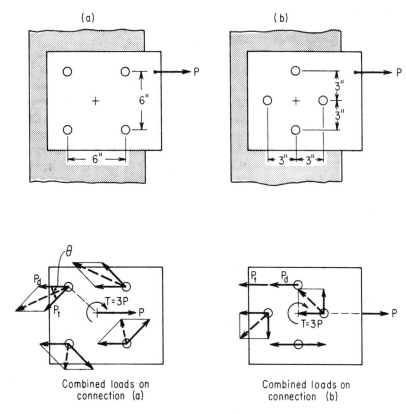

Combined loads on
connection (a)

Combined loads on
connection (b)

Figure 15-11

Solution: Direct loads and coupling loads are the same for both connections; the difference in load-carrying capacity will depend, therefore, on geometry alone.

Connection (a). The direct load P_d on each of the rivets acts horizontally to the left and is equal to $P/4$. The torsional load P_t on the rivets is the same for all, since the rivets are equidistant from the centroid. P_t can be found by a moment summation:

$$\sum M = 0$$

$$4P_t\sqrt{3^2 + 3^2} = 3P$$

$$P_t = \frac{P}{4\sqrt{2}}$$

By observing the vectors, the two top rivets seem to be the most severely loaded in the group. The resultant force, by the cosine law, for either of these rivets is

$$R = \sqrt{(P_d)^2 + (P_t)^2 + 2P_dP_t\cos\theta}$$

$$= \sqrt{\left(\frac{P}{4}\right)^2 + \left(\frac{P}{4\sqrt{2}}\right)^2 + 2\left(\frac{P}{4}\right)\left(\frac{P}{4\sqrt{2}}\right)\cos 45°}$$

$$= 0.395\,P$$

The permissible load, based on the safe working stress, is

$$0.395P = 15,000\left(\frac{\pi}{4}\right)\left(\frac{1}{2}\right)^2$$

$$P = 7460 \text{ lb}$$

Joint (b). The direct force on each of the rivets is $P/4$, and the torsional load is

$$4(3)P_t = 3P$$

$$P_t = \frac{P}{4}$$

The uppermost rivet is the most severely loaded in this arrangement; the forces, in this case, are colinear and add directly:

$$R = P_d + P_t = \frac{P}{4} + \frac{P}{4} = \frac{P}{2}$$

$$\frac{P}{2} = 15,000\left(\frac{\pi}{4}\right)\left(\frac{1}{2}\right)^2$$

$$P = 5890 \text{ lb}$$

Rivet arrangement (a) is the stronger of the two.

EXAMPLE 7. A steel gusset plate is bolted to a machine frame, as shown in Fig. 15-12. Find the greatest load that acts on any one bolt.

Solution: The centroid of the bolt array is found by taking first moments of the bolt areas about the left connector.

$$3A\bar{x} = A(0) + A(3) + A(9)$$

$$\bar{x} = 4 \text{ in.}$$

By observation, bolt C is seen to be the most severely loaded connector. The direct load on C is $P/3 = 1.67$ kips, and the torsional load can be found by summing moments of the torsional loads about the centroid.

$$4P_{At} + P_{Bt} + 5P_{Ct} = 6P = 30$$

By similar triangles,

$$P_{At} = \tfrac{4}{5}P_{Ct}$$

and

$$P_{Bt} = \tfrac{1}{5}P_{Ct}$$

Therefore,

$$\tfrac{16}{5}P_{Ct} + \tfrac{1}{5}P_{Ct} + 5P_{Ct} = 30$$

$$P_{Ct}(3.2 + 0.2 + 5) = 30$$

$$P_{Ct} = \frac{30}{8.4} = 3.57 \text{ kips}$$

The resultant force on bolt C is the vector sum of the direct load and the torsional load.

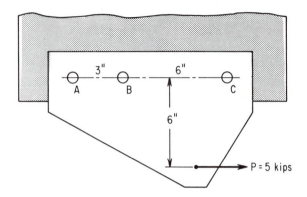

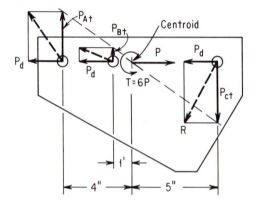

Figure 15-12

$$R = \sqrt{(1.67)^2 + (3.57)^2} = 3.94 \text{ kips}$$

In this example the torsion load is far greater than the direct load; this clearly indicates the severe effects of eccentricity.

QUESTIONS, PROBLEMS, AND ANSWERS

15-1. Two steel plates are double-butt welded as shown. Determine the safe load *P* and the efficiency of the joint if the permissible stress in the steel plate is 20,000 psi.

Ans. 80 per cent.

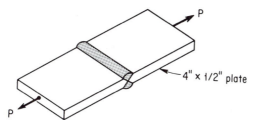

Problem 15-1

15-2. Determine the strength and efficiency of the welded connection shown. The working stress of the steel plate is 18,000 psi.

Ans. 83.3 per cent.

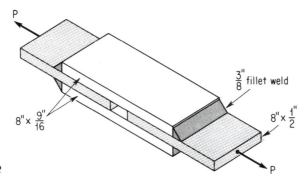

Problem 15-2

15-3. The ideal efficiency of 100 per cent in a welded joint is difficult to achieve. Which is more desirable: to have the weld stronger than the base metal or the base metal stronger than the weld?

Ans. Welding is expensive; why "overweld"?

15-4. Tanks and boilers operating under high pressure are fabricated by welding "dished" heads to cylindrical shells by the various methods shown on p. 406. Determine safe working pressures based on the strength of the weld for each of the attachments. Assume that the vessels, which are capable of sustaining these pressures, have diameters of 72 in. and have shell and head thicknesses of $\frac{3}{8}$ in.

Ans. (a) 333 psi; (b) 417 psi; (c) 209 psi; (d) 209 psi.

15-5. A 40-ft-diameter spherical gas tank is fabricated by welding together two $\frac{1}{2}$-in.-thick steel hemispheres. Find the safe working pressure for the tank and the efficiency of the joint based on a working stress of 20,000 psi for the steel plate.

Ans. 66.7 psi; 80 per cent.

15-6. The structural joint illustrated is fillet-welded and has strength equal to that

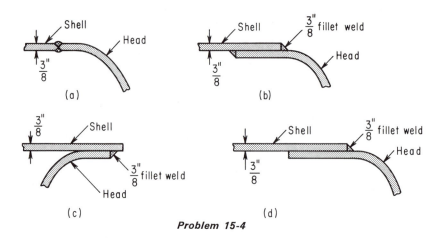

Problem 15-4

of the angle. Determine the proper length L_1 and L_2 if the permissible stress in the steel member is 20,000 psi. Assume that the load is applied at the centroid of the angle.

Ans. $L_1 = 7.13$ in.; $L_2 = 13.43$ in.

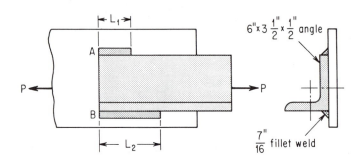

Problem 15-6

15-7. An additional transverse fillet weld 6 in. long is added along AB to the structural joint of Prob. 15-6. Determine the appropriate lengths L_1 and L_2.

Ans. $L_1 = 5$ in.; $L_2 = 9.5$ in. These are approximate answers; why?

15-8. Cover plates are used to join two sections of wide-flange beam as shown. Determine the required length L of $\frac{5}{8}$-in. fillet weld required if the bending moment in the beam at section A-A is 100,000 lb ft. Assume that vertical shear is negligible in this section.

Ans. 6.86 in.

15-9. Telescoping tubes are welded to form the joint shown. Determine the safe

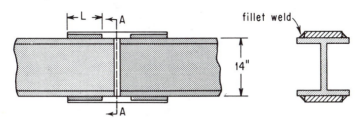

Problem 15-8   Section A-A

load P and the efficiency of the connection based on a working stress of 20,000 psi in the tubing.

Ans. 11,780 lb; 52.2 per cent.

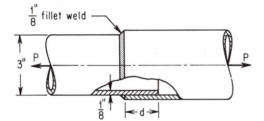

Problem 15-9

15-10. Assume that the joint of Prob. 15-9 is sweat soldered instead of welded and that the permissible shear stress between the tubes is 800 psi. Determine the distance d to achieve 90 per cent efficiency.

Ans. 2.69 in.

15-11. To increase the strength in the welded joint, the larger of the square tubes is "fish-tailed" on four sides as shown. What is the efficiency of the joint if the permissible stress in the base metal is 20,000 psi?

Ans. 95.2 per cent.

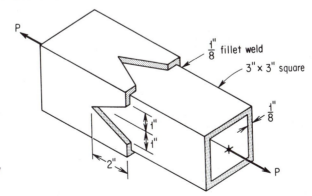

Problem 15-11

15-12. Determine the permissible torque that can be applied to the welded steel shaft illustrated. Assume a working stress in shear of 15,000 psi in the weld.

Ans. 25,500 lb in.

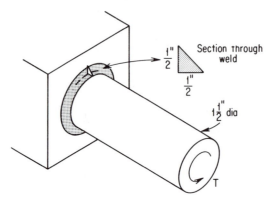

Problem 15-12

15-13. Four legs are used to support the water storage tank shown. The tank is 10 ft in diameter and has a capacity of 40,000 gallons. Determine the appropriate length L of $\frac{5}{16}$-in. fillet weld required on each side of the four legs. Water weighs 62.4 lb per ft³.

Ans. 13.3 in.

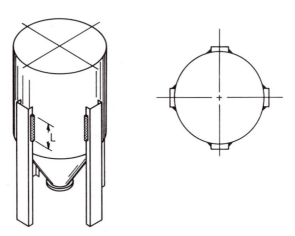

Problem 15-13

15-14. Two plates are connected by a single rivet as shown. Determine the strength and efficiency of the joint if the following working stresses apply: $\tau = 15,000$ psi, $\sigma_b = 30,000$ psi, $\sigma_t = 18,000$ psi.

Ans. 32.7 per cent; shear governs.

15-15. Two pieces of steel are spot-welded in six places as shown. Assume that

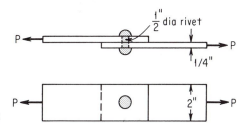

Problem 15-14

each "spot" supports $\frac{1}{6}$ of the total load. What is the tensile stress in the top plate at A, B, and C?

Ans. A: $5P/6wt$; B: $P/3wt$; C: $P/6wt$.

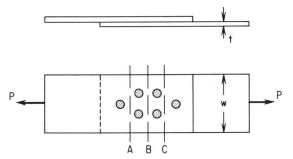

Problem 15-15   A   B   C

15-16. Two cover plates are employed in the connection shown. The $\frac{5}{8}$-in. bolts have a cross-sectional area of 0.3068 in.^{2}. Determine the strength and efficiency of the joint if the following working stresses apply: $\tau = 10,000$ psi, $\sigma_b = 20,000$ psi, $\sigma_t = 15,000$ psi. *Hint:* The bolts are in double shear.

Ans. 33.3 per cent.

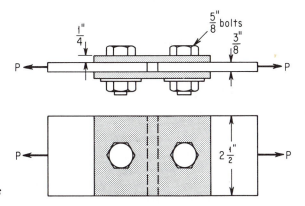

Problem 15-16

15-17. Determine the proper bolt diameter d (nearest standard size by $\frac{1}{8}$-in. increments), width w, and thickness t of the connection shown if it is to withstand a pull of $P = 10{,}000$ lb. The working stresses are $\tau = 15{,}000$ psi, $\sigma_b = 32{,}000$ psi, $\sigma_t = 18{,}000$ psi.

Ans. $d = 1$ in. dia.; $t = 0.3125$ in.; $w = 2.78$ in.

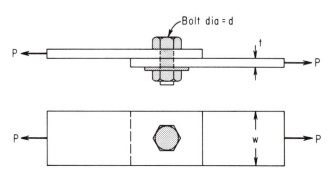

Problem 15-17

15-18. Find the critical dimensions d, w, t_1, and t_2 for the coupling shown. Base the computation on the following working stresses: $\tau = 15{,}000$ psi, $\sigma_b = 30{,}000$ psi, $\sigma_t = 20{,}000$ psi.

Ans. $d = 0.461$ in.; $t_1 = 0.362$ in.; $t_2 = 0.724$ in.; $w = 1.15$ in.

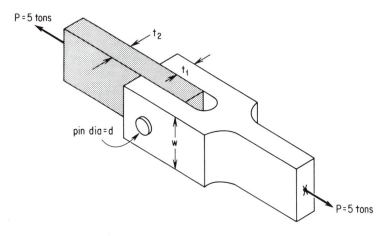

Problem 15-18

15-19. Determine the safe load P that can be supported by the riveted lap joint shown. The rivets have $\frac{5}{8}$-in. diameters, and the permissible stresses are $\tau = 15{,}000$ psi, $\sigma_b = 32{,}000$ psi, and $\sigma_t = 15{,}000$ psi, What is the efficiency of the joint?

Ans. 23 kips; 51.1 per cent.

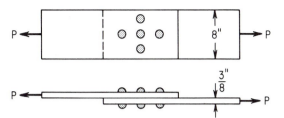

Problem 15-19

15-20. Nine $\frac{7}{8}$-in.-diameter rivets are used to secure the three plates illustrated. Find the design load P and the efficiency of the joint if the following stresses apply: $\tau = 15,000$ psi, $\sigma_b = 32,000$ psi for single shear, $\sigma_b = 40,000$ psi for double shear, $\sigma_t = 18,000$ psi.

Ans. 79.1 kips; 87.9 per cent.

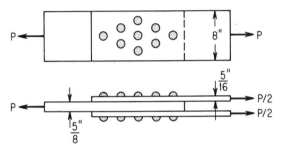

Problem 15-20

15-21. List several factors that are not accounted for in computing the strength of riveted and bolted connections, and indicate if these factors help or hinder.

Ans. Friction is not accounted for—it works in favor of the designer.

15-22. Two steel plates 10 in. wide by $\frac{3}{8}$ in. thick are to be lapped and connected by using $\frac{3}{4}$-in.-diameter rivets. Determine the number of rivets required and their arrangement if the joint is to have maximum efficiency The following stresses apply: $\tau = 15,000$ psi, $\sigma_b = 32,000$ psi, $\sigma_t = 20,000$ psi.

Ans. 69.4 kips; 92.5 per cent; 12 rivets arranged in rows of $1:2:3:3:2:1$.

15-23. Tension member B consists of two 4-in. by 3-in. by $\frac{3}{8}$-in. angles bolted back to

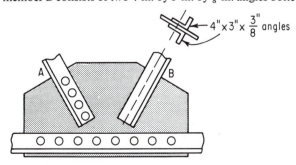

Problem 15-23

back to a $\frac{1}{2}$-in. gusset plate as shown. The member is to carry a maximum load based on a tensile stress of 18,000 psi. Determine the required number of $\frac{7}{8}$-in.-diameter *in-line* bolts. The allowable stresses are: $\tau = 15,000$ psi, $\sigma_b = 32,000$ psi, $\sigma_t = 20,000$ psi.

Ans. 5 bolts.

15-24. A 21 WF 62 beam is attached to two 14 WF 68 columns by means of the standard beam connection shown. Two 4-in. by $3\frac{1}{2}$-in. by $\frac{7}{16}$-in. angles and ten $\frac{7}{8}$-in.-diameter rivets are used at each column. Determine the safe load in kips per ft as governed by bearing and shear. Use the permissible stresses of Prob. 15-23.

Ans. 5.6 kips/ft; bearing on web of beam governs.

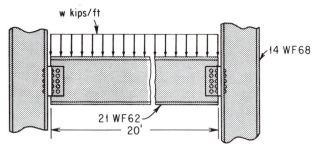

Problem 15-24

15-25. Determine the safe load P that may be carried by the eccentrically loaded bolted connection shown. The working stress in shear is 15,000 psi, and the bolt diameters are $\frac{7}{8}$ in.

Ans. 10.8 kips.

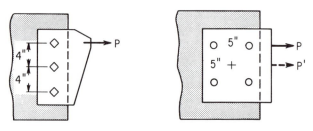

Problem 15-25   *Problem 15-26*

15-26. The riveted connection shown can carry either a centric load P' or an eccentric load of P. Find the ratio of P/P'. Assume the stress in the most severely loaded rivet to be the same for both methods of loading.

Ans. 0.633.

15-27. Find the permissible eccentricity *e* in the connection illustrated if the maximum load any one rivet can support is 5 kips.

Ans. 6 in.

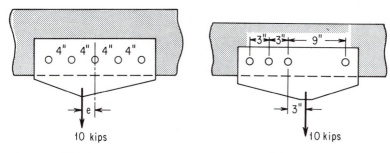

Problem 15-27   Problem 15-28

15-28. A load acts on a riveted connection shown. Find the maximum and minimum rivet loads.

Ans. 4.65 kips; 1.07 kips.

15-29. Find *P* for the rivet arrangement shown. The maximum load for any one rivet is 4000 lb.

Ans. 8000 lb.

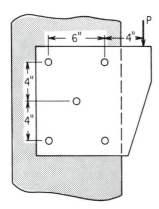

Problem 15-29

15-30. Riveting and bolting is not nearly as popular as welding as a means of structural assembly—it does, however, have some advantages. List some of these.

Ans. Disassembly is one; another advantage is that alignment is easier.

Chapter 16

COLUMNS

A member that fails in compression by *buckling*, or *collapsing*, is defined as a column. The term buckling refers to an *unstable state*, and the force causing this instability is called the *critical force*. Little concern is given to the degree of buckling: once instability is reached, the member is presumed to have failed. Stress that accompanies buckling failure is always less than that required for a direct compressive failure.

A rather forcible illustration of the effects of buckling appears in Fig. 16-1. Critical (a better word would be crippling) loads on the three flagpole columns, each a 1-in.-diameter steel bar, are indicated, together with the stress at a point of instability. The shortest member, in the true sense of the word, is not a column, but rather a *post;* failure here is due to direct compression. The illustration shows the relationship between length and instability.

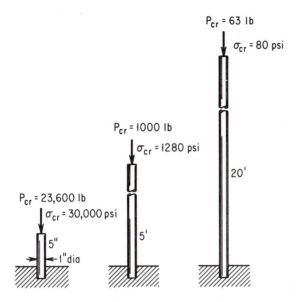

Critical loads and critical stresses on flag-pole columns

Figure 16-1

16-1 Long Columns: Euler's Equation

In 1744 Leonard Euler[1] published a lengthy paper covering a variety of cases of elastic bending. Created by the studies is the now-famous "Euler formula," an expression for the buckling strength of hinge-ended columns like that of Fig. 16-2. The equation is a statement of the critical buckling load P_{cr} in terms of the elastic and gemoetric properties of the column.

$$P_{cr} = \frac{\pi^2 EI}{l^2} \qquad (16\text{-}1)$$

where

E = modulus of elasticity.

I = least moment of inertia.

l = length.

It is rather interesting to observe that the critical buckling load does not depend upon the strength of the column material, but rather on elasticity and geometry. High-strength alloy steels and

Figure 16-2

[1] Leonard Euler (1707–1783), has been called the most prolific mathematician in history: he is probably the greatest man of science that Switzerland has produced.

common structural steels used as similar columns both fail under the same critical load.

The stress at buckling can be found by dividing both sides of the Euler equation by the cross-sectional area A of the column.

$$\sigma_{cr} = \frac{P_{cr}}{A} = \frac{\pi^2 EI}{l^2 A}$$

Since the square of the radius of gyration[2] r is equal to the ratio I/A, the critical stress becomes

$$\sigma_{cr} = \frac{\pi^2 E r^2}{l^2} = \frac{\pi^2 E}{\left(\dfrac{l}{r}\right)^2} \tag{16-2}$$

When expressed in this form, the term l/r is called the *slenderness ratio*.

In terms of pure mathematics, the slenderness ratio can assume any value other than zero; from the practical point of view, however, severe restrictions must be placed on the term. Imagine, for example, that a steel column has a slenderness ratio equal to 1. This would mean that the critical stress at buckling would be roughly

$$\sigma_{cr} = \frac{\pi^2 E}{\left(\dfrac{l}{r}\right)^2} = \frac{\pi^2 (30)10^6}{1} = 300,000,000 \text{ psi}$$

a ridiculous number.

For any given material, then, there must be a limit to the least value of l/r: a limit that can be found by allowing the critical stress to equal the yield stress. For steel, the slenderness ratio must be greater than 100, if a yield stress of approximately 30,000 psi is assumed.

$$\left(\frac{l}{r}\right)^2_{\min} = \frac{\pi^2 E}{\sigma_{yp}} = \frac{100(30)10^6}{30,000}$$

$$\left(\frac{l}{r}\right)_{\min} = 100$$

For high-strength steels, the slenderness ratio can be considerably smaller; for low-strength steels, higher. Good design practice generally governs the upper limit of the slenderness ratio. For structural-steel design, a rule of thumb places the greatest permissible l/r at 200.

Another way of determining the validity of Euler's equation is to divide the critical load, as found in the equation, by the cross-sectional area. If the compressive stress is less than the yield stress, the column is *long* and will fail by buckling. If the stress is greater than the yield stress, the member will fail by fracture or by plastic deformation.

An important point that must be remembered when using Euler's equation

---

[2]In mechanics of materials the r is generally used to denote radius of gyration, whereas in engineering mechanics k is the usual symbol.

is that the moment of inertia (or radius of gyration) must be the least value for the section.

EXAMPLE 1. An aluminum alloy has a modulus of elasticity of 10×10^6 psi and a yield strength of 6000 psi. Determine the least value of the slenderness ratio for which Euler's equation applies.

Solution: The least value of the slenderness ratio is found by substituting numerical values into Eq. 16-2.

$$\sigma_{cr} = \frac{\pi^2 E}{\left(\dfrac{l}{r}\right)^2}$$

$$\left(\frac{l}{r}\right)_{min} = \sqrt{\frac{\pi^2 E}{\sigma_{yp}}} = \sqrt{\frac{(\pi)^2 10(10)^6}{6000}} = 128$$

EXAMPLE 2. A structural-steel column 20 ft long is to support an axial load of 50 kips. Use a factor of safety of 3 and find the lightest wide-flanged section that can support this load.

Solution: To satisfy the required factor of safety, the critical design load is

$$P_{cr} = 3(50) = 150 \text{ kips}$$

Numerical data are substituted into Euler's equation, and the least moment of inertia is computed:

$$I = \frac{P_{cr}l^2}{\pi^2 E} = \frac{150(10)^3(20 \times 12)^2}{\pi^2 30(10)^6}$$
$$= 29.2 \text{ in.}^4$$

A suitable column must, therefore, have a moment of inertia in its weak direction of at least 29.2 in.^{4}. Tables in the Appendix give the lightest beam as an 8 WF 31. The least radius of gyration of this section is 2.01; therefore,

$$\frac{l}{r} = \frac{20(12)}{2.01} = 119$$

This section is satisfactory, since its slenderness ratio is greater than 100. If the ratio l/r were less than 100, the next lightest beam would be checked, and so on, until a satisfactory section could be found.

16-2 Effects of Bracing and End Restraints

The braced column of Fig. 16-3(a) is, in effect, two columns, one on top of the other; to fail by buckling each half must collapse. The *effective length* of this member is $l/2$, thereby making the column four times stronger than when unbraced; thus

$$P_{cr} = \frac{\pi^2 EI}{\left(\dfrac{l}{2}\right)^2} = \frac{4\pi^2 EI}{l^2} \tag{16-3}$$

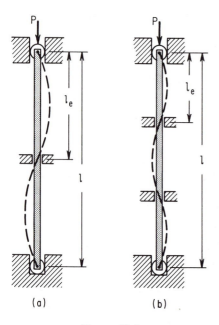

(a)   (b)

Figure 16-3

By similar reasoning, bracing at the third points, Fig. 16-3(b) increases the column strength nine times.

$$P_{cr} = \frac{\pi^2 EI}{(l/3)^2} = \frac{9\pi^2 EI}{l^2} \quad (16\text{-}4)$$

Experience has shown that Euler's formula can be corrected to apply to the variety of end conditions illustrated in Fig. 16-4. In the second and third cases, the end restraints tend to strengthen the column, whereas the lack of restraints in the fourth case weakens the column. It is usual practice to consider columns "hinge-ended" when in doubt as to the degree of restraint offered by the supports. This simply increases the factor of safety in the design.

EXAMPLE 3. A 5-in. I-beam weighing

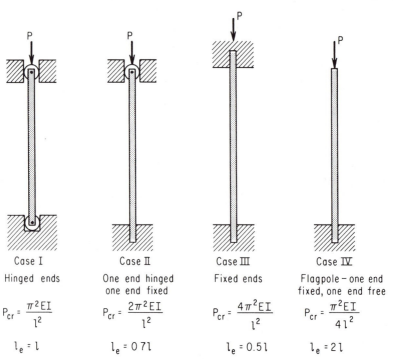

Case I	Case II	Case III	Case IV
Hinged ends	One end hinged one end fixed	Fixed ends	Flagpole – one end fixed, one end free
$P_{cr} = \dfrac{\pi^2 EI}{l^2}$	$P_{cr} = \dfrac{2\pi^2 EI}{l^2}$	$P_{cr} = \dfrac{4\pi^2 EI}{l^2}$	$P_{cr} = \dfrac{\pi^2 EI}{4l^2}$
$l_e = l$	$l_e = 0.7l$	$l_e = 0.5l$	$l_e = 2l$

Figure 16-4

10 lb per ft is to be used as a hinge-ended column 18 ft long. The beam is braced in its weakest direction at the midpoint. Determine the safe load P based on a factor of safety of 2.5.

Solution: Two possibilities exist: The column can fail in its strongest direction, where its length is considered to be the full 18 ft, or it can fail in its weakest direction, where its effective length is 9 ft. Both possibilities must be investigated, as well as the slenderness ratio for each case.

As an 18-*ft column in the strong direction:*

The slenderness ratio must be checked to ensure that Euler's equation applies. Data are obtained from the Appendix.

$$\frac{l}{r} = \frac{18(12)}{2.05} = 105$$

Since the slenderness ratio is greater than 100, Euler's equation is valid:

$$P_{cr} = \frac{\pi^2 EI}{l^2} = \frac{\pi^2(30)10^6(12.1)}{(18 \times 12)^2} = 76,700 \text{ lb}$$

The design load is

$$P = \frac{76,700}{2.5} = 30,700 \text{ lb}$$

As a 9-*ft column in the weak direction:*

$$\frac{l}{r} = \frac{9(12)}{0.65} = 166$$

Euler's equation is valid; hence,

$$P_{cr} = \frac{\pi^2 EI}{l^2} = \frac{\pi^2(30)10^6(1.2)}{(9 \times 12)^2} = 30,500 \text{ lb}$$

$$P = \frac{30,500}{2.5} = 12,200 \text{ lb}$$

The weak direction governs the design. This conclusion could have been reached by merely comparing the two slenderness ratios; the configuration which gives the greatest slenderness ratio governs the design.

EXAMPLE 4. Two 12-in., 25-lb channels are latticed together to form a column. The distance between the channels is designed so that the moments of inertia about the principal axes are equal. Find the critical (minimum) length of this column, assuming one end to be fixed and the other hinged. $E = 30 \times 10^6$ psi; $\sigma_{yp} = 30,000$ psi.

Solution: The critical load for this column is given in Fig. 16-4:

$$P_{cr} = \frac{2\pi^2 EI}{l^2}$$

where $P_{cr} = \sigma_{yp} A$

$$I = 2(143.5) = 287 \text{ in.}^4$$

$$A = 2(7.32) = 14.64 \text{ in.}^2$$

Substitution of numerical data gives

$$l^2 = \frac{2\pi^2 EI}{\sigma_{yp}A} = \frac{2\pi^2(30 \times 10^6)287}{30,000(14.64)} = 38.7 \times 10^4$$

$$l = 622 \text{ in. or } 51.8 \text{ ft}$$

Alternate Solution: If the column were hinge-ended, the critical slenderness ratio would be approximately 100.

$$l/r \approx 100$$

$$l = 100\sqrt{\frac{I}{A}} = 100\sqrt{\frac{287}{14.64}} = 443 \text{ in.}$$

With one end fixed and the other hinged, the effective length, given in Fig. 16-4, is

$$l_e = 0.7\,l$$

Therefore

$$l = \frac{l_e}{0.7} = \frac{443}{0.7} = 634 \text{ in. or } 52.8 \text{ ft}$$

which is a fair approximation to the exact answer.

16-3 Long Columns Loaded Between Supports

When long columns sustain concentric loads applied at points other than the ends, Fig. 16-5, Euler's formula becomes

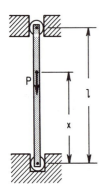

$$P_{cr} = = \frac{\pi^2 EI}{l_r^2} \qquad (16\text{-}5)$$

where l_r is called the *reduced length* and is similar to the *effective length* employed in correcting for various end conditions. In other words, l_r is the length of a hinged column which has the same critical buckling load as the column loaded between supports.

Values of l_r may be obtained from the graph[3] of Fig. 16-6, which relates the ratios x/l and l_r/l. The former is obtained from known data and the latter from the graph.

For validity, Euler's equation must give a critical stress less than the yield stress, which means that a failure will occur through buckling rather than compression.

This point is easily checked by simply computing the value of the critical stress at the buckling load:

Figure 16-5

$$\sigma_a = \frac{P_{cr}}{A}$$

[3]James Dow, "Columns Loaded Between Supports," *Machine Design*, February, 1961, pp. 167–168.

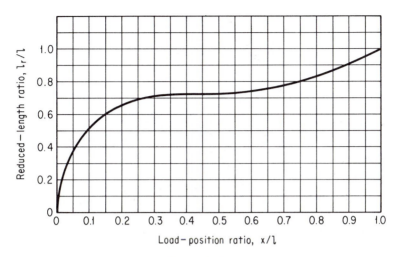

Figure 16-6

EXAMPLE 5. The control linkage shown in Fig. 16-7 consists of a high-strength steel rod and a pivot arm. The load is transmitted to the rod through pin A. Determine (a) the critical value of F based on the buckling strength of the rod, (b) the stress in the rod at the critical load. $E = 30 \times 10^6$ psi; $\sigma_{yp} = 60,000$ psi.

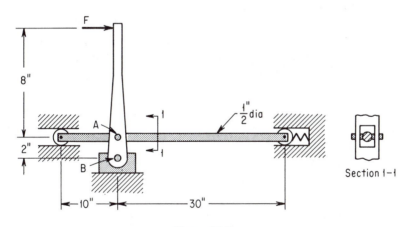

Figure 16-7

Solution: The ratio of $x/l = 30/40 = 0.75$; l_r/l from the graph of Fig. 16-6 is 0.8; hence,

$$l_r = 0.8(40) = 32 \text{ in.}$$

Pin A can tolerate a critical force of

$$P_{cr} = \frac{\pi^2 EI}{l_r^2}$$

where

$$I = \frac{\pi d^4}{64} = \frac{\pi(0.5)^4}{64} = 0.00307 \text{ in.}^4$$

Thus

$$P_A = P_{cr} = \frac{\pi^2(30 \times 10^6)(0.00307)}{(32)^2} = 887 \text{ lb}$$

Stress, at the critical load, is

$$\sigma_{cr} = \frac{P_{cr}}{A} = \frac{887}{\pi(0.25)^2} = 4520 \text{ psi}$$

which is well below the yield stress of 60,000 psi.

Moments taken about the pin B will give the critical force F:

$$10F = 2(887)$$
$$F = 177 \text{ lb}$$

16-4 Intermediate Columns: Empirical Formulas

A graph of Euler's equation, Fig. 16-8, shows the critical stress to decrease as the slenderness ratio increases, the useful portion of the curve being below the proportional limit. The range of columns whose slenderness ratios are less than Euler's critical value are called *short columns* and *intermediate columns*. Limiting values of compressive stress govern the design of the former class, and empirical formulas based on observation and experience govern the latter. The newest and most practical of the intermediate column formulas[4] is based on studies made by the American Insitute of Steel Con-

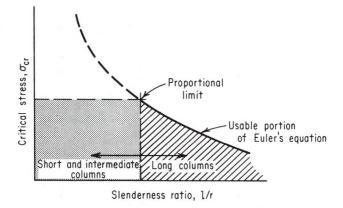

Figure 16-8

[4]Adopted November 30, 1961, by the American Institute of Steel Construction and reprinted by permission.

struction. The formula defines the permissible stress as

$$\sigma_a = \frac{\left[1 - \frac{(l/r)^2}{2C_c^2}\right]\sigma_{yp}}{\text{f.s.}} \tag{16-6}$$

where

$$\text{f.s.} = \text{factor of safety} = \frac{5}{3} + \frac{3\left(\dfrac{l}{r}\right)}{8C_c} - \frac{\left(\dfrac{l}{r}\right)^3}{8C_c^3} \tag{a}$$

and

$$C_c = \sqrt{\frac{2\pi^2 E}{\sigma_{yp}}} \tag{b}$$

The term σ_{yp} is the stress at the yield point for a given grade of steel; the term σ_a is the permissible axial stress. Equation (16-6) is valid for any axial-loaded hinge-ended column whose slenderness ratio is less than the constant C_c.

The AISC equation is interesting from a very practical point of view in that it provides a *variable factor of safety*—high, where the column is most vulnerable. Similar provisions have been included in the British and German design standards for some time and can be justified by the insensitivity of columns to accidental eccentricities.

Allowable stresses, as calculated by Eq. (16-6), for five grades of steel are presented in the graph of Fig. 16-9. It is interesting to see how the values of σ_a converge to those permitted by Euler's equation for high values of l/r.

EXAMPLE 6. Determine the permissible load that can be carried by a 10-ft 10 WF 45 steel column. The stress at the yield point of this particular steel is 33,000 psi.

Solution: The Appendix gives the least radius of gyration of this section as $r = 2$ in. The slenderness ratio therefore, is

$$\frac{l}{r} = \frac{10(12)}{2} = 60$$

The constant C_e and the factor of safety are next computed:

$$C_c = \sqrt{\frac{2\pi^2 E}{\sigma_{yp}}} = \sqrt{\frac{2\pi^2(30 \times 10^6)}{33,000}} = 134$$

$$\text{f.s.} = \frac{5}{3} + \frac{3(l/r)}{8C_c} - \frac{(l/r)^3}{8C_c^3}$$

$$= \frac{5}{3} + \frac{3(60)}{8(134)} - \frac{(60)^3}{8(134)^3} = 1.83$$

Numerical values are substituted into Eq. (16-6).

$$\sigma_a = \frac{\left[1 - \frac{(l/r)^2}{2C_c^2}\right]\sigma_{yp}}{\text{f.s.}}$$

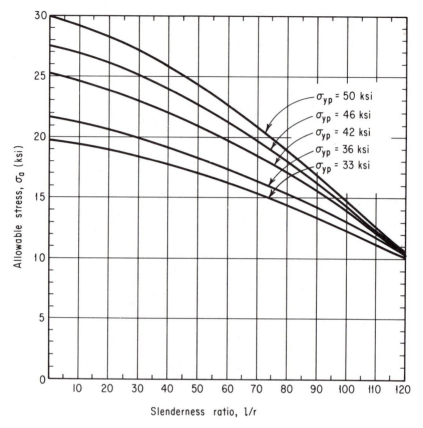

Figure 16-9

$$= \frac{\left[1 - \frac{(60)^2}{2(134)^2}\right] 33,000}{1.83} = 16,200 \text{ psi}$$

Since this column has a cross-sectional area of 13.24 in.², the allowable load is

$$P_a = \sigma_a A = 16,200(13.24) = 214,000 \text{ lb or } 214 \text{ kips}$$

Alternate Solution: The graph of Fig. 16-9 gives an allowable stress of 16.2 ksi at $l/r = 60$ for 33,000-psi-grade steel; therefore,

$$P_a = 16.2(13.24) = 214 \text{ kips}$$

16-5 Eccentrically Loaded Columns: Conservative Approach

When the line of action of the axial force does not coincide with the centroidal axis of the column, bending stresses as well as compressive stresses are induced in the member. This is illustrated in Fig. 16-10, where the moment

M is the product of the force P and the eccentricity e; the combined stress has a maximum magnitude of

$$\sigma_{max} = \frac{P}{A} + \frac{Pe}{Z} \qquad (16\text{-}7)$$

where A and Z are the area and section modulus, respectively.

For a column acted upon by an axial load P_o and an eccentric load P, a more realistic situation, the maximum stress is

$$\sigma_{max} = \frac{P_o + P}{A} + \frac{Pe}{Z} \qquad (16\text{-}8)$$

In these equations the maximum stress is actually the working stress σ_a obtained through Eq. (16-6) or by the graph of Fig. 16-9, and is always based on the section's least radius of gyration.

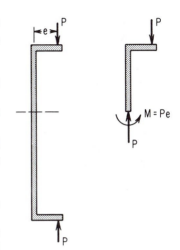

Figure 16-10

EXAMPLE 7. A 12 WF 40 steel column 12 ft long supports an axial load of 100 kips. Determine the additional load that may be applied in the weak direction at an eccentricity $e = 2$ ft. The steel has a yield-point stress of 42,000 psi.

Solution: The properties of the section are

$$A = 11.77 \text{ in.}^2$$
$$r_{min} = 195 \text{ in.}$$
$$Z_{min} = 11.0 \text{ in.}^3$$
$$\frac{l}{r} = \frac{12(12)}{19.4} = 74.2$$

The working stress is estimated from the graph of Fig. 16-9.

$$\sigma_a = 18 \text{ ksi}$$

Numerical values are substituted into Eq. (16-8):

$$\sigma_{max} = \frac{P_o + P}{A} + \frac{Pe}{Z}$$

$$18 = \frac{100 + P}{11.77} + \frac{P(2 \times 12)}{11.0}$$

$$P = 4.19 \text{ kips}$$

EXAMPLE 8. Select a wide-flanged section to act as a column 15 ft long. The axial compressive load is 80 kips, and the eccentric load is 60 kips applied in the strong direction, as shown in Fig. 16-11. The steel has a yield-point stress of 33 ksi.

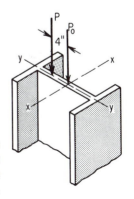

Figure 16-11

Solution: This problem is more difficult than the previous example, since the properties of the section are not known; trial-and-error methods must be used.

If the total load were applied concentrically and if the slenderness ratio were zero, the maximum stress from the graph of Fig. 16-9 would be approximately 20 ksi. The column would require an area of at least

$$A = \frac{P}{A} = \frac{80 + 60}{20} = 7 \text{ in.}^2$$

Column sections having less cross-sectional area need not be considered.

Several members must be tried until an agreement between the left- and right-hand portions of Eq. (16-8) can be found.

Trial 1. 12 WF 40

$$A = 11.77 \text{ in.}^2$$

$$r_{min} = 1.94 \text{ in.}$$

$$Z_{xx} = 51.9 \text{ in.}^3$$

$$\frac{l}{r} = \frac{15(12)}{1.94} = 92.8$$

$$\sigma_{max} = \frac{P + P_o}{A} + \frac{Pe}{Z}$$

$$\sigma_{max} = \frac{80 + 60}{11.77} + \frac{60(4)}{51.9} = 16.5 \text{ ksi}$$

The graph of Fig. 16-9 gives a maximum stress of 13.1 ksi for this section. A heavier member is, therefore, required.

Trial 2. 10 WF 49

$$A = 14.4 \text{ in.}^2$$

$$r_{min} = 2.54 \text{ in.}$$

$$Z_{xx} = 54.6 \text{ in.}^3$$

$$\frac{l}{r} = \frac{15(12)}{2.54} = 70.9$$

$$\sigma_{max} = \frac{80 + 60}{14.4} + \frac{60(4)}{54.6} = 14.1 \text{ ksi}$$

This section is satisfactory, since the allowable stress from the graph of Fig. 16-9 is 15.3 ksi, a greater value.

Further trials may result in finding a lighter, and, therefore, more economical section.

QUESTIONS, PROBLEMS, AND ANSWERS

16-1. How would you differentiate between a column and a post and how would each be treated in a design situation?

Ans. A column fails by buckling or collapsing; a post fails by compression. Special column formulas are needed for the former, while the simple stress-force-area approach is used in the design of posts.

16-2. Find the greatest value of the slenderness ratio for each of the following vertical columns: (a) square cross section, 1 in. on edge and 10 ft long; (b) a plate 5 ft wide, $\frac{1}{4}$ in. thick, and 3 ft high; (c) a solid circular section, 2 in. in diameter and 5 ft long; (d) an 8 WF 31 section, 20 ft long; (e) two 9-in., 20-lb channels, 6 ft long placed back-to-back; (f) four 3-in. by 3-in. by $\frac{1}{4}$-in. angles welded together to form a box section 10 ft long; (g) a 6-in. by 4-in. by $\frac{1}{2}$-in. angle, 3 ft long.

Ans. (a) 416; (b) 499; (c) 120; (d) 119; (e) 82.7; (f) 51.2; (g) 41.4.

16-3. Two 15-in., 50-lb channels are latticed together as shown to form a column section having equal moments of inertia I_x and I_y. Find the length of this column if it has a slenderness ratio of 110.

Ans. 48.0 ft.

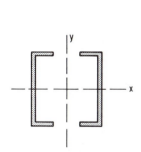

Problem 16-3

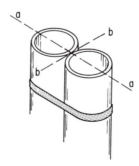

Problem 16-4

16-4. Two sections of $3\frac{1}{2}$-in.-diameter standard pipe are strapped together to form the column section shown. Each pipe has an outside diameter of 4 in., a cross-sectional area of 2.68 in.^{2}, and a diametral moment of inertia of 4.79 in.^{4}. Find the radius of gyration of the section about axis *a-a* and about axis *b-b*.

Ans. 1.34 in.; 2.41 in.

16-5. A certain high-tensile-strength steel has a modulus of elasticity of 30×10^6 psi and a yield point stress of 90,000 psi. Find the limiting value of the slenderness ratio for which Euler's equation is valid. Is this a minimum or maximum value?

Ans. $l/r \geq 57.3$.

16-6. Douglas fir has a modulus of elasticity of 1.80×10^6 and a yield stress parallel to the grain of 4500 psi. Find the limiting value of the slenderness ratio for which Euler's equation is valid.

Ans. $l/r \geq 62.8$.

16-7. Determine the safe axial load that may act on a 15-ft hinge-ended 8 WF 24 steel column. Use a factor of safety of 2.5.

Ans. 66.5 kips.

16-8. Determine the critical load F that can be applied as indicated to the steel column shown. The member has rounded ends and is braced at the midpoint in the weak direction. Use Euler's equation and a yield-point stress of 45,000 psi.

Ans. 2050 lb.

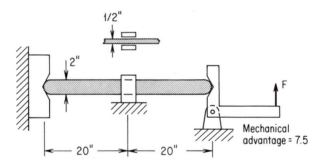

Problem 16-8

16-9. Two steel columns with square cross sections support a rigid beam of negligible weight as shown. Column A is fixed at one end and pinned at the other, and column B is pinned at both ends. Find the magnitude and location of the critical load P that can be supported by the assembly. Use Euler's equation.

Ans. 167 kips.

16-10. Four aluminum tubes are latticed together to form the antenna shown. Guy wires with equal tensions T support the antenna. (a) Determine the permissible cable tension, using Euler's equation with a factor of safety of 2, and (b) find the stress in the tubes. $E = 10 \times 10^6$ psi.

Ans. 3780 lb; 2790 psi.

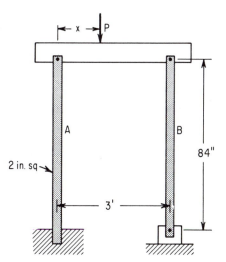

Problem 16-9

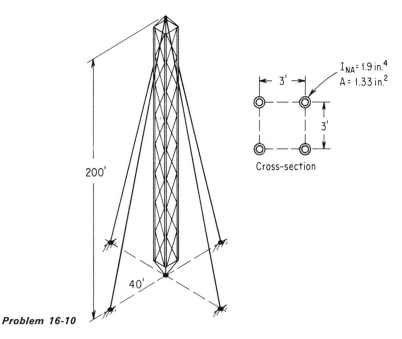

Problem 16-10

16-11. A steel rod, rigidly supported at its base as shown on p. 430, is free to expand 0.04 in. before making contact with the rigid support. What is the minimum temperature change necessary to cause buckling?

Ans. 129°F.

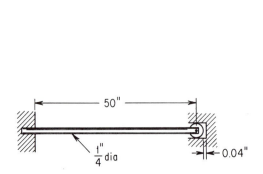

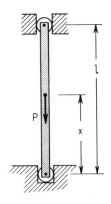

Problem 16-11   **Problem 16-12**

16-12. Use the graph of Fig. 16-6 to find the reduced length for each of the following conditions for the column shown: (a) $l = 10$ ft, $x = 6$ ft; (b) $l = 12$ ft, $x = 4$ ft; (c) $l = 20$ ft, $x = 18$ ft; (d) $l = 60$ in., $x = 10$ in.

Ans. (a) 7.4 ft; (b) 8.6 ft; (c) 18.2 ft; (d) 37.5 in.

16-13. A 3-in., 5-lb-per-ft steel channel is used as a 10-ft-long hinge-ended column. Find (a) the critical axial load that may be applied at a point 6 ft from the base, and (b) the critical stress for this loading arrangement.

Ans. 9380 lb; 6420 psi.

16-14. Pressure applied at C is used to operate the push-rod switch mechanism shown. Find the safe value of F, based on a factor of safety of 2, and the resulting contact force at A. The rod is a 1-in.-diameter steel tube and has the properties shown in the figure.

Ans. $F = 724$ lb; $F_a = 28,900$ lb.

16-15. A 12-ft long 10 WF 77 steel section is used as a hinge-ended column. Use the AISC formulas to determine the load capacity of the column. The steel has a yield-point stress of 36 ksi.

Ans. 405 kips.

16-16. A column has a slenderness ratio of 70 and is made of steel that has a yield-point stress of 40,000 psi. Use the AISC formulas to determine the permissible stress and the appropriate factor of safety.

Ans. 18 ksi; 1.86.

16-17. What are the maximum and minimum safety factors given by the AISC formulas for the permissible range of l/r?

Ans. 1.67 for $l/r = 0$; 1.92 for $l/r = C_c$.

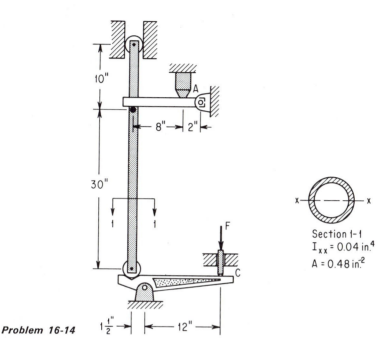

Section 1-1
$I_{xx} = 0.04$ in.^{4}
$A = 0.48$ in.^{2}

Problem 16-14

16-18. What load can be carried by two 12-in., 25-lb steel channels that are 8 ft long: (a) if they are fastened back to back; (b) if they are latticed together to form a section having equal moments of inertia in the two principal directions? The steel has a yield-point stress of 42 ksi.

Ans. (a) 222 kips; (b) 348 kips.

16-19. Find a suitable wide-flanged steel beam (33-ksi stress grade) to be used as a 20-ft column capable of supporting an axial load of 150 kips. Use the AISC formulas.

Ans. 8 WF 48.

16-20. An 8-ft long steel pipe with outside and inside diameters of 6 in. and 4 in., respectively, is used as a column. Determine the safe axial load based on the AISC formulas. The steel has a yield-point stress of 36 ksi.

Ans. 283 kips.

16-21. A 12-ft boxed column is fabricated by welding four 3-in. by $\frac{1}{2}$-in. steel angles together. Determine, by means of the AISC formula, the safe concentric load that may be carried. The steel has a yield-point stress of 50 ksi.

Ans. 242 kips.

16-22. Determine the permissible column load that may be supported by a 6-in.-long, $\frac{1}{2}$-in.-diameter alloy steel rod. The yield-point stress is 90,000 psi, and the modulus of elasticity is 30×10^6 psi. Use either the AISC formulas or the Euler equation, whichever applies.

Ans. 8000 lb.

16-23. What is the maximum beam reaction that may be carried by a 14 WF 34 steel column that is 12 ft long? Use the AISC formulas and Eq. (16-7). The beam reaction occurs at the outside flange of the column as shown. Assume that this steel has a yield-point stress of 33 ksi.

Ans. 51.2 kips.

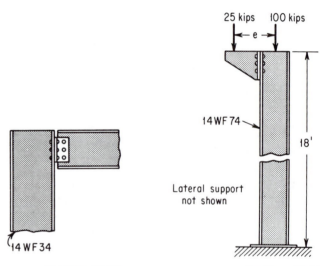

Problem 16-23   **Problem 16-24**

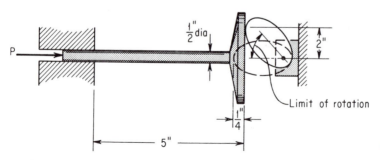

Problem 16-25

16-24. Determine the maximum permissible eccentricity, e, for the column illustrated. The section, a 14 WF 74 steel beam, has a yield-point stress of 36 ksi. Use Eq. (16-8) and the AISC formulas and assume that the ends of the column are hinge-supported.

Ans. 39.8 in.

16-25. A cam-operated steel plunger bar is illustrated. The bar, a "flag-pole" column, has a yield-point stress of 70,000 psi. What are the maximum and minimum critical values of P for the limits of cam rotation shown? The greatest eccentricity is 2 in.

Ans. $P_{max} = 8190$ lb; $P_{min} = 275$ lb.

Chapter 17

EXPERIMENTAL
STRESS ANALYSIS

Mathematics is the most exact science, and its conclusions are capable of absolute proof. But this is so only because mathematics does not attempt to draw absolute conclusions. All mathematical truths are relative, conditional—Charles Proteus Steinmetz (1923)[1]

This chapter is concerned with the more significant experimental methods used in the study of mechanics of materials. Although the pure theorist will cringe at the importance about to be attributed to the experimental method, he cannot deny its many accomplishments. In truth, a theory or a formula is only useful when it can predict with reasonable accuracy what will happen physically. To a degree, the equations of torsion, bending, and axial loading —theoretical equations all—give reasonable and accurate answers; they fail, however, to account for the variance in the physical properties of materials,

[1]From *Men of Mathematics*, copyright 1937 by E. T. Bell, by permission of Simon and Schuster, Inc.

the stress and strain in the region of the applied load, the abrupt change in geometry, and the effects of dynamic loading.

The designer is, more often than not, in these regions of the unknown, and he frequently relies on intuition based on past experience. A difficulty that accompanies this approach is the high cost of both success and failure. The high cost of success is measured by the waste of raw materials caused by overdesigning; the high cost of failure is usually represented in newspaper headlines.

Experimental stress analysis has one great claim to fame: it provides a *nondestructive* and *nondisruptive* method of examining a part in actual service.

17-1 The Bonded Resistance Strain Gage

One of the most versatile tools available for the measurement of strain is the bonded resistance strain gage, Fig. 17-1, a device which undergoes a minute change in electrical resistance as its length changes. The physical bases of the gage were observed by Lord Kelvin over 100 years ago; it took some 70 years before ingenious thinking put the concept to work in the form of the strain gage.

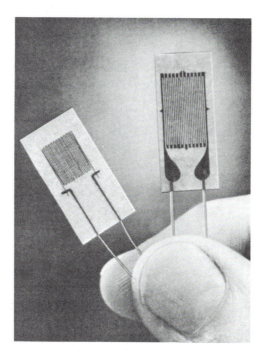

Figure 17-1. Wire filament and foil filament strain gages. Courtesy Baldwin-Lima-Hamilton Corporation.

There are two essential and inseparable parts to a bonded resistance strain gage: a resistive filament, which may take the form of a fine wire or thin foil, and a backing or carrier to support and electrically insulate the filament.

Gages are available in a variety of shapes and sizes designed to meet specific requirements. There are *general-purpose gages*, Fig. 17-2, with grid

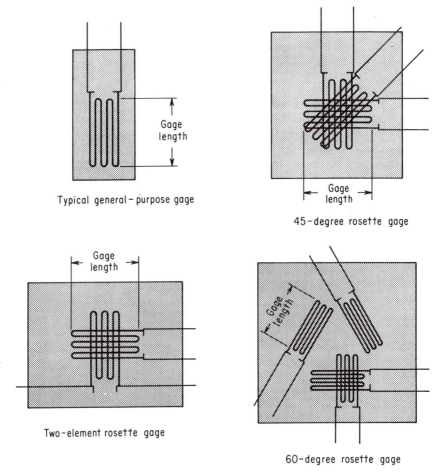

Figure 17-2. *Typical general-purpose bonded wire strain gages. Courtesy Baldwin-Lima-Hamilton Corporation.*

lengths from $\frac{3}{8}$ to 6 in., and *rosettes*, which consist of two or more gages mounted on a single backing. Etched-foil gages, a rather recent development, are also available in a variety of shapes and sizes, some as small as $\frac{1}{64}$-in. gage length. These appear as shown in Fig. 17-3.

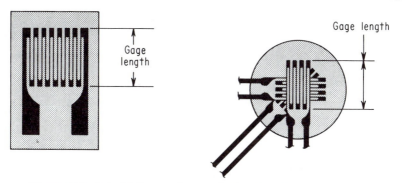

Figure 17-3. Etched-foil gages. Courtesy Baldwin-Lima-Hamilton Corporation.

A number of different materials are used as carriers of the resistive filaments, and the selection of a backing depends upon the temperature at which the gage is to be used. Paper- or epoxy-backed gages are limited to strain measurements below 180°F, whereas Bakelite-backed gages can be used successfully up to 450°F. At elevated temperatures ceramics are used to hold and insulate the filaments. Strain gages are sensitive devices which can give true measures of strain only if properly bonded to a clean, scale-free surface. The backing material used in the gage generally dictates the appropriate cement and cementing procedure to use. For paper-backed gages, Duco

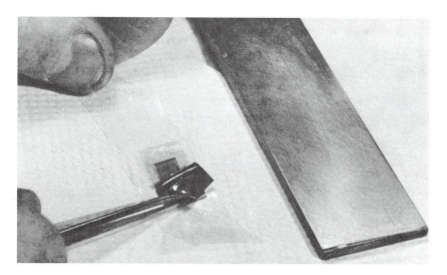

(a) Gage has been picked up with transparent tape, plastic protective backing removed, and cement catalyst applied to gage.

Figure 17-4. Rapid method of installing Budd Metal Film strain gage with catalytic adhesives. Courtesy Instruments Division of the Budd Company.

Household Cement is universally used, and Bakelite cements (phenol-resins) are used with Bakelite-backed gages.

There are remarkable adhesives now available which transform, with the addition of catalysts and moderate pressure, from free-flowing liquids to rigid plastics. These adhesives, which are particularly compatible with

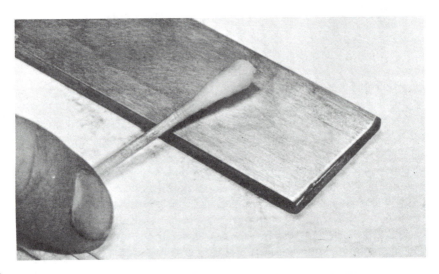

(b) Mounting surface has been cleaned and neutralized. Meanwhile catalyst has dried.

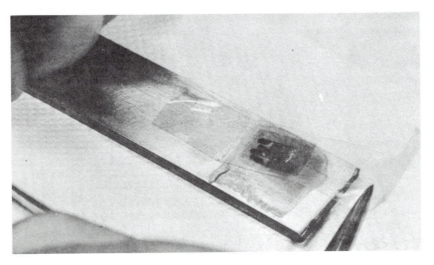

(c) Gage has been oriented on surface applied with cement and smoothed out.

(d) While cement sets, lead wires have been tinned. Transparent tape has been peeled from gage.

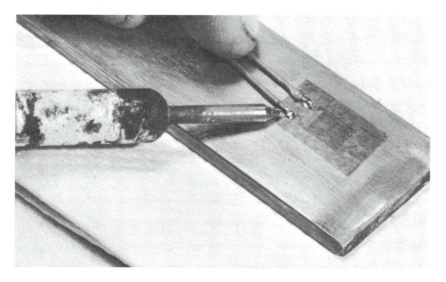

(e) Lead wires have been attached and supported and water-proofing compound applied.

epoxy-backed strain gages, provide a simple and rapid method of gage installation. The five simple steps outlined in Fig. 17-4 illustrate the bonding procedure generally used with these adhesives; Fig. 17-5 shows two typical applications of the epoxy-backed foil gage in the automotive field.

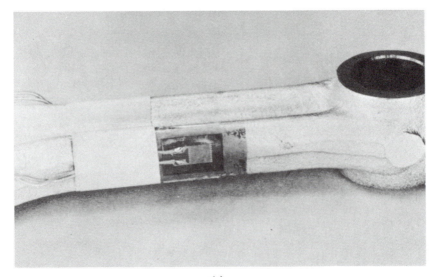

(a)

(b)

Figure 17-5. *Application of etched-foil gages for the analysis of strain in (a) an automotive connecting rod, (b) a critical section of a crankshaft journal. Courtesy General Motors Engineering Staff.*

440

17-2 Strain-Gage Instrumentation

A relationship between two physical quantities, resistance and length, forms the basis of strain-gage analysis. The sensitivity of a strain gage is dependent upon its *gage factor G*, a constant defined by the ratio

$$G = \frac{\frac{\Delta R}{R}}{\frac{\Delta L}{L}} \tag{17-1}$$

where R and L are the initial values of resistance and length; ΔR and ΔL are the change in electrical resistance and the accompanying change in length, respectively. Gage factor, initial resistance, and initial length are fixed quantities, established in the manufacture of the gage; the variables in Eq. (17-1) are the change in resistance ΔR and the change in length ΔL. Many American-made gages have nominal resistances of $R = 120$ ohms and gage factors of $G = 2$. The magnitude of ΔR in terms of strain can be found by substituting these two numerical values into Eq. (17-1):

$$\Delta R = G \times \frac{\Delta L}{L} \times R$$
$$\Delta = GR\epsilon$$
$$= 2 \times 120\epsilon$$

For these particular constants, a strain ϵ of 1 microinch (one-millionth of an inch) would be equivalent to a change in resistance of

$$\Delta R = 240 \times 10^{-6} = 0.000240 \ \Omega$$

hardly a very large value, yet a measurable one.

A basic circuit capable of detecting and measuring these small change in resistance is the *Wheatstone bridge:* for the most part, instrumentation, both

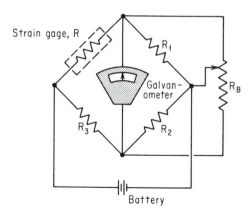

Figure 17-6   Battery

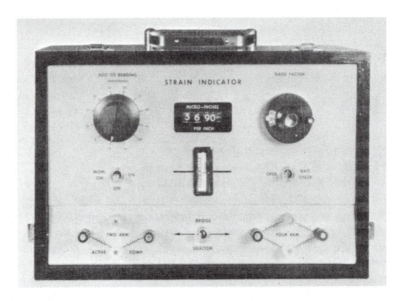

Figure 17-7. *Baldwin Model 120 strain indicator. This instrument, which features digital readout, may be used with strain gages in one-, two-, or four-arm networks. Dynamic strains up to 50 cps and 5000 microinches per inch may be seen by coupling the indicator to a standard cathode-ray oscilloscope. Courtesy Baldwin-Lima-Hamilton Corporation.*

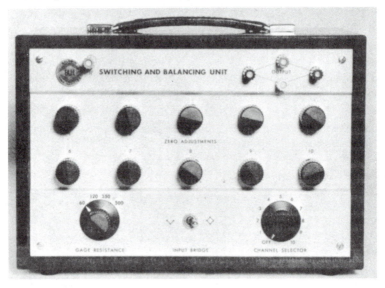

Figure 17-8. *Baldwin Model 225 Switching and Balancing Unit. The unit provides a means for initially balancing each of several bridges to zero and quickly switching active and compensating gages into a strain indicator. Courtesy Baldwin-Lima-Hamilton Corporation.*

442

simple and elaborate, makes use of this circuit. A diagram of a typical arrangement is shown in Fig. 17-6. The strain gage, suitably mounted on a structure to be tested, is used as one leg of the bridge. Fixed resistors R_1, R_2, and R_3, which might be non-acting strain gages, are used as the remaining three legs. A galvanometer, power supply, and balancing resistor R_B complete the circuit. R_B is a very high resistance, usually 50,000 or 100,000 Ω, used to establish a *zero* or *null* balance of the galvanometer. Changes in resistance of the strain gage appear in the circuit as a galvanometer deflection. If the balancing resistor is calibrated, a measured dial rotation can restore the galvanometer to null balance, and strain can be read directly.

Since most materials expand when heated, strain gages are extremely sensitive to temperature variation. A change of one degree Fahrenheit in

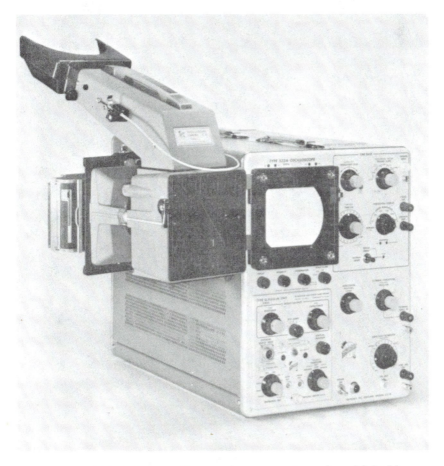

Figure 17-9. Tektronix oscilloscope, strain gage plug-in unit, and Polaroid Land Camera. The unit is self-contained and requires no external equipment other than strain gages. Courtesy Tektronix, Inc.

steel, for example, is equivalent to 6.5 microinches of strain, which, in turn, could be misinterpreted as a stress of 195 psi:

$$\sigma = \epsilon E = (6.5 \times 10^{-6})(30 \times 10^6) = 195 \text{ psi}$$

To eliminate temperature errors, a *dummy*, or *compensating gage*, alike in all respects to the measuring gage, is mounted on a material similar to that to be tested. This gage replaces R_3 in the circuit of Fig. 17-6, the hope being that the temperature of the measuring gage and the dummy gage will undergo equal changes and, therefore, will balance out expansion and contraction strains.

Figure 17-10. *Sanborn Channel 2 Carrier Recorder. This instrument provides the power for two separate strain-gage bridges and records the outputs of these bridges. Courtesy Sanborn Company.*

Instruments are available, like the strain indicator shown in Fig. 17-7, which feature *digital readout* of strain in microinches per inch. This instrument can also be used with a standard cathode-ray oscilloscope, and dynamic strains up to 50 cps and 5000 microinches per inch can be seen and measured. A switching and balancing network, like that shown in Fig. 17-8, allows as many as ten gages to be read with a single strain indicator.

A variety of strain-gage instrumentation is used for amplifying and recording strain measurement; oscilloscopes with camera attachments, like that shown in Fig. 17-9, are capable of recording high-frequency dynamic strain. Oscillographs, Fig. 17-10, are available which will simultaneously record the output of several gages.

17-3 Interpretation of Strain-Gage Data

For simple stress, like that in Fig. 17-11(a), a single gage mounted in the direction of the load is used. Strain readings are readily converted to stress by multiplying the indicated strain by the modulus of elasticity.

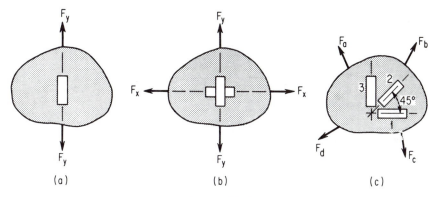

Figure 17-11

$$\sigma = \epsilon E \tag{17-2}$$

For biaxial stress, Fig. 17-11(b), principle strains in terms of principal stresses are

$$\epsilon_x = \frac{\sigma_x}{E} - \mu\frac{\sigma_y}{E}$$

and $\tag{17-3}$

$$\epsilon_y = \frac{\sigma_y}{E} - \mu\frac{\sigma_x}{E}$$

where μ is Poisson's ratio.

Since the measured quantity is strain, it is convenient to express stress in terms of strain in Eq. (17-3). By solving simultaneously for σ_x and σ_y,

Eq. (17-3) can be rewritten as

$$\sigma_x = \frac{E}{1 - \mu^2}(\epsilon_x + \mu\epsilon_y)$$

$$\sigma_y = \frac{E}{1 - \mu^2}(\epsilon_y + \mu\epsilon_x)$$

(17-4)

Numerical values of measured strain ϵ_x and ϵ_y can be substituted into Eq. (17-4), and the values of stress can be computed.

When directions of principal stresses are not known, as is frequently the case, the problem becomes somewhat more complicated. A minimum of three strain readings in known directions is required for the analysis. A variety of rosette gages are available for this purpose, and each has a characteristic set of equations for stress based on indicated strain. Principal stresses in terms of indicated strains ϵ_1, ϵ_2, and ϵ_3 for a rectangular rosette, Fig. 17-11(c), are

$$\sigma_{\substack{max \\ min}} = \frac{E}{2}\left\{\frac{\epsilon_1 + \epsilon_3}{1 - \mu} \pm \frac{1}{1 + \mu}\sqrt{(\epsilon_1 - \epsilon_3)^2 + [2\epsilon_2 - (\epsilon_1 - \epsilon_3)]^2}\right\}$$

$$\tau_{max} = \frac{E}{2(1 + \mu)}\sqrt{(\epsilon_1 - \epsilon_3)^2 + [2\epsilon_2 - (\epsilon_1 + \epsilon_3)]^2}$$

(17-5)

θ_p (the angle from gage 1 to σ_{max} axis) is equal to

$$\frac{1}{2}\tan^{-1}\left[\frac{2\epsilon_2 - (\epsilon_1 + \epsilon_3)}{\epsilon_1 - \epsilon_3}\right]$$

(17-6)

The slide rule, scratch paper, and eraser approach to these equations can be rather tedious; fortunately, however, electronic computers can be programmed to handle hundreds of these and similar calculations in a matter of seconds.

17-4 Brittle Coatings

The crazing and cracking in the glazes on antique pottery is caused by unequal expansion rates. The glaze, which is a thin, brittle, glass-like substance with little tensile strength, cracks as the body of the pottery expands with temperature. A simple observation like this led to the brittle coating techniques used in stress analysis. These methods involve coating a part to be studied with a brittle lacquer-like substance.[2] Calibration strips are always processed along with the part to be tested, and when dry, these strips are subjected to a known strain in a test fixture, like that shown in Fig. 17-12(a). The brittle

[2]These coatings are sold under the trade name of Stresscoat and are manufactured by Magnaflux Corporation, Chicago, Illinois.

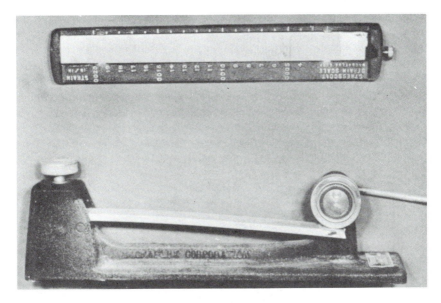

(a)

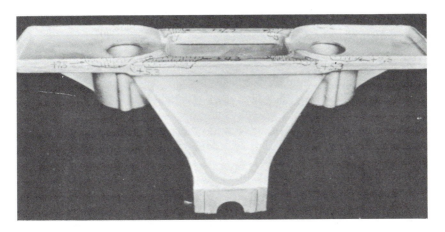

(b)

Figure 17-12. (a) *View of Stresscoat calibrator and strain scale, with test strip in place.* (b) *Stresscoat crack patterns on a heavily loaded aluminum beam saddle for the double rear-axle tandem suspension of a truck. Courtesy Magnaflux Corporation.*

coating fractures under the strain, and note is made of the minimum value of stress required to produce a failure in the coating.

Brittle coatings respond to both static and dynamic loading, and the usual procedure in static testing is to observe the start and the spread of the cracks as measured loads are applied. Since it is difficult to apply measured dynamic loads, several coatings, each with a different strain sensitivity, are used. This technique gives a quantitative picture of the magnitude as well as the distribution of strain.

With careful handling, brittle coatings are capable of a quantitative accuracy of 10 per cent. Where more precise results are desired, strain gages can be applied at critical regions indicated by the coating.

Ceramic-base coatings, a rather recent development, are available for the analysis of strain at temperatures up to 700°F. These coatings are, in reality, glazes suspended in a vehicle that is sprayed on the part, dried, and then fired at 1000°F, a process similar to porcelain enameling. Success of the ceramic coating depends upon a controlled difference between the coefficient of expansion of the coating and that of the base metal. If the coefficient of expansion of the coating is greater than that of the base metal, residual tension will be *locked in* the coating when it cools from the firing temperature. This develops a coating which cracks at low strain levels. If, on the other hand, the coefficient of expansion of the coating is less than that of the base metal, residual compression will be locked in the coating, which then requires higher strain levels to initiate crack patterns.

17-5 Photoelastic Stress Analysis

The photoelastic approach requires exacting techniques and is the most mathematical of the methods discussed. Basis of the method is the fact that certain transparent materials, like glass and Bakelite, become *birefringent*, or *doubly refracting*, when stressed. This simply means that these substances will divide an incident ray of light into two beams, which travel at different velocities through the material. In addition, the two beams transmitted

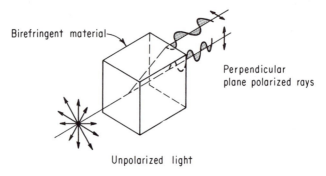

Birefringent material

Perpendicular
plane polarized rays

Unpolarized light

Figure 17-13

through the stressed body are *polarized* at right angles to one another, as shown in Fig. 17-13.

In photoelastic stress analysis, polarized light is used to reveal the presence of strains in a transparent *model* of a part to be studied. The light is split into two components, which vibrate in the directions of the principal stresses. A second polarizing screen, called an *analyzer*, permits the passage of only the component of light in its plane of polarization. A pattern of alternate bright and dark lines indicates the variation in refraction produced in the strained model. Figure 17-14 is a photoelastic stress pattern of an "eye-bar" in tension.

Figure 17-14. Photoelastic stress pattern of an "eyebar" in tension. Courtesy Professor W. W. Murray, M.I.T.

The dark bands, called *stress fringes*, or *isochromatics*, pass through points of constant difference in principal stress. A variety of techniques, some borrowed from the field of physical optics, some developed especially for photoelasticity, are used to translate fringe patterns into stress magnitudes. In general, optical observation, mechanical measurements, and mathematical methods are involved in the interpretation of data. Also involved is the mechanical ability to machine accurate, stress-free models and calibration bars.

The photoleastic method has made an enormous contribution to the theories of mechanics of materials, and in particular to the development of stress-concentration factors. It does not, however, easily lend itself as a method to be used by the untrained or novice in the field of stress analysis.[3]

17-6 Photoelastic Coatings

New techniques have recently been developed in which the *actual structure* to be stress-analyzed is covered with a photoelastic coating. When loads are applied to the structure, strains are transmitted to the plastic coating, which then becomes birefringent. The strain distribution, which is visible when the plastic is illuminated with suitable polarizing instruments, appears as a colored fringe pattern. These fringes, called *isochromatics*, are the loci of points where the difference in principal strains is constant; an area of uniform color denotes an area of uniform shear strain. Since the *birefringence* (or color) is directly proportional to the difference in principal strains, the method can produce quantitative results. A typical fringe pattern on an actual part is shown in Fig. 17-15.

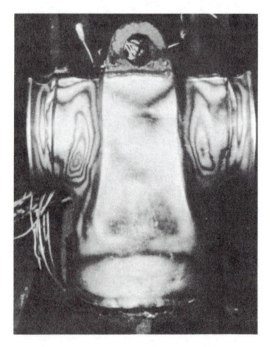

Figure 17-15. Photostress pattern on a portion of the main landing gear of a supersonic bomber. High stress concentrations, particularly in the fillet area, are indicated by the close fringe pattern. Courtesy Instruments Division of the Budd Company.

[3] For a more detailed, nontechnical discussion of the photoelastic method, see *Photoelastic Stress Analysis*, Eastman Kodak Co.

The photoelastic coating, in effect, acts as an infinite number of strain gages uniformly distributed over the surface being studied. In many respects, the technique combines the functions of photoelasticity, brittle lacquers, and electrical strain gages. The process does have its limitations, however, since stresses cannot be measured in areas of an assembly which are inaccessible to light. Correction factors must also be introduced for measurement at temperatures above 100°F, and errors in measurement are inherent when the thickness of the plastic is comparable to the thickness of the part under study.

The photoelastic coating, sold under the trade name of Photostress,[4] can be applied in either of two ways, depending upon the contour of the structure. If flat surfaces are to be studied, the plastic in sheet form can be bonded directly to the part. When the surfaces to be analyzed are curved, the plastic can be brushed on and allowed to polymerize directly on the part. An alternate method involves the pouring of a flat sheet of plastic and allowing it to polymerize partially; the flexible sheet can then be formed about the complex surface and allowed to cure.

17-7 The Measurement of Fatigue Strength

At one time or another everyone has bent a paper clip, a wire, or a nail back and forth until it broke. The mechanism of failure, called *fatigue*, is the result of a cyclic reversal of stress from tension to compression. The exact mechanism of fatigue failure is not completely understood; it seems reasonable, however, to assume the failure to progress in three stages: (1) the start of a surface crack; (2) the propagation of the crack under repeated stress; (3) final rupture of the weakened section. There is little argument concerning the second and third stages; the question debated is how a crack can start under cyclic stresses lower than the static rupture stress. Attempts have been made to correlate fatigue characteristics of materials with other engineering properties, such as tensile strength, yield strength, impact strength, and so on; for the most part, these attempts have been unsuccessful. Some general observations, based on a mountainous amount of experimental data, indicate that (a) most structural materials, like metals, wood, and some plastics, will fail by cracking under cyclic loads at stresses lower than the ultimate tensile stress; (b) fatigue in metals generally depends upon the number of cycles of load change in a given stress range rather than the rate of loading; (c) most metals have a safe range of stress, called the *fatigue limit*, or *endurance limit*, below which failure will not occur even after a large number of cycles; (d) notches, grooves, stamp marks, and machining marks appreciably decrease the fatigue strength.

[4]Photostress, a photoelastic coating, is produced by Instruments Division of The Budd Company, Phoenixville, Pa.

The knowledge of fatigue characteristics of materials is largely dependent upon experimental data obtained by repeated loading in either bending, tension, or torsion. Data are usually presented in graphic form in which fatigue life in terms of cycles N of load reversal is plotted against a stress σ to cause failure. Because of the large number of cycles involved, N is plotted on a logarithmic scale. The curves, referred to as S-N diagrams, appear as shown in Fig. 17-16. For most metals, a limiting stress exists, below which a material will endure what appears to be an infinite number of stress cycles. This stress is called the *fatigue limit*, or *endurance limit*.

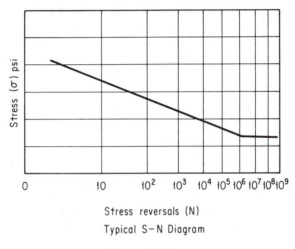

Stress reversals (N)

Typical S–N Diagram

Figure 17-16

The S-N diagram can only be of value in designing against fatigue if the part in question is always subjected to a constant fatigue cycle. This condition is rarely the case, since most structural members, for instance, the automobile spring, are stressed at a variety of levels under normal operations; this variety of levels has a decided effect on fatigue life. Experience and experimentation have disclosed that the two most important factors affecting fatigue life are surface roughness and geometric discontinuity. Carefully finished surfaces and liberal fillets are common features of a good design when dynamic loads are involved.

REVIEW
PROBLEMS

Each problem in this section uses one or more of the basic concepts discussed in the text. No attempt has been made to present this review in a preferred order of topic sequence or difficulty, but rather to present the subjects of statics and strength of materials in its most general form.

More than twenty state engineering-registration boards have generously supplied the problems in this section. Therefore, the problems represent the types of questions which will probably be encountered in the portion of the examination devoted to statics and strength of materials.

A-1. A wood-stave tank filled with water is bound together with threaded steel rods that are $\frac{3}{4}$ in. in diameter, as shown on p. 454. If the allowable tensile stress for the steel is 20 ksi, how many rods are required to resist the water pressure?

Ans. 12 rods.

A-2. Two plates are welded together, as shown on p. 454, to carry a concrete-block wall over a doorway. If the simple effective span length is 6 ft and the load per

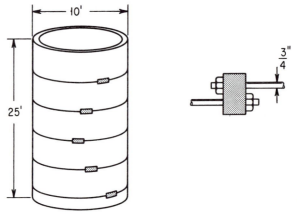

Problem A-1

foot is 400 lb, find: (a) the maximum tensile stress, (b) the maximum compressive stress, (c) the shearing stress between the plates.

Ans. (a) 1700 psi; (b) 5270 psi; (c) 664 psi.

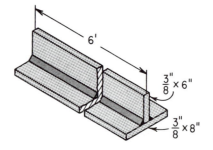

Problem A-2

A-3. A perfectly straight column, rigidly fixed at the bottom and free at the top, must support a concentric load of 35 kips. The column is a 4-in.-o.d. steel pipe with $\frac{1}{4}$-in.-thick walls. The steel has a modulus of elasticity of 30×10^6 psi and an elastic limit of 40,000 psi. How long can the column be and still have a factor of safety against buckling of 1.5?

Ans. 85 in.

A-4. Two 4-in.-diameter solid shafts are connected through a coupling with six $\frac{3}{4}$-in. bolts located on a 10-in. bolt circle. If the shafts rotate at 150 rpm, what horsepower can be transmitted without exceeding a shearing stress of 10,000 psi in the bolts or in the shafts?

Ans. 299 hp.

A-5. A block of wood and a block of aluminum, each 2 in. by 4 in. by 8 in., support a 20-lb beam and a concentrated load of 500 lb, as shown. If the

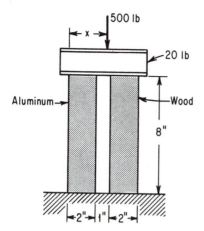

Problem A-5

beam is to remain rigid and level after the 500-lb load is applied, where must the load be placed? $E_a = 10 \times 10^6$ psi; $E_w = 2 \times 10^6$ psi.

Ans. 1.46 in.

A-6. A timber cantilever beam, 10 in. wide by 2 in. deep and 10 ft long, supports a load of 150 lb at a point 3 ft from the free end. Find: (a) the slope at the free end, (b) the deflection at the free end, (c) the maximum bending stress. $E = 1.5 \times 10^6$ psi.

Ans. (a) 0.053 rad; (b) 4.86 in.; (c) 1890 psi.

A-7. A short timber compression member has a cross-section of 6 in. by 8 in. It is reinforced by the addition of two $\frac{1}{2}$-in. by 6-in. steel plates placed to form a composite post. What is the maximum load this member can carry? The limiting stresses for timber and steel are 1000 psi and 18,000 psi, respectively. $E_t = 1.5 \times 10^6$ psi; $E_s = 30 \times 10^6$ psi.

Ans. 151,000 lb.

A-8. A circular open-link chain is made of a $\frac{3}{8}$-in.-round bar, and the links have an outside diameter of $1\frac{3}{4}$ in. If the maximum tensile stress is limited to 30,000 psi, what is the greatest load that can be safely supported by the chain?

Ans. 211 lb.

A-9. What size circular steel shaft can transmit 100 hp at 1000 rpm with an allowable extreme fiber stress in shear of 10,000 psi? The angle of twist must not exceed 1 degree per foot of length.

Ans. 1.48 in.; use 1.50 in. material.

A-10. A square steel bar of 1 in. by 1 in. cross-section and 6 ft long is to be used as a column. The ends are perfectly free to rotate, but may not be displaced.

What is the maximum load this column can support if the allowable stress is 30,000 psi?

Ans. 4770 lb.

A-11. A steel punch has a diameter of 0.750 in. When a hole is punched in a plate, a total compressive force of 35,000 lb acts on the punch. What is the actual diameter when the load is applied? $\mu = 0.25$.

Ans. 0.7505 in.

A-12. A cantilever beam, 8 ft long, is fixed at the right end. It carries a uniformly distributed load of 100 lb per ft, including its own weight, and a concentrated load of 1000 lb at the free end. Write the bending-moment equation for any point along the beam as a function of the length x measured from the free end.

Ans. $M = -1000x - 50x^2$.

A-13. An 8-in., 40-lb, wide-flange beam, used as a column, is 30 ft long. It is supported at the middle in a direction normal to the web but is unsupported in the direction parallel to the web. Find the safe load for this column by means of Euler's equation. Assume a factor of safety of 3 and "round-end" conditions. $\sigma_{pl} = 30,000$ psi.

Ans. 111,500 lb.

A-14. A vertical, cylindrical steel standpipe is 20 ft in diameter and 50 ft high. Compute the maximum circumferential stress exerted within the walls when the tank is filled with water. The walls are $\frac{3}{8}$ in. thick.

Ans. 6930 psi.

A-15. An aluminum bar, 10 in. long, is placed on a 12-in.-long steel bar, as shown. Both bars have the same cross-section. A gap of 0.01 in. exists between the top of the aluminum bar and the rigid support. What stress is produced in the aluminum when the temperature is raised by 100°F?

Ans. 7360 psi.

A-16. An I-beam that is 14 in. deep and 20 ft long is simply supported at its ends. The beam has a moment of inertia of 440 in.^{4}. (a) What load may be placed at midspan if the deflection is limited to $\frac{1}{4}$ in.? (b) What is the maximum bending stress for this value of load? (c) What distributed load would produce the same deflection? (d) What is the maximum bending stress associated with the distributed load of part (c)? Neglect the weight of the beam.

Ans. (a) 11,500 lb; (b) 10,900 psi; (c) 917 lb/ft; (d) 8750 psi.

A-17. A 1-in.-diameter steel pipe that is 4 ft long acts as a spreader bar, as illustrated. What pull P may be applied through the cables and connectors? Use Euler's formula for pinned-end conditions and assume a factor of safety of 3. $I = 0.087$ in.^{4}; $A = 0.494$ in.^{2}.

Ans. 9310 lb.

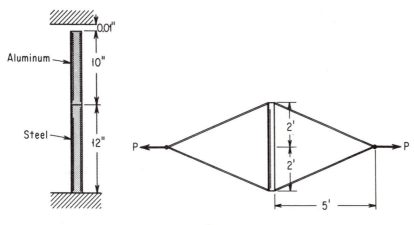

Problem A-15   **Problem A-17**

A-18. A short length of 7-in.-diameter steel propeller shaft, turning at 160 rpm, is subjected to an axial compressive load of 231 tons. What is the greatest horsepower that can be transmitted if the maximum shearing stress must not exceed 10,000 psi and the maximum normal stress must not exceed 13,500 psi?

Ans. 770 hp.

A-19. A 16-ft-long steel beam is fixed against rotation at one end and simply supported at the other. A load of 1000 lb per ft is uniformly distributed over the entire length of the beam. The moment of inertia of the beam section is 100 in.^{4}. What is the reaction at the simply supported end if the end settles $\frac{1}{2}$ in. when the load is applied?

Ans. 5680 lb.

A-20. Calculate the spacing of $\frac{7}{8}$-in.-diameter rivets required to resist a girder shear of 70 kips in the composite section shown on p. 458. The permissible shearing stress is 8000 psi.

Ans. 4.8 in.

A-21. Select the most economical wide-flange steel beam capable of supporting the loads shown on p. 458. The limiting bending stress is 20 ksi. The beam is braced, and it is not necessary to check for deflection or shear. Neglect the weight of the beam.

Ans. 14 WF 30.

A-22. A steel column section consists of a 12-in. by $\frac{1}{2}$-in. web plate, four 6-in. by 4-in. by $\frac{1}{2}$-in. angles, with the 4-in. leg attached to the web, and two 14-in. by $\frac{1}{2}$-in. cover plates. What is the smallest radius of gyration of the section?

Ans. 3.17 in.

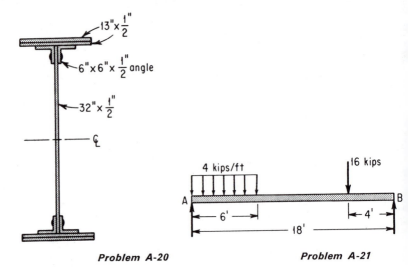

Problem A-20

Problem A-21

A-23. A 10-in., 25.4-lb standard I-beam is supported as shown. What is the deflection at *C*? Neglect the weight of the beam.

Ans. 0.018 in.

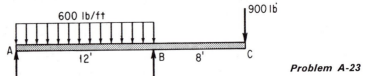

Problem A-23

A-24. If the allowable bending stress is 15,000 psi, what is the required section modulus for the beam illustrated?

Ans. 29.8 in.^{3}.

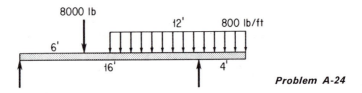

Problem A-24

A-25. A flanged and bolted coupling is to connect two solid shafts, each 3 in. in diameter. How many $\frac{3}{4}$-in. bolts must be used in a 6-in. bolt circle, if the shearing stresses in the shaft and in the bolts are not to exceed 12,000 psi and 8000 psi, respectively?

Ans. Six bolts.

A-26. The frame of the riveting machine illustrated is acted upon by the forces shown. Find the maximum tensile and compressive stresses at section *a-a*.

Ans. 4920 psi; −6600 psi.

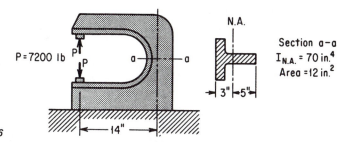

Problem A-26

A-27. The composite compression member shown in the figure consists of a block, 4 in. by 4 in. by 6 in., of each of the following materials: steel, copper, bronze, and aluminum. A load of 800,000 lb is applied in a way that causes each block to decrease an equal amount in height. Find the stress in each block. $E_s = 30 \times 10^6$ psi, $E_c = 15 \times 10^6$ psi, $E_b = 12 \times 10^6$ psi, $E_a = 10 \times 10^6$ psi.

Ans. $\sigma_s = 22,400$ psi; $\sigma_c = 11,200$ psi; $\sigma_b = 8960$ psi; $\sigma_a = 7470$ psi.

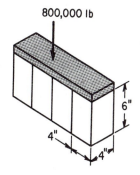

Problem A-27

A-28. A Douglas fir column, $4\frac{1}{2}$ in. by $9\frac{1}{2}$ in., with a modulus of $E = 1.6 \times 10^6$ psi, and a spruce column of the same size with a modulus of $E = 1.2 \times 10^6$ psi are bolted together to form a short column. If lateral bending is prevented, what portion of a load of 70,000 lb will each carry?

Ans. $F_{\text{fir}} = 40,000$ lb; $F_{\text{spruce}} = 30,000$ lb.

A-29. A load of 12 kips is supported on a bracket, which, in turn, is fastened to a column by a single vertical row of five rivets spaced on 3-in. centers. The load is 9 in. out from a vertical center line that passes through the rivets. What is the maximum shearing force developed in the rivets?

Ans. 7.59 kips.

A-30. The rungs of a ladder are $1\frac{1}{8}$ in. in diameter and are carried by side rails spaced a maximum of 12 in. apart. The limiting bending stress in the rungs is

2400 psi. What is the maximum centrally placed concentrated load that can be applied?

Ans. 224 lb.

A-31. A 6-in. by 12-in. Douglas fir beam (1600-psi-stress grade) is reinforced by the addition of two 12-in., 20.7-lb channels properly bolted to the wood beam. What is the relative strength of the reinforced beam to the original wood beam? The allowable stress in the steel is 20,000 psi, and the modulus of the wood is $E = 1.5 \times 10^6$ psi.

Ans. $M_{\text{reinforced}}/M_{\text{wood}} = 4.34$.

A-32. A series of Douglas fir joists, 16 ft long, carry a uniformly distributed load of 45 lb per ft^2 plus a cross-partition 4 ft from one end which carries a load of 300 lb per running ft. If the joists are spaced 16 in. on center, what size joists ($1\frac{5}{8}$ in. wide) are needed? The stresses are limited to 1600 psi in bending and 100 psi in shear.

Ans. $1\frac{5}{8} \times 8.82$ in. (use standard 2 in. $\times$ 10 in. timber).

A-33. A horizontal beam is 22 ft long. It is supported at one point that is 4 ft from the left end and at another that is 6 ft from the right. The beam supports a concentrated load of 6000 lb at its center and a distributed load of 1000 lb per ft from the left support to the extreme right end. Draw the shear and moment diagrams and determine maximum values in each.

Ans. $V_{\text{max}} = -11,000$ lb; $M_{\text{max}} = 24,500$ lb ft.

TERMS USED IN MECHANICS OF MATERIALS[1]

AGE HARDENING. A process of aging that increases hardness and strength and ordinarily decreases ductility. Age hardening usually follows rapid cooling or cold working.

AIR-HARDENING STEEL. An alloy steel that is hardened by cooling in air from a temperature higher than the transformation range. Also called *self-hardening steel*.

ALLOYING ELEMENTS. Chemical elements constituting an alloy; in steels, usually limited to the metallic elements added to modify the properties of the steel.

ANNEALING. A process involving heating and cooling, usually applied to induce softening. The term also refers to treatments intended to alter mechanical or physical properties, produce a definite microstructure, or remove gases. When applicable, the following more specific terms should be used:

[1]Reprinted from the Metals Handbook by permission of the American Society for Metals.

black annealing   isothermal annealing
blue annealing   malleablizing
box annealing   process annealing
bright annealing   spheroidizing
full annealing   stabilizing annealing
graphitizing

ARC WELDING. Welding accomplished by using an electric arc that may be formed between a metal or carbon electrode and the metal being welded; between two separate electrodes, as in *atomic hydrogen welding;* or between the two separate pieces being welded, as in *flash welding.*

AUSTEMPERING. A trade name for a patented heat treating process that consists in quenching a ferrous alloy from a temperature above the transformation range, in a medium having a rate of heat abstraction sufficiently high to prevent the formation of high-temperature transformation products; and in maintaining the alloy, until transformation is complete, at a temperature below that of pearlite formation and above that of martensite formation.

AUTOGENOUS WELDING. A method of uniting two pieces of metal by melting their edges together without solder or any added welding metal.

BASIC STEEL. Steel melted in a furnace that has a basic bottom and lining, and under a slag that is dominantly basic.

BEADING. Raising a ridge on sheet metal.

BEND TESTS. Various tests used to determine the ductility of sheet or plate that is subjected to bending. These tests may include determination of the minimum radius or diameter required to make a satisfactory bend and the number of repeated bends that the material can withstand without failure when it is bent through a given angle and over a definite radius.

BLANKING. Shearing out a piece of sheet metal in preparation for deep drawing.

BLUE ANNEALING. A process of softening ferrous alloys in the form of hot rolled sheet, by heating in the open furnace to a temperature within the transformation range and then cooling in air. The formation of a bluish oxide on the surface is incidental.

BOWING. Lack of flatness in sheet or strip metal in which the longitudinal or transverse section forms an arc.

BRITTLENESS. A tendency to fracture without appreciable deformation.

BUTT WELDING. Joining two edges or ends by placing one against the other and welding them.

CARBON STEEL. Steel that owes its properties chiefly to the presence of carbon, without substantial amounts of other alloying elements; also termed "ordinary steel," "straight carbon steel," "plain carbon steel."

CARBURIZING. A process that introduces carbon into a solid ferrous alloy by heating the metal in contact with a carbonaceous material—solid, liquid, or gas—to a temperature above the transformation range and holding at that temperature. Carburizing is generally followed by quenching to produce a hardened case.

CASE. In a ferrous alloy, the surface layer that has been made substantially harder than the interior or *core* by a process of *case hardening*.

CASE HARDENING. A process of hardening a ferrous alloy so that the surface layer or *case* is made substantially harder than the interior or *core*. Typical case-hardening processes are *carburizing* and *quenching, cyaniding, carbonitriding, nitriding, induction hardening*, and *flame hardening*.

CENTRIFUGAL CASTING. A casting technique in which the mold is rotated during solidification of the casting. Unusually sound castings may be produced by the action of centrifugal force pressing toward the periphery.

CHARPY TEST. A pendulum type of impace test in which a specimen, supported at both ends as a simple beam, is broken by the impact of the falling pendulum. The energy absorbed in breaking the specimen, as determined by the decreased rise of the pendulum, is a measure of the impact strength of the metal.

CHIPPING. A method for removing seams and other surface defects with chisel or gouge so that such defects will not be worked into the finished product. Chipping is often employed also to remove metal that is excessive but not defective. Removal of defects by gas cutting is known as "deseaming" or "scarfing."

COERCIVE FORCE. The magnetizing force that must be applied in the direction opposite to that of the previous magnetizing force in order to remove residual magnetism; thus, an indicator of the "strength" of magnetically hard materials.

COHESIVE STRENGTH. A term used with one of the following meanings: (1) The maximum stress required in order to cause tensile fracture in the absence of any deformation, when two of the three principal stresses equal zero; that is, with an unnotched bar. Sometimes called "initial cohesive strength." (2) The maximum principal stress required in order to cause tensile fracture when triaxial stresses are present; that is, by using a notched test bar. This is frequently called the "technical cohesive strength" and is variable, depending on the relative magnitude of the three principal stresses, the amount of plastic deformation preceding fracture, and the temperature and rate of straining.

COINING. A process of impressing images or characters of the die and punch onto a plane metal surface.

COLD WORK. Plastic deformation at such temperatures and rates that substantial increases occur in the strength and hardness of the metal. Visible structural changes include changes in grain shape and, in some instances, mechanical twinning or banding.

COLD WORKING. Deforming a metal plastically at such a temperature and rate that strain hardening occurs. The upper limit of temperature for this process is the *recrystallization temperature*.

COMPRESSIVE STREINGTH. *Yield.* The maximum stress that a metal, subjected to compression, can withstand without a predefined amount of deformation. *Ultimate.* The maximum stress that a brittle material can withstand without fracture when subjected to compression.

CONSTANT-DEFLECTION TEST. A stress-corrosion-cracking or mechanical test in which the specimen is stressed by bending to a definite and constant curvature.

COOLING STRESSES. Stresses developed by uneven contraction or external constraint of metal during cooling; also those stresses resulting from localized plastic deformation during cooling, and retained.

CORROSION FATIGUE. The repeated cyclic stressing of a metal in a corrosive medium, resulting in more rapid deterioration of properties than would be encountered as a result of either cyclic stressing or of corrosion alone.

CREEP. The flow of plastic deformation of metals held for long periods of time at stresses lower than the normal yield strength. The effect is particularly important if the temperature of stressing is in the vicinity of the recrystallization temperature of the metal.

CREEP LIMIT. The maximum stress that will result in creep at a rate lower than an assigned rate.

CUPPING. The breaking of wire with a cup fracture accompanied by very little reduction of area; observed during cold drawing. Also the forming of sheet into cuplike objects such as shells, by deep drawing.

DAMPING CAPACITY. The ability of a metal to absorb vibrations, changing the mechanical energy into heat.

DEEP DRAWING. Forming cup-shaped articles or shells by using a punch to force sheet metal into a die.

DILATOMETER. An instrument for measuring the expansion or contraction caused in a metal by changes in temperature of structure.

DISRUPTIVE STRENGTH. The maximum strength of a metal when subjected to three principal tensile stresses at right angles to one another and of equal magnitude.

DUCTILITY. The property that permits permanent deformation before fracture by stress in tension.

ELASTIC AFTEREFFECT. A slight contraction that occurs slowly while metal is standing with no load, subsequent to plastic tensile flow and immediate elastic recovery. *Microscopic stresses*, acting in compression, are responsible for this as well as for the Bauschinger effect.

ELASTIC CONSTANTS. In the general form of Hooke's law, the elastic moduli, which vary in individual crystals with the direction of test.

ELASTIC DEFORMATION. Temporary changes caused in dimensions by stress. The material returns to the original dimensions after removal of the stress.

ELASTIC HYSTERESIS. Energy absorbed by reversed deformation, represented by the closed loop of stress-strain curves in the elastic range, formed by curves for loading and unloading.

ELASTIC LIMIT. The maximum stress that a material will withstand without permanent deformation. (Almost never determined experimentally; *yield strength* is customarily determined.)

ELONGATION. The amount of permanent extension in the vicinity of the fracture in the tension test; usually expressed as a percentage of the original gage length, as 25 per cent in 2 in. Elongation may also refer to the amount of extension at any stage in any process that elongates a body continuously, as in rolling.

ENDURANCE LIMIT. The maximum stress that a metal will withstand without failure during a specified large number of cycles of stress. If the term is employed without qualification, the cycles of stress are usually such as to produce complete reversal of flexural stress.

ENDURANCE RATIO. The ratio of the endurance limit for cycles of reversed flexural stress to the tensile strength.

FATIGUE. The tendency for a metal to break under conditions of repeated cyclic stressing considerably below the ultimate tensile strength.

FATIGUE CRACK OR FAILURE. A fracture starting from a nucleus where there is an abnormal concentration of cyclic stress and propagating through the metal. The surface is smooth and frequently shows concentric (sea shell) markings with a nucleus as a center.

FIBER STRESS. Local stress at a point or line on a section over which stress is not uniform, such as on the cross section of a beam under a bending load.

FLOW STRESS. The shear stress required to cause plastic deformation of solid metals.

FRACTURE STRESS. The maximum principal true stress (fracture load divided by fracture area).

FRACTURE TEST. Breaking a piece of metal for the purpose of examining the fractured surface to determine the structure of carbon content of the metal or to detect the presence of internal defects.

FREE MACHINING. The property that makes machining easy because of the forming of small chips, a characteristic imparted to steel by sulfur, to brass by lead, to aluminum alloys by lead and bismuth, to nickel alloys by sulfur or carbon, and so on.

GAS WELDING. Manual or automatic torch welding of metals, in which the heating is accomplished by an oxyacetylene or oxyhydrogen flame.

HARD DRAWN. A temper produced in wire, rod, or tube by cold drawing.

HARDENABILITY. In a ferrous alloy, the property that determines the depth and distribution of hardness induced by quenching.

HARDNESS. Defined in terms of the method of measurement. (1) Usually the resistance to indentation. (2) Stiffness or temper of wrought products. (3) Machinability characteristics.

HEAT TREATMENT. A combination of heating and cooling operations, timed and applied to a metal or alloy in the solid state in a way that will produce desired properties. Heating for the sole purpose of hot working is excluded from the meaning of this definition.

HOOKE'S LAW. Stress is proportional to strain in the elastic range.

HOT WORKING. Plastic deformation of metal at such a temperature and rate that strain hardening does not occur. The lower limit of temperature for this process is the recrystallization temperature.

IMPACT TEST. A test to determine the energy absorbed in fracturing a test bar at high velocity. The test may be in tension or in bending, or it may properly be a *notch* test if a notch is present, creating multiaxial stresses.

INDENTATION HARDNESS. The resistance of a material to indentation. This is the usual type of hardness test, in which a pointed or rounded indenter is pressed into a surface under a substantially static load.

INITIAL CREEP. The early part of the time-elongation curve for creep, in which extension increases at a rapid rate.

INTERNAL FRICTION. Ability of a metal to transform vibratory energy into heat. Internal friction generally refers to low stress levels of vibration; *damping* has a broader connotation since it may refer to stresses approaching or exceeding the yield strength.

IZOD TEST. A pendulum type of impact test, in which the specimen is supported at one end as a cantilever beam and the energy required to break off the free end is used as a measure of impact strength.

KNOOP HARDNESS. Microhardness determined from the resistance of the metal or of individual microconstituents to indentation by a diamond indenter making a rhombohedral impression with one long and one short diagonal.

LAP WELD. A term applied to a weld formed by lapping two pieces of metal and then pressing or hammering, and applied particularly to the longitudinal joint produced by a welding process for tubes or pipe, in which the edges of the skelp are beveled or scarfed so that when they are overlapped they can be welded together.

LONGITUDINAL DIRECTION. The direction in a wrought metal product parallel to direction of working (drawing, extruding, rolling).

MACROSCOPIC STRESSES. Residual stresses of such scope that relatively large areas of the material or the whole specimen are involved. These stresses are accompanied by strain measurable with ordinary extensometers under appropriate test conditions.

MALLEABILITY. The property that determines the ease of deforming a metal when the metal is subjected to rolling or hammering. The more malleable metals can be hammered or rolled into thin sheet more easily than others.

MECHANICAL TESTING. Methods of determining *mechanical properties.*

MECHANICAL WORKING. Subjecting metal to pressure exerted by rolls, dies, presses, or hammers, to change its form or to affect the structure and consequently the mechanical and physical properties.

MICROHARDNESS. The hardness of microscopic points within an alloy. The hardness of individual constituent particles and of localized areas of the solid solution matrix can be determined.

MODULUS OF ELASTICITY. The slope of the elastic portion of the stress-strain curve in mechanical testing. The stress is divided by the unit elongation. The tensile or compressive elastic modulus is called "Young's modulus"; the torsional elastic modulus is known as the "shear modulus" or "modulus of rigidity."

MODULUS OF RIGIDITY. In a torsion test, the ratio of the unit shear stress to the displacement caused by it per unit length in the elastic range. This modulus corresponds to the modulus of elasticity in the tension test.

MODULUS OF RUPTURE. The ultimate strength of the breaking load per unit area of a specimen tested in torsion or in bending (flexure). In tension, it is the tensile strength.

MODULUS OF STRAIN HARDENING. When the true stress (σ) is plotted against true strain (δ) in tensile testing, the slope of the tangent to the curve in the plastic range is sometimes called the "strain hardening modulus." A better expression would be "rate of strain hardening," defined as $d\sigma/d\delta$.

NECKING DOWN. Reduction in area concentrated at the subsequent fracture when a ductile metal is tested in tension.

NOTCH BRITTLENESS. Susceptibility of a material to brittleness in areas containing a groove, scratch, sharp fillet, or notch.

NOTCH FATIGUE FACTOR. The reduction caused in fatigue strength by the presence of a sharp notch in the stressed test section.

NOTCH SENSITIVITY. The reduction caused in nominal strength, impact or static, by the presence of a stress concentration, usually expressed as the ratio of the notched to the unnotched strength.

OLSEN DUCTILITY TEST. A cupping test in which a piece of sheet, restrained except at the center, is deformed by a standard steel punch with a hemispherical end. This test measures the depth of the impression required to fracture the metal.

OPERATING STRESS. The stress to which a structural unit is subjected during service.

OVERBENDING. Allowance for spring-back when bending metal to a desired angle.

OVERSTRESSING. Permanently deforming a metal by subjecting it to stresses that exceed the elastic limit.

PERCUSSION WELDING. An electric-resistance butt-welding process in which the weld area is upset or forged at the moment of welding.

PHYSICAL PROPERTIES. Those properties familiarly discussed in physics, exclusive of those described under *mechanical properties;* for example, density, electrical conductivity, coefficient of thermal expansion. This term has often been used to describe mechanical properties, but this usage is not recommended.

PHYSICAL TESTING. Testing methods by which physical properties are determined. This term is also used inadvisedly to mean the determination of the mechanical properties.

PLASTIC DEFORMATION. Permanent distortion of a material under the action of applied stresses.

PLASTICITY. The ability of a metal to be deformed extensively without rupture.

POISSON'S RATIO. The ratio of the transverse contraction of a strained test specimen to its longitudinal elongation; essentially a statement of constancy of volume during deformation.

PRINCIPAL STRESSES. Normal stresses along rectilinear co-ordinates that are so chosen in direction that shearing stresses are zero.

PROOF STRESS. In a test, stress that will cause a specified permanent deformation in a material, usually 0.01 per cent or less.

PROPORTIONAL LIMIT. The greatest stress that the material is capable of sustaining without a deviation from the law of proportionality of stress to strain (*Hooke's law*).

REDUCTION IN AREA. The difference between the original cross-sectional area and that of the smallest area at the point of rupture; usually stated as a percentage of the original area; also called "contraction of area."

RELAXATION. Relief of stress as the result of creep. Some types of tests are designed to provide diminution of stress by relaxation at constant strain, as frequently occurs in service.

RESIDUAL STRESS. Macroscopic stresses that are set up within a metal as the result of nonuniform plastic deformation. This deformation may be caused by cold working or by drastic gradients of temperature from quenching or welding.

RESILIENCE. The tendency of a material to return to its original shape after the removal of a stress that has produced elastic strain.

RESISTANCE WELDING. A type of welding process in which the work pieces are heated by the passage of an electric current through the contact. Such processes include *spot welding, seam* or *line welding*, and *percussion welding. Flash* and *butt welding* are sometimes considered as resistance welding processes.

RESOLVED SHEAR STRESS. The vectorial component of load acting on the shear plane, divided by the shear plane area.

RUPTURE STRESS. The stress given by dividing the load at the moment of incipient fracture, by the area supporting that load.

SCLEROSCOPE TEST. A hardness test where the loss in kinetic energy of a falling metal "tup," absorbed by indentation upon impact of the tup on the metal being tested, is indicated by the height of rebound.

SEAM WELDING. An electric-resistance type of welding process, in which the lapped sheet is passed between electrodes of the roller type while a series of overlapping *spot welds* is made by the intermittent application of electric current.

SECONDARY CREEP. The second portion of the creep curve following the *initial creep* stage and in which the rate of creep has reached a rather constant value.

SHEAR. (1) A type of cutting operation in which the metal object (sheet, wire, rod, or such) is cut by means of a moving blade and fixed edge or by a pair of moving blades that may be either flat or curved. (2) A type of deformation in which parallel planes in metal crystals slide so as to retain their parallel relation to one another, resulting in block movement.

SHEAR MODULUS. See *modulus of rigidity*.

SHEAR STRESS. The component of stress acting on the plane of shear.

SHIELDED-ARC WELDING. Electric-arc welding in which the metal is protected from the air atmosphere. An inert gaseous atmosphere may be used or *flux-coated electrodes*.

SLENDERNESS RATIO. Length of a test specimen divided by the square root of the cross-sectional area.

SPOT WELDING. An electric-resistance welding process in which the fusion is limited to a small area. The pieces being welded are pressed together between a pair of water-cooled electrodes through which an electrical current is passed during a very short interval so that fusion occurs over a small area at the interface between the pieces.

STRAIN. Deformation expressed as a pure number or ratio. (1) Ordinarily expressed as epsilon (ϵ), equivalent to the change in length divided by original length. (2) *True* strain (δ or ϵ) is the logarithm of the ratio of the length at the moment of observation, to the original length. True strain δ does not differ much from ϵ until above 20 per cent.

STRAIN HARDENING. An increase in hardness and strength caused by plastic deformation at temperatures lower than the recrystalization range.

STRESS. The load per unit of area. Ordinarily stress-strain curves do not show the true stress (load divided by area at that moment) but a fictitious value obtained by using always the original area.

STRESS RELIEVING. A process of reducing residual stresses in a metal object by heating the object to a suitable temperature and holding for a sufficient time. This treatment may be applied to relieve stresses induced by casting, quenching, normalizing, machining, cold working, or welding.

TANGENT MODULUS. The slope of the stress-strain curve of a metal at any point along the curve in the plastic region. In the elastic region the tangent modulus is equivalent to *Young's modulus*.

TENSILE STRENGTH. The value obtained by dividing the maximum load observed during tensile straining by the specimen cross-sectional area before straining. Also called "ultimate strength."

THERMAL STRESSES. Stresses in metal, resulting from non-uniform distribution of temperature.

TORSION. Strain created in a material by a twisting action. Correspondingly, the stress within the material resisting the twisting.

TOUGHNESS. Property of absorbing considerable energy before fracture; usually represented by the area under a stress-strain curve, and therefore involving both ductility and strength.

TUKON HARDNESS TEST. A method for determining microhardness by using a Knoop diamond indenter or Vickers square-base pyramid indenter.

ULTIMATE STRENGTH. See *tensile strength*.

VICKERS HARDNESS TEST. An indentation hardness test employing a 136-degree diamond pyramid indenter and variable loads enabling the use of one hardness scale from very soft lead to tungsten carbide.

WELD BEAD. The built-up portion of a fusion weld, formed either from the filler metal or from the melting of the parent metal.

WELDING. A process used to join metals by the application of heat. *Fusion welding*, which includes *gas*, *arc*, and *resistance welding*, requires that the parent metals be melted. This distinguishes fusion welding from brazing. In *pressure welding* joining is accomplished by the use of heat and pressure without melting. The parts that are being welded are pressed together and heated simultaneously, so that recrystallization occurs across the interface.

WELDING STRESS. The stress resulting from localized heating and cooling of metal during welding.

WORK HARDNESS. Hardness developed in metal as a result of cold working.

YIELD STRENGTH. The stress at which a material exhibits a specified limiting deviation from proportionality of stress to strain. An offset of 0.2 per cent is used for many metals such as aluminim-base and magnesium-base alloys, while a 0.5 per cent total elongation under load is frequently used for copper alloys.

YOUNG'S MODULUS. The *modulus of elasticity* in tension of compression.

Appendix **C**

TABLES

1. Natural Trigonometric Functions
2. Exponentials
3. Typical Physical Properties of Metals
4. Typical Properties of Structural Timber
5. American Standard Steel I-Beams, Properties for Designing
6. Steel Wide-Flange Beams, Properties for Designing
7. American Standard Steel Channels, Properties for Designing
8. Steel Angles with Equal Legs, Properties for Designing
9. Steel Angles with Unequal Legs, Properties for Designing
10. Properties of Areas
11. Beam Deflection Equations

Tables 5 through 9 are taken from the *AISC Manual of Steel Construction* and are reproduced by permission of The American Institute of Steel Construction.

Table 1. Natural Trigonometric Functions

sin

	.0	.1	.2	.3	.4	.5	.6	.7	.8	.9	
0°	.0000	.0017	.0035	.0052	.0070	.0087	.0105	.0122	.0140	.0157	.0175	89°
1°	.0175	.0192	.0209	.0227	.0244	.0262	.0279	.0297	.0314	.0332	.0349	88°
2°	.0349	.0366	.0384	.0401	.0419	.0436	.0454	.0471	.0488	.0506	.0523	87°
3°	.0523	.0541	.0558	.0576	.0593	.0610	.0628	.0645	.0663	.0680	.0698	86°
4°	.0698	.0715	.0732	.0750	.0767	.0785	.0802	.0819	.0837	.0854	.0872	85°
5°	.0872	.0889	.0906	.0924	.0941	.0958	.0976	.0993	.1011	.1028	.1045	84°
6°	.1045	.1063	.1080	.1097	.1115	.1132	.1149	.1167	.1184	.1201	.1219	83°
7°	.1219	.1236	.1253	.1271	.1288	.1305	.1323	.1340	.1357	.1374	.1392	82°
8°	.1392	.1409	.1426	.1444	.1461	.1478	.1495	.1513	.1530	.1547	.1564	81°
9°	.1564	.1582	.1599	.1616	.1633	.1650	.1668	.1685	.1702	.1719	.1736	80°
10°	.1736	.1754	.1771	.1788	.1805	.1822	.1840	.1857	.1874	.1891	.1908	79°
11°	.1908	.1925	.1942	.1959	.1977	.1994	.2011	.2028	.2045	.2062	.2079	78°
12°	.2079	.2096	.2113	.2130	.2147	.2164	.2181	.2198	.2215	.2233	.2250	77°
13°	.2250	.2267	.2284	.2300	.2317	.2334	.2351	.2368	.2385	.2402	.2419	76°
14°	.2419	.2436	.2453	.2470	.2487	.2504	.2521	.2538	.2554	.2571	.2588	75°
15°	.2588	.2605	.2622	.2639	.2656	.2672	.2689	.2706	.2723	.2740	.2756	74°
16°	.2756	.2773	.2790	.2807	.2823	.2840	.2857	.2874	.2890	.2907	.2924	73°
17°	.2924	.2940	.2957	.2974	.2990	.3007	.3024	.3040	.3057	.3074	.3090	72°
18°	.3090	.3107	.3123	.3140	.3156	.3173	.3190	.3206	.3223	.3239	.3256	71°
19°	.3256	.3272	.3289	.3305	.3322	.3338	.3355	.3371	.3387	.3404	.3420	70°
20°	.3420	.3437	.3453	.3469	.3486	.3502	.3518	.3535	.3551	.3567	.3584	69°
21°	.3584	.3600	.3616	.3633	.3649	.3665	.3681	.3697	.3714	.3730	.3746	68°
22°	.3746	.3762	.3778	.3795	.3811	.3827	.3843	.3859	.3875	.3891	.3907	67°
23°	.3907	.3923	.3939	.3955	.3971	.3987	.4003	.4019	.4035	.4051	.4067	66°
24°	.4067	.4083	.4099	.4115	.4131	.4147	.4163	.4179	.4195	.4210	.4226	65°
25°	.4226	.4242	.4258	.4274	.4289	.4305	.4321	.4337	.4352	.4368	.4384	64°
26°	.4384	.4399	.4415	.4431	.4446	.4462	.4478	.4493	.4509	.4524	.4540	63°
27°	.4540	.4555	.4571	.4586	.4602	.4617	.4633	.4648	.4664	.4679	.4695	62°
28°	.4695	.4710	.4726	.4741	.4756	.4772	.4787	.4802	.4818	.4833	.4848	61°
29°	.4848	.4863	.4879	.4894	.4909	.4924	.4939	.4955	.4970	.4985	.5000	60°
30°	.5000	.5015	.5030	.5045	.5060	.5075	.5090	.5105	.5120	.5135	.5150	59°
31°	.5150	.5165	.5180	.5195	.5210	.5225	.5240	.5255	.5270	.5284	.5299	58°
32°	.5299	.5314	.5329	.5344	.5358	.5373	.5388	.5402	.5417	.5432	.5446	57°
33°	.5446	.5461	.5476	.5490	.5505	.5519	.5534	.5548	.5563	.5577	.5592	56°
34°	.5592	.5606	.5621	.5635	.5650	.5664	.5678	.5693	.5707	.5721	.5736	55°
35°	.5736	.5750	.5764	.5779	.5793	.5807	.5821	.5835	.5850	.5864	.5878	54°
36°	.5878	.5892	.5906	.5920	.5934	.5948	.5962	.5976	.5990	.6004	.6018	53°
37°	.6018	.6032	.6046	.6060	.6074	.6088	.6101	.6115	.6129	.6143	.6157	52°
38°	.6157	.6170	.6184	.6198	.6211	.6225	.6239	.6252	.6266	.6280	.6293	51°
39°	.6293	.6307	.6320	.6334	.6347	.6361	.6374	.6388	.6401	.6414	.6428	50°
40°	.6428	.6441	.6455	.6468	.6481	.6494	.6508	.6521	.6534	.6547	.6561	49°
41°	.6561	.6574	.6587	.6600	.6613	.6626	.6639	.6652	.6665	.6678	.6691	48°
42°	.6691	.6704	.6717	.6730	.6743	.6756	.6769	.6782	.6794	.6807	.6820	47°
43°	.6820	.6833	.6845	.6858	.6871	.6884	.6896	.6909	.6921	.6934	.6947	46°
44°	.6947	.6959	.6972	.6984	.6997	.7009	.7022	.7034	.7046	.7059	.7071	45°
	.9	.8	.7	.6	.5	.4	.3	.2	.1	.0	

cos

Table 1. Natural Trigonometric Functions

sin

	.0	.1	.2	.3	.4	.5	.6	.7	.8	.9	
45°	.7071	.7083	.7096	.7108	.7120	.7133	.7145	.7157	.7169	.7181	.7193	44°
46°	.7193	.7206	.7218	.7230	.7242	.7254	.7266	.7278	.7290	.7302	.7314	43°
47°	.7314	.7325	.7337	.7349	.7361	.7373	.7385	.7396	.7408	.7420	.7431	42°
48°	.7431	.7443	.7455	.7466	.7478	.7490	.7501	.7513	.7524	.7536	.7547	41°
49°	.7547	.7559	.7570	.7581	.7593	.7604	.7615	.7627	.7638	.7649	.7660	40°
50°	.7660	.7672	.7683	.7694	.7705	.7716	.7727	.7738	.7749	.7760	.7771	39°
51°	.7771	.7782	.7793	.7804	.7815	.7826	.7837	.7848	.7859	.7869	.7880	38°
52°	.7880	.7891	.7902	.7912	.7923	.7934	.7944	.7955	.7965	.7976	.7986	37°
53°	.7986	.7997	.8007	.8018	.8028	.8039	.8049	.8059	.8070	.8080	.8090	36°
54°	.8090	.8100	.8111	.8121	.8131	.8141	.8151	.8161	.8171	.8181	.8192	35°
55°	.8192	.8202	.8211	.8221	.8231	.8241	.8251	.8261	.8271	.8281	.8290	34°
56°	.8290	.8300	.8310	.8320	.8329	.8339	.8348	.8358	.8368	.8377	.8387	33°
57°	.8387	.8396	.8406	.8415	.8425	.8434	.8443	.8453	.8462	.8471	.8480	32°
58°	.8480	.8490	.8499	.8508	.8517	.8526	.8536	.8545	.8554	.8563	.8572	31°
59°	.8572	.8581	.8590	.8599	.8607	.8616	.8625	.8634	.8643	.8652	.8660	30°
60°	.8660	.8669	.8678	.8686	.8695	.8704	.8712	.8721	.8729	.8738	.8746	29°
61°	.8746	.8755	.8763	.8771	.8780	.8788	.8796	.8805	.8813	.8821	.8829	28°
62°	.8829	.8838	.8846	.8854	.8862	.8870	.8878	.8886	.8894	.8902	.8910	27°
63°	.8910	.8918	.8926	.8934	.8942	.8949	.8957	.8965	.8973	.8980	.8988	26°
64°	.8988	.8996	.9003	.9011	.9018	.9026	.9033	.9041	.9048	.9056	.9063	25°
65°	.9063	.9070	.9078	.9085	.9092	.9100	.9107	.9114	.9121	.9128	.9135	24°
66°	.9135	.9143	.9150	.9157	.9164	.9171	.9178	.9184	.9191	.9198	.9205	23°
67°	.9205	.9212	.9219	.9225	.9232	.9239	.9245	.9252	.9259	.9265	.9272	22°
68°	.9272	.9278	.9285	.9291	.9298	.9304	.9311	.9317	.9323	.9330	.9336	21°
69°	.9336	.9342	.9348	.9354	.9361	.9367	.9373	.9379	.9385	.9391	.9397	20°
70°	.9397	.9403	.9409	.9415	.9421	.9426	.9432	.9438	.9444	.9449	.9455	19°
71°	.9455	.9461	.9466	.9472	.9478	.9483	.9489	.9494	.9500	.9505	.9511	18°
72°	.9511	.9516	.9521	.9527	.9532	.9537	.9542	.9548	.9553	.9558	.9563	17°
73°	.9563	.9568	.9573	.9578	.9583	.9588	.9593	.9598	.9603	.9608	.9613	16°
74°	.9613	.9617	.9622	.9627	.9632	.9636	.9641	.9646	.9650	.9655	.9659	15°
75°	.9659	.9664	.9668	.9673	.9677	.9681	.9686	.9690	.9694	.9699	.9703	14°
76°	.9703	.9707	.9711	.9715	.9720	.9724	.9728	.9732	.9736	.9740	.9744	13°
77°	.9744	.9748	.9751	.9755	.9759	.9763	.9767	.9770	.9774	.9778	.9781	12°
78°	.9781	.9785	.9789	.9792	.9796	.9799	.9803	.9806	.9810	.9813	.9816	11°
79°	.9816	.9820	.9823	.9826	.9829	.9833	.9836	.9839	.9842	.9845	.9848	10°
80°	.9848	.9851	.9854	.9857	.9860	.9863	.9866	.9869	.9871	.9874	.9877	9°
81°	.9877	.9880	.9882	.9885	.9888	.9890	.9893	.9895	.9898	.9900	.9903	8°
82°	.9903	.9905	.9907	.9910	.9912	.9914	.9917	.9919	.9921	.9923	.9925	7°
83°	.9925	.9928	.9930	.9932	.9934	.9936	.9938	.9940	.9942	.9943	.9945	6°
84°	.9945	.9947	.9949	.9951	.9952	.9954	.9956	.9957	.9959	.9960	.9962	5°
85°	.9962	.9963	.9965	.9966	.9968	.9969	.9971	.9972	.9973	.9974	.9976	4°
86°	.9976	.9977	.9978	.9979	.9980	.9981	.9982	.9983	.9984	.9985	.9986	3°
87°	.9986	.9987	.9988	.9989	.9990	.9990	.9991	.9992	.9993	.9993	.9994	2°
88°	.9994	.9995	.9995	.9996	.9996	.9997	.9997	.9997	.9998	.9998	.9998	1°
89°	.9998	.9999	.9999	.9999	.9999	1.000	1.000	1.000	1.000	1.000	1.000	0°
		.9	.8	.7	.6	.5	.4	.3	.2	.1	.0	

cos

473

Table 1. Natural Trigonometric Functions

tan

	.0	.1	.2	.3	.4	.5	.6	.7	.8	.9	
0°	.0000	.0017	.0035	.0052	.0070	.0087	.0105	.0122	.0140	.0157	.0175	89°
1°	.0175	.0192	.0209	.0227	.0244	.0262	.0279	.0297	.0314	.0332	.0349	88°
2°	.0349	.0367	.0384	.0402	.0419	.0437	.0454	.0472	.0489	.0507	.0524	87°
3°	.0524	.0542	.0559	.0577	.0594	.0612	.0629	.0647	.0664	.0682	.0699	86°
4°	.0699	.0717	.0734	.0752	.0769	.0787	.0805	.0822	.0840	.0857	.0875	85°
5°	.0875	.0892	.0910	.0928	.0945	.0963	.0981	.0998	.1016	.1033	.1051	84°
6°	.1051	.1069	.1086	.1104	.1122	.1139	.1157	.1175	.1192	.1210	.1228	83°
7°	.1228	.1246	.1263	.1281	.1299	.1317	.1334	.1352	.1370	.1388	.1405	82°
8°	.1405	.1423	.1441	.1459	.1477	.1495	.1512	.1530	.1548	.1566	.1584	81°
9°	.1584	.1602	.1620	.1638	.1655	.1673	.1691	.1709	.1727	.1745	.1763	80°
10°	.1763	.1781	.1799	.1817	.1835	.1853	.1871	.1890	.1908	.1926	.1944	79°
11°	.1944	.1962	.1980	.1998	.2016	.2035	.2053	.2071	.2089	.2107	.2126	78°
12°	.2126	.2144	.2162	.2180	.2199	.2217	.2235	.2254	.2272	.2290	.2309	77°
13°	.2309	.2327	.2345	.2364	.2382	.2401	.2419	.2438	.2456	.2475	.2493	76°
14°	.2493	.2512	.2530	.2549	.2568	.2586	.2605	.2623	.2642	.2661	.2679	75°
15°	.2679	.2698	.2717	.2736	.2754	.2773	.2792	.2811	.2830	.2849	.2867	74°
16°	.2867	.2886	.2905	.2924	.2943	.2962	.2981	.3000	.3019	.3038	.3057	73°
17°	.3057	.3076	.3096	.3115	.3134	.3153	.3172	.3191	.3211	.3230	.3249	72°
18°	.3249	.3269	.3288	.3307	.3327	.3346	.3365	.3385	.3404	.3424	.3443	71°
19°	.3443	.3463	.3482	.3502	.3522	.3541	.3561	.3581	.3600	.3620	.3640	70°
20°	.3640	.3659	.3679	.3699	.3719	.3739	.3759	.3779	.3799	.3819	.3839	69°
21°	.3839	.3859	.3879	.3899	.3919	.3939	.3959	.3979	.4000	.4020	.4040	68°
22°	.4040	.4061	.4081	.4101	.4122	.4142	.4163	.4183	.4204	.4224	.4245	67°
23°	.4245	.4265	.4286	.4307	.4327	.4348	.4369	.4390	.4411	.4431	.4452	66°
24°	.4452	.4473	.4494	.4515	.4536	.4557	.4578	.4599	.4621	.4642	.4663	65°
25°	.4663	.4684	.4706	.4727	.4748	.4770	.4791	.4813	.4834	.4856	.4877	64°
26°	.4877	.4899	.4921	.4942	.4964	.4986	.5008	.5029	.5051	.5073	.5095	63°
27°	.5095	.5117	.5139	.5161	.5184	.5206	.5228	.5250	.5272	.5295	.5317	62°
28°	.5317	.5340	.5362	.5384	.5407	.5430	.5452	.5475	.5498	.5520	.5543	61°
29°	.5543	.5566	.5589	.5612	.5635	.5658	.5681	.5704	.5727	.5750	.5774	60°
30°	.5774	.5797	.5820	.5844	.5867	.5890	.5914	.5938	.5961	.5985	.6009	59°
31°	.6009	.6032	.6056	.6080	.6104	.6128	.6152	.6176	.6200	.6224	.6249	58°
32°	.6249	.6273	.6297	.6322	.6346	.6371	.6395	.6420	.6445	.6469	.6494	57°
33°	.6494	.6519	.6544	.6569	.6594	.6619	.6644	.6669	.6694	.6720	.6745	56°
34°	.6745	.6771	.6796	.6822	.6847	.6873	.6899	.6924	.6950	.6976	.7002	55°
35°	.7002	.7028	.7054	.7080	.7107	.7133	.7159	.7186	.7212	.7239	.7265	54°
36°	.7265	.7292	.7319	.7346	.7373	.7400	.7427	.7454	.7481	.7508	.7536	53°
37°	.7536	.7563	.7590	.7618	.7646	.7673	.7701	.7729	.7757	.7785	.7813	52°
38°	.7813	.7841	.7869	.7898	.7926	.7954	.7983	.8012	.8040	.8069	.8098	51°
39°	.8098	.8127	.8156	.8185	.8214	.8243	.8273	.8302	.8332	.8361	.8391	50°
40°	.8391	.8421	.8451	.8481	.8511	.8541	.8571	.8601	.8632	.8662	.8693	49°
41°	.8693	.8724	.8754	.8785	.8816	.8847	.8878	.8910	.8941	.8972	.9004	48°
42°	.9004	.9036	.9067	.9099	.9131	.9163	.9195	.9228	.9260	.9293	.9325	47°
43°	.9325	.9358	.9391	.9424	.9457	.9490	.9523	.9556	.9590	.9623	.9657	46°
44°	.9657	.9691	.9725	.9759	.9793	.9827	.9861	.9896	.9930	.9965	1.000	45°
	.9	.8	.7	.6	.5	.4	.3	.2	.1	.0		

cot

Table 1. Natural Trigonometric Functions

tan

	.0	.1	.2	.3	.4	.5	.6	.7	.8	.9	
45°	1.000	1.003	1.007	1.011	1.014	1.018	1.021	1.025	1.028	1.032	1.036	**44°**
46°	1.036	1.039	1.043	1.046	1.050	1.054	1.057	1.061	1.065	1.069	1.072	**43°**
47°	1.072	1.076	1.080	1.084	1.087	1.091	1.095	1.099	1.103	1.107	1.111	**42°**
48°	1.111	1.115	1.118	1.122	1.126	1.130	1.134	1.138	1.142	1.146	1.150	**41°**
49°	1.150	1.154	1.159	1.163	1.167	1.171	1.175	1.179	1.183	1.188	1.192	**40°**
50°	1.192	1.196	1.200	1.205	1.209	1.213	1.217	1.222	1.226	1.230	1.235	**39°**
51°	1.235	1.239	1.244	1.248	1.253	1.257	1.262	1.266	1.271	1.275	1.280	**38°**
52°	1.280	1.285	1.289	1.294	1.299	1.303	1.308	1.313	1.317	1.322	1.327	**37°**
53°	1.327	1.332	1.337	1.342	1.347	1.351	1.356	1.361	1.366	1.371	1.376	**36°**
54°	1.376	1.381	1.387	1.392	1.397	1.402	1.407	1.412	1.418	1.423	1.428	**35°**
55°	1.428	1.433	1.439	1.444	1.450	1.455	1.460	1.466	1.471	1.477	1.483	**34°**
56°	1.483	1.488	1.494	1.499	1.505	1.511	1.517	1.522	1.528	1.534	1.540	**33°**
57°	1.540	1.546	1.552	1.558	1.564	1.570	1.576	1.582	1.588	1.594	1.600	**32°**
58°	1.600	1.607	1.613	1.619	1.625	1.632	1.638	1.645	1.651	1.658	1.664	**31°**
59°	1.664	1.671	1.678	1.684	1.691	1.698	1.704	1.711	1.718	1.725	1.732	**30°**
60°	1.732	1.739	1.746	1.753	1.760	1.767	1.775	1.782	1.789	1.797	1.804	**29°**
61°	1.804	1.811	1.819	1.827	1.834	1.842	1.849	1.857	1.865	1.873	1.881	**28°**
62°	1.881	1.889	1.897	1.905	1.913	1.921	1.929	1.937	1.946	1.954	1.963	**27°**
63°	1.963	1.971	1.980	1.988	1.997	2.006	2.014	2.023	2.032	2.041	2.050	**26°**
64°	2.050	2.059	2.069	2.078	2.087	2.097	2.106	2.116	2.125	2.135	2.145	**25°**
65°	2.145	2.154	2.164	2.174	2.184	2.194	2.204	2.215	2.225	2.236	2.246	**24°**
66°	2.246	2.257	2.267	2.278	2.289	2.300	2.311	2.322	2.333	2.344	2.356	**23°**
67°	2.356	2.367	2.379	2.391	2.402	2.414	2.426	2.438	2.450	2.463	2.475	**22°**
68°	2.475	2.488	2.500	2.513	2.526	2.539	2.552	2.565	2.578	2.592	2.605	**21°**
69°	2.605	2.619	2.633	2.646	2.660	2.675	2.689	2.703	2.718	2.733	2.747	**20°**
70°	2.747	2.762	2.778	2.793	2.808	2.824	2.840	2.856	2.872	2.888	2.904	**19°**
71°	2.904	2.921	2.937	2.954	2.971	2.989	3.006	3.024	3.042	3.060	3.078	**18°**
72°	3.078	3.096	3.115	3.133	3.152	3.172	3.191	3.211	3.230	3.251	3.271	**17°**
73°	3.271	3.291	3.312	3.333	3.354	3.376	3.398	3.420	3.442	3.465	3.487	**16°**
74°	3.487	3.511	3.534	3.558	3.582	3.606	3.630	3.655	3.681	3.706	3.732	**15°**
75°	3.732	3.758	3.785	3.812	3.839	3.867	3.895	3.923	3.952	3.981	4.011	**14°**
76°	4.011	4.041	4.071	4.102	4.134	4.165	4.198	4.230	4.264	4.297	4.331	**13°**
77°	4.331	4.366	4.402	4.437	4.474	4.511	4.548	4.586	4.625	4.665	4.705	**12°**
78°	4.705	4.745	4.787	4.829	4.872	4.915	4.959	5.005	5.050	5.097	5.145	**11°**
79°	5.145	5.193	5.242	5.292	5.343	5.396	5.449	5.503	5.558	5.614	5.671	**10°**
80°	5.671	5.730	5.789	5.850	5.912	5.976	6.041	6.107	6.174	6.243	6.314	**9°**
81°	6.314	6.386	6.460	6.535	6.612	6.691	6.772	6.855	6.940	7.026	7.115	**8°**
82°	7.115	7.207	7.300	7.396	7.495	7.596	7.700	7.806	7.916	8.028	8.144	**7°**
83°	8.144	8.264	8.386	8.513	8.643	8.777	8.915	9.058	9.205	9.357	9.514	**6°**
84°	9.514	9.677	9.845	10.02	10.20	10.39	10.58	10.78	10.99	11.20	11.43	**5°**
85°	11.43	11.66	11.91	12.16	12.43	12.71	13.00	13.30	13.62	13.95	14.30	**4°**
86°	14.30	14.67	15.06	15.46	15.89	16.35	16.83	17.34	17.89	18.46	19.08	**3°**
87°	19.08	19.74	20.45	21.20	22.02	22.90	23.86	24.90	26.03	27.27	28.64	**2°**
88°	28.64	30.14	31.82	33.69	35.80	38.19	40.92	44.07	47.74	52.08	57.29	**1°**
89°	57.29	63.66	71.62	81.85	95.49	114.6	143.2	191.0	286.5	573.0		**0°**
	.9	.8	.7	.6	.5	.4	.3	.2	.1	.0		

cot

Table 2. Exponentials

x	e^x	x	e^x	x	e^x	x	e^x	x	e^x
0.00	1.000	1.00	2.718	2.00	7.389	3.00	20.086	4.00	54.598
0.05	1.051	1.05	2.858	2.05	7.768	3.05	21.115	40.5	57.397
0.10	1.105	1.10	3.004	2.10	8.166	3.10	22.198	4.10	60.340
0.15	1.162	1.15	3.158	2.15	8.585	3.15	23.336	4.15	63.434
0.20	1.221	1.20	3.320	2.20	9.025	3.20	24.533	4.20	66.686
0.25	1.284	1.25	3.490	2.25	9.488	3.25	25.790	4.25	70.105
0.30	1.350	1.30	3.669	2.30	9.974	3.30	27.113	4.30	73.700
0.35	1.419	1.35	3.857	2.35	10.486	3.35	28.503	4.35	77.487
0.40	1.492	1.40	4.005	2.40	11.023	3.40	29.964	4.40	81.451
0.45	1.568	1.45	4.263	2.45	11.588	3.45	31.500	4.45	85.627
0.50	1.649	1.50	4.482	2.50	12.182	3.50	33.115	4.50	90.017
0.55	1.733	1.55	4.712	2.55	12.807	3.55	34.813	4.55	94.632
0.60	1.822	1.60	4.953	2.60	13.464	3.60	36.598	4.60	99.484
0.65	1.916	1.65	5.207	2.65	14.154	3.65	38.475	4.65	104.58
0.70	2.014	1.70	5.474	2.70	14.880	3.70	40.447	4.70	109.95
0.75	2.117	1.75	5.755	2.75	15.643	3.75	42.521	4.75	115.58
0.80	2.226	1.80	6.050	2.80	16.445	3.80	44.701	4.80	121.51
0.85	2.340	1.85	6.360	2.85	17.288	3.85	46.993	4.85	127.74
0.90	2.460	1.90	6.686	2.90	18.174	3.90	49.902	4.90	134.29
0.95	2.586	1.95	7.029	2.95	19.106	3.95	51.935	4.95	141.17
1.00	2.718	2.00	7.389	3.00	20.086	4.00	54.598	5.00	148.41

Table 3. Typical Physical Properties of Metals

Wrought iron	Fe —Bal. Slag—2.5	Hot-rolled	30	48	30	100	0.278	7.70	6.35	29	
Aluminum alloy	Al —99 plus	Annealed–0	5	13	45	23	0.098	2.71	13.1	10	3.8
		Cold-rolled–H14	17	18	20	32					
		Cold-rolled–H18	22	24	15	44					
Copper	Cu —99.9 plus	Annealed	10	32	45	42	0.322	8.91	9.3	17	6.4
		Cold-drawn	40	45	15	90					
		Cold-rolled	40	46	5	100					
Magnesium alloy	Mg—Bal. Al —9.0 Zn —2.0 Mn —0.10 min	Sand-cast	14	24	6	50	0.066	1.83	14.5	6.5	2.4
Manganese bronze	Cu —58.5 Zn —39.2 Fe —1.0 Sn —1.0 Mn —0.3	Annealed	30	60	30	95	0.302	8.36	11.2	15	5.6
		Cold-drawn	50	80	20	180					
Nickel (pure)	Ni —99.99	Annealed	8.5	46	30		0.322	8.91	7.4	30	11
Red brass (wrought)	Cu —85 Zn —15	Annealed	15	40	50	50	0.316	8.75	9.8	17	6.4
		Cold-drawn	55	70	15	120					
		Cold-rolled	60	75	7	135					
Titanium (commercially pure)	Ti —Bal. Fe —0.2 max N_2 —0.05 max C —0.08 max H_2 —0.015 max	Annealed	70	90	23	200	0.163	4.54	5.0	16.5	6.6

Table 3. (continued)

Material	Nominal composition (essential elements) per cent	Form and condition	Typical mechanical properties				Typical physical constants				
			Yield strength (0.2% offset) 1000 psi	Tensile strength 1000 psi	Elongation in 2 in., per cent	Hardness brinell	Density lb/cu in.	Specific gravity	Thermal expansion coefficient (32°-212°F) x10⁻⁶ in./in./°F	Tensile modulus of elasticity x10⁶ psi	Torsional modulus of elasticity x10⁶ psi
Carbon steel AISI–SAE 1020	Fe —Bal. Mn—0.45 Si —0.25 C —0.20	Annealed	38	65	30	130	0.284	7.86	6.5	30	11.6
		Hot-rolled	42	68	32	135					
		Hardened	62	90	25	179					
Cast gray iron	C —3.4 Si —1.8 Mn—0.5 Fe —Bal.	Cast		25 min	0.5 max	180	0.260	7.20	6.7	13	6.7
Ingot iron	Fe —99.9 plus	Hot-rolled	29	45	26	90	0.284	7.86	6.8	30.1	11.8
		Annealed	19	38	45	67					
Malleable iron	C —2.5 Si —1.0 Mn—0.55 max	Cast	33	52	12	130	0.264	7.32	6.6	25	
Stainless steel type 431	Fe —Bal. Cr —16 Ni — 2	Annealed	85	120	25	250	0.280	7.75	6.5	29	10.5
		Heat-treated	150	195	20	400					

478

Table 4. Typical Properties of Structural Timber

Species	Allowable unit stresses (psi)				Modulus of elasticity (psi)
	Tension	Horizontal shear	Compression (perpendicular to grain)	Compression (parallel to grain)	
Douglas fir	2000	120	450	1500	1.8×10^6
Hemlock	1300	80	360	850	1.2×10^6
Pine (southern)	3000	160	470	2300	1.8×10^6
Pine (Norway)	1200	75	360	900	1.3×10^6
Redwood	1700	110	320	1500	1.3×10^6

Table 5

AMERICAN STANDARD
STEEL I BEAMS
PROPERTIES FOR DESIGNING

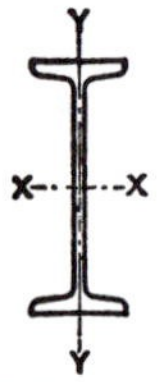

Nominal Size *	Weight per Foot	Area	Depth	Flange Width	Flange Thickness	Web Thickness	AXIS X-X I	AXIS X-X $\frac{I}{c}$	AXIS X-X r	AXIS Y-Y I	AXIS Y-Y $\frac{I}{c}$	AXIS Y-Y r
In.	Lb.	In.²	In.	In.	In.	In.	In.⁴	In.³	In.	In.⁴	In.³	In.
24 x 7⅞	120.0	35.13	24.00	8.048	1.102	.798	3010.8	250.9	9.26	84.9	21.1	1.56
	105.9	30.98	24.00	7.875	1.102	.625	2811.5	234.3	9.53	78.9	20.0	1.60
24 x 7	100.0	29.25	24.00	7.247	.871	.747	2371.8	197.6	9.05	48.4	13.4	1.29
	90.0	26.30	24.00	7.124	.871	.624	2230.1	185.8	9.21	45.5	12.8	1.32
	79.9	23.33	24.00	7.000	.871	.500	2087.2	173.9	9.46	42.9	12.2	1.36
20 x 7	95.0	27.74	20.00	7.200	.916	.800	1599.7	160.0	7.59	50.5	14.0	1.35
	85.0	24.80	20.00	7.053	.916	.653	1501.7	150.2	7.78	47.0	13.3	1.38
20 x 6¼	75.0	21.90	20.00	6.391	.789	.641	1263.5	126.3	7.60	30.1	9.4	1.17
	65.4	19.08	20.00	6.250	.789	.500	1169.5	116.9	7.83	27.9	8.9	1.21
18 x 6	70.0	20.46	18.00	6.251	.691	.711	917.5	101.9	6.70	24.5	7.8	1.09
	54.7	15.94	18.00	6.000	.691	.460	795.5	88.4	7.07	21.2	7.1	1.15
15 x 5½	50.0	14.59	15.00	5.640	.622	.550	481.1	64.2	5.74	16.0	5.7	1.05
	42.9	12.49	15.00	5.500	.622	.410	441.8	58.9	5.95	14.6	5.3	1.08
12 x 5¼	50.0	14.57	12.00	5.477	.659	.687	301.6	50.3	4.55	16.0	5.8	1.05
	40.8	11.84	12.00	5.250	.659	.460	268.9	44.8	4.77	13.8	5.3	1.08
12 x 5	35.0	10.20	12.00	5.078	.544	.428	227.0	37.8	4.72	10.0	3.9	.99
	31.8	9.26	12.00	5.000	.544	.350	215.8	36.0	4.83	9.5	3.8	1.01
10 x 4⅝	35.0	10.22	10.00	4.944	.491	.594	145.8	29.2	3.78	8.5	3.4	.91
	25.4	7.38	10.00	4.660	.491	.310	122.1	24.4	4.07	6.9	3.0	.97
8 x 4	23.0	6.71	8.00	4.171	.425	.441	64.2	16.0	3.09	4.4	2.1	.81
	18.4	5.34	8.00	4.000	.425	.270	56.9	14.2	3.26	3.8	1.9	.84
7 x 3⅝	20.0	5.83	7.00	3.860	.392	.450	41.9	12.0	2.68	3.1	1.6	.74
	15.3	4.43	7.00	3.660	.392	.250	36.2	10.4	2.86	2.7	1.5	.78
6 x 3⅜	17.25	5.02	6.00	3.565	.359	.465	26.0	8.7	2.28	2.3	1.3	.68
	12.5	3.61	6.00	3.330	.359	.230	21.8	7.3	2.46	1.8	1.1	.72
5 x 3	14.75	4.29	5.00	3.284	.326	.494	15.0	6.0	1.87	1.7	1.0	.63
	10.0	2.87	5.00	3.000	.326	.210	12.1	4.8	2.05	1.2	.82	.65
4 x 2⅝	9.5	2.76	4.00	2.796	.293	.326	6.7	3.3	1.56	.91	.65	.58
	7.7	2.21	4.00	2.660	.293	.190	6.0	3.0	1.64	.77	.58	.59
3 x 2⅜	7.5	2.17	3.00	2.509	.260	.349	2.9	1.9	1.15	.59	.47	.52
	5.7	1.64	3.00	2.330	.260	.170	2.5	1.7	1.23	.46	.40	.53

*Steel I-beams are designated by giving their depth in inches first; then the letter I to designate an I-beam; then the weight in pounds per linear foot. For example, 24 I 120.0.

Table 6

I   STEEL WIDE FLANGE BEAMS

PROPERTIES FOR DESIGNING

(ABRIDGED LIST)

Nominal* Size	Weight per Foot	Area	Depth	Flange Width	Flange Thickness	Web Thickness	AXIS X-X I	AXIS X-X $\frac{I}{c}$	AXIS X-X r	AXIS Y-Y I	AXIS Y-Y $\frac{I}{c}$	AXIS Y-Y r
In.	Lb.	In.²	In.	In.	In.	In.	In.⁴	In.³	In.	In.⁴	In.³	In.
36 x 16½	230	67.73	35.88	16.475	1.260	.765	14988.4	835.5	14.88	870.9	105.7	3.59
36 x 12	150	44.16	35.84	11.972	.940	.625	9012.1	502.9	14.29	250.4	41.8	2.38
33 x 15¾	200	58.79	33.00	15.750	1.150	.715	11048.2	669.6	13.71	691.7	87.8	3.43
33 x 11½	130	38.26	33.10	11.510	.855	.580	6699.0	404.8	13.23	201.4	35.0	2.29
30 x 15	172	50.65	29.88	14.985	1.065	.655	7891.5	528.2	12.48	550.1	73.4	3.30
30 x 10½	108	31.77	29.82	10.484	.760	.548	4461.0	299.2	11.85	135.1	25.8	2.06
27 x 14	145	42.68	26.88	13.965	.975	.600	5414.3	402.9	11.26	406.9	58.3	3.09
27 x 10	94	27.65	26.91	9.990	.747	.490	3266.7	242.8	10.87	115.1	23.0	2.04
24 x 14	130	38.21	24.25	14.000	.900	.565	4009.5	330.7	10.24	375.2	53.6	3.13
24 x 12	100	29.43	24.00	12.000	.775	.468	2987.3	248.9	10.08	203.5	33.9	2.63
24 x 9	76	22.37	23.91	8.985	.682	.440	2096.4	175.4	9.68	76.5	17.0	1.85
21 x 13	112	32.93	21.00	13.000	.865	.527	2620.6	249.6	8.92	289.7	44.6	2.96
21 x 9	82	24.10	20.86	8.962	.795	.499	1752.4	168.0	8.53	89.6	20.0	1.93
21 x 8¼	62	18.23	20.99	8.240	.615	.400	1326.8	126.4	8.53	53.1	12.9	1.71
18 x 11¾	96	28.22	18.16	11.750	.831	.512	1674.7	184.4	7.70	206.8	35.2	2.71
18 x 8¾	64	18.80	17.87	8.715	.686	.403	1045.8	117.0	7.46	70.3	16.1	1.93
18 x 7½	50	14.71	18.00	7.500	.570	.358	800.6	89.0	7.38	37.2	9.9	1.59
16 x 11½	88	25.87	16.16	11.502	.795	.504	1222.6	151.3	6.87	185.2	32.2	2.67
16 x 8½	58	17.04	15.86	8.464	.645	.407	746.4	94.1	6.62	60.5	14.3	1.88
16 x 7	50	14.70	16.25	7.073	.628	.380	655.4	80.7	6.68	34.8	9.8	1.54
	36	10.59	15.85	6.992	.428	.299	446.3	56.3	6.49	22.1	6.3	1.45
14 x 16	142	41.85	14.75	15.500	1.063	.680	1672.2	226.7	6.32	660.1	85.2	3.97
	†320	94.12	16.81	16.710	2.093	1.890	4141.7	492.8	6.63	1635.1	195.7	4.17
14 x 14½	87	25.56	14.00	14.500	.688	.420	966.9	138.1	6.15	349.7	48.2	3.70
14 x 12	84	24.71	14.18	12.023	.778	.451	928.4	130.9	6.13	225.5	37.5	3.02
	78	22.94	14.06	12.000	.718	.428	851.2	121.1	6.09	206.9	34.5	3.00

*Steel WF beams are designated by giving their nominal depth in inches first; then the letters WF to designate a wide-flange beam; then the weight in pounds per linear foot. For example, 36 WF 230.

†Column core section.

Table 6. (continued)

Nominal Size	Weight per Foot	Area	Depth	Flange Width	Flange Thickness	Web Thickness	AXIS X-X I	AXIS X-X $\frac{I}{c}$	AXIS X-X r	AXIS Y-Y I	AXIS Y-Y $\frac{I}{c}$	AXIS Y-Y r
In.	Lb.	In.²	In.	In.	In.	In.	In.⁴	In.³	In.	In.⁴	In.³	In.
14 x 10	74	21.76	14.19	10.072	.783	.450	796.8	112.3	6.05	133.5	26.5	2.48
	68	20.00	14.06	10.040	.718	.418	724.1	103.0	6.02	121.2	24.1	2.46
	61	17.94	13.91	10.000	.643	.378	641.5	92.2	5.98	107.3	21.5	2.45
14 x 8	53	15.59	13.94	8.062	.658	.370	542.1	77.8	5.90	57.5	14.3	1.92
	43	12.65	13.68	8.000	.528	.308	429.0	62.7	5.82	45.1	11.3	1.89
14 x 6¾	38	11.17	14.12	6.776	.513	.313	385.3	54.6	5.87	24.6	7.3	1.49
	34	10.00	14.00	6.750	.453	.287	339.2	48.5	5.83	21.3	6.3	1.46
	30	8.81	13.86	6.733	.383	.270	289.6	41.8	5.73	17.5	5.2	1.41
12 x 12	85	24.98	12.50	12.105	.796	.495	723.3	115.7	5.38	235.5	38.9	3.07
	65	19.11	12.12	12.000	.606	.390	533.4	88.0	5.28	174.6	29.1	3.02
12 x 10	53	15.59	12.06	10.000	.576	.345	426.2	70.7	5.23	96.1	19.2	2.48
12 x 8	40	11.77	11.94	8.000	.516	.294	310.1	51.9	5.13	44.1	11.0	1.94
12 x 6½	36	10.59	12.24	6.565	.540	.305	280.8	45.9	5.15	23.7	7.2	1.50
	31	9.12	12.09	6.525	.465	.265	238.4	39.4	5.11	19.8	6.1	1.47
	27	7.97	11.95	6.500	.400	.240	204.1	34.1	5.06	16.6	5.1	1.44
10 x 10	112	32.92	11.38	10.415	1.248	.755	718.7	126.3	4.67	235.4	45.2	2.67
	100	29.43	11.12	10.345	1.118	.685	625.0	112.4	4.61	206.6	39.9	2.65
	89	26.19	10.88	10.275	.998	.615	542.4	99.7	4.55	180.6	35.2	2.63
	77	22.67	10.62	10.195	.868	.535	457.2	86.1	4.49	153.4	30.1	2.60
	49	14.40	10.00	10.000	.558	.340	272.9	54.6	4.35	93.0	18.6	2.54
10 x 8	45	13.24	10.12	8.022	.618	.350	248.6	49.1	4.33	53.2	13.3	2.00
	39	11.48	9.94	7.990	.528	.318	209.7	42.2	4.27	44.9	11.2	1.98
	33	9.71	9.75	7.964	.433	.292	170.9	35.0	4.20	36.5	9.2	1.94
10 x 5¾	29	8.53	10.22	5.799	.500	.289	157.3	30.8	4.29	15.2	5.2	1.34
	21	6.19	9.90	5.750	.340	.240	106.3	21.5	4.14	9.7	3.4	1.25
8 x 8	67	19.70	9.00	8.287	.933	.575	271.8	60.4	3.71	88.6	21.4	2.12
	58	17.06	8.75	8.222	.808	.510	227.3	52.0	3.65	74.9	18.2	2.10
	48	14.11	8.50	8.117	.683	.405	183.7	43.2	3.61	60.9	15.0	2.08
	40	11.76	8.25	8.077	.558	.365	146.3	35.5	3.53	49.0	12.1	2.04
	35	10.30	8.12	8.027	.493	.315	126.5	31.1	3.50	42.5	10.6	2.03
	31	9.12	8.00	8.000	.433	.288	109.7	27.4	3.47	37.0	9.2	2.01
8 x 6½	28	8.23	8.06	6.540	.463	.285	97.8	24.3	3.45	21.6	6.6	1.62
	24	7.06	7.93	6.500	.398	.245	82.5	20.8	3.42	18.2	5.6	1.61
8 x 5¼	20	5.88	8.14	5.268	.378	.248	69.2	17.0	3.43	8.5	3.2	1.20
	17	5.00	8.00	5.250	.308	.230	56.4	14.1	3.36	6.7	2.6	1.16

Table 7

AMERICAN STANDARD STEEL CHANNELS PROPERTIES FOR DESIGNING

Nominal* Size	Weight per Foot	Area	Depth	Flange		Web Thickness	AXIS X-X			AXIS Y-Y			
				Width	Average Thickness		I	$\frac{I}{c}$	r	I	$\frac{I}{c}$	r	x
In.	Lb.	In.²	In.	In.	In.	In.	In.⁴	In.³	In.	In.⁴	In.³	In.	In.
†18 x 4	58.0	16.98	18.00	4.200	.625	.700	670.7	74.5	6.29	18.5	5.6	1.04	.88
	51.9	15.18	18.00	4.100	.625	.600	622.1	69.1	6.40	17.1	5.3	1.06	.87
	45.8	13.38	18.00	4.000	.625	.500	573.5	63.7	6.55	15.8	5.1	1.09	.89
	42.7	12.48	18.00	3.950	.625	.450	549.2	61.0	6.64	15.0	4.9	1.10	.90
15 x 3⅜	50.0	14.64	15.00	3.716	.650	.716	401.4	53.6	5.24	11.2	3.8	.87	.80
	40.0	11.70	15.00	3.520	.650	.520	346.3	46.2	5.44	9.3	3.4	.89	.78
	33.9	9.90	15.00	3.400	.650	.400	312.6	41.7	5.62	8.2	3.2	.91	.79
12 x 3	30.0	8.79	12.00	3.170	.501	.510	161.2	26.9	4.28	5.2	2.1	.77	.68
	25.0	7.32	12.00	3.047	.501	.387	143.5	23.9	4.43	4.5	1.9	.79	.68
	20.7	6.03	12.00	2.940	.501	.280	128.1	21.4	4.61	3.9	1.7	.81	.70
10 x 2⅝	30.0	8.80	10.00	3.033	.436	.673	103.0	20.6	3.42	4.0	1.7	.67	.65
	25.0	7.33	10.00	2.886	.436	.526	90.7	18.1	3.52	3.4	1.5	.68	.62
	20.0	5.86	10.00	2.739	.436	.379	78.5	15.7	3.66	2.8	1.3	.70	.61
	15.3	4.47	10.00	2.600	.436	.240	66.9	13.4	3.87	2.3	1.2	.72	.64
9 x 2½	20.0	5.86	9.00	2.648	.413	.448	60.6	13.5	3.22	2.4	1.2	.65	.59
	15.0	4.39	9.00	2.485	.413	.285	50.7	11.3	3.40	1.9	1.0	.67	.59
	13.4	3.89	9.00	2.430	.413	.230	47.3	10.5	3.49	1.8	.97	.67	.61
8 x 2¼	18.75	5.49	8.00	2.527	.390	.487	43.7	10.9	2.82	2.0	1.0	.60	.57
	13.75	4.02	8.00	2.343	.390	.303	35.8	9.0	2.99	1.5	.86	.62	.56
	11.5	3.36	8.00	2.260	.390	.220	32.3	8.1	3.10	1.3	.79	.63	.58
7 x 2⅛	14.75	4.32	7.00	2.299	.366	.419	27.1	7.7	2.51	1.4	.79	.57	.53
	12.25	3.58	7.00	2.194	.366	.314	24.1	6.9	2.59	1.2	.71	.58	.53
	9.8	2.85	7.00	2.090	.366	.210	21.1	6.0	2.72	.98	.63	.59	.55
6 x 2	13.0	3.81	6.00	2.157	.343	.437	17.3	5.8	2.13	1.1	.65	.53	.52
	10.5	3.07	6.00	2.034	.343	.314	15.1	5.0	2.22	.87	.57	.53	.50
	8.2	2.39	6.00	1.920	.343	.200	13.0	4.3	2.34	.70	.50	.54	.52
5 x 1¾	9.0	2.63	5.00	1.885	.320	.325	8.8	3.5	1.83	.64	.45	.49	.48
	6.7	1.95	5.00	1.750	.320	.190	7.4	3.0	1.95	.48	.38	.50	.49
4 x 1⅝	7.25	2.12	4.00	1.720	.296	.320	4.5	2.3	1.47	.44	.35	.46	.46
	5.4	1.56	4.00	1.580	.296	.180	3.8	1.9	1.56	.32	.29	.45	.46
3 x 1½	6.0	1.75	3.00	1.596	.273	.356	2.1	1.4	1.08	.31	.27	.42	.46
	5.0	1.46	3.00	1.498	.273	.258	1.8	1.2	1.12	.25	.24	.41	.44
	4.1	1.19	3.00	1.410	.273	.170	1.6	1.1	1.17	.20	.21	.41	.44

*Steel channels are designated by giving their depth in inches first; then the symbol ⌐ to designate a channel; then the weight in pounds per linear foot. For example. 15 ⌐ 50.0.

†Car and shipbuilding channel; not an American Standard.

Table 8

	STEEL ANGLES							
	EQUAL LEGS							
	PROPERTIES FOR DESIGNING							

| Size | Thickness | Weight per Foot | Area | AXIS X-X AND AXIS Y-Y | | | | AXIS Z-Z |
				I	$\frac{I}{c}$	r	x or y	r
In.	In.	Lb.	In.²	In.⁴	In.³	In.	In.	In.
8 x 8	1⅛	56.9	16.73	98.0	17.5	2.42	2.41	1.56
	1	51.0	15.00	89.0	15.8	2.44	2.37	1.56
	⅞	45.0	13.23	79.6	14.0	2.45	2.32	1.57
	¾	38.9	11.44	69.7	12.2	2.47	2.28	1.57
	⅝	32.7	9.61	59.4	10.3	2.49	2.23	1.58
	9⁄16	29.6	8.68	54.1	9.3	2.50	2.21	1.58
	½	26.4	7.75	48.6	8.4	2.50	2.19	1.59
6 x 6	1	37.4	11.00	35.5	8.6	1.80	1.86	1.17
	⅞	33.1	9.73	31.9	7.6	1.81	1.82	1.17
	¾	28.7	8.44	28.2	6.7	1.83	1.78	1.17
	⅝	24.2	7.11	24.2	5.7	1.84	1.73	1.18
	9⁄16	21.9	6.43	22.1	5.1	1.85	1.71	1.18
	½	19.6	5.75	19.9	4.6	1.86	1.68	1.18
	7⁄16	17.2	5.06	17.7	4.1	1.87	1.66	1.19
	⅜	14.9	4.36	15.4	3.5	1.88	1.64	1.19
	5⁄16	12.5	3.66	13.0	3.0	1.89	1.61	1.19
5 x 5	⅞	27.2	7.98	17.8	5.2	1.49	1.57	.97
	¾	23.6	6.94	15.7	4.5	1.51	1.52	.97
	⅝	20.0	5.86	13.6	3.9	1.52	1.48	.98
	½	16.2	4.75	11.3	3.2	1.54	1.43	.98
	7⁄16	14.3	4.18	10.0	2.8	1.55	1.41	.98
	⅜	12.3	3.61	8.7	2.4	1.56	1.39	.99
	5⁄16	10.3	3.03	7.4	2.0	1.57	1.37	.99
4 x 4	¾	18.5	5.44	7.7	2.8	1.19	1.27	.78
	⅝	15.7	4.61	6.7	2.4	1.20	1.23	.78
	½	12.8	3.75	5.6	2.0	1.22	1.18	.78
	7⁄16	11.3	3.31	5.0	1.8	1.23	1.16	.78
	⅜	9.8	2.86	4.4	1.5	1.23	1.14	.79
	5⁄16	8.2	2.40	3.7	1.3	1.24	1.12	.79
	¼	6.6	1.94	3.0	1.1	1.25	1.09	.80
3½ x 3½	½	11.1	3.25	3.6	1.5	1.06	1.06	.68
	7⁄16	9.8	2.87	3.3	1.3	1.07	1.04	.68
	⅜	8.5	2.48	2.9	1.2	1.07	1.01	.69
	5⁄16	7.2	2.09	2.5	.98	1.08	.99	.69
	¼	5.8	1.69	2.0	.79	1.09	.97	.69
3 x 3	½	9.4	2.75	2.2	1.1	.90	.93	.58
	7⁄16	8.3	2.43	2.0	.95	.91	.91	.58
	⅜	7.2	2.11	1.8	.83	.91	.89	.58
	5⁄16	6.1	1.78	1.5	.71	.92	.87	.59
	¼	4.9	1.44	1.2	.58	.93	.84	.59
	3⁄16	3.71	1.09	.96	.44	.94	.82	.59
2½ x 2½	½	7.7	2.25	1.2	.72	.74	.81	.49
	⅜	5.9	1.73	.98	.57	.75	.76	.49
	5⁄16	5.0	1.47	.85	.48	.76	.74	.49
	¼	4.1	1.19	.70	.39	.77	.72	.49
	3⁄16	3.07	.90	.55	.30	.78	.69	.49

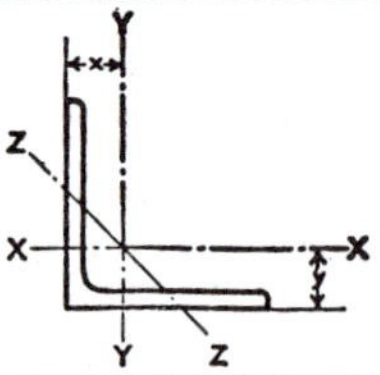

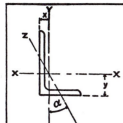

Table 7

STEEL ANGLES
UNEQUAL LEGS

PROPERTIES FOR DESIGNING

Size	Thick- ness	Weight per Foot	Area	AXIS X-X				AXIS Y-Y				AXIS Z-Z	
				I	$\frac{I}{c}$	r	y	I	$\frac{I}{c}$	r	x	r	Tan α
In.	In.	Lb.	In.^{2}	In.^{4}	In.^{3}	In.	In.	In.^{4}	In.^{3}	In.	In.	In.	
8 x 6	1	44.2	13.00	80.8	15.1	2.49	2.65	38.8	8.9	1.73	1.65	1.28	.543
	⅞	39.1	11.48	72.3	13.4	2.51	2.61	34.9	7.9	1.74	1:61	1.28	.547
	¾	33.8	9.94	63.4	11.7	2.53	2.56	30.7	6.9	1.76	1.56	1.29	.551
	⅝	28.5	8.36	54.1	9.9	2.54	2.52	26.3	5.9	1.77	1.52	1.29	.554
	⁹⁄₁₆	25.7	7.56	49.3	9.0	2.55	2.50	24.0	5.3	1.78	1.50	1.30	.556
	½	23.0	6.75	44.3	8.0	2.56	2.47	21.7	4.8	1.79	1.47	1.30	.558
	⁷⁄₁₆	20.2	5.93	39.2	7.1	2.57	2.45	19.3	4.2	1.80	1.45	1.31	.560
8 x 4	1	37.4	11.00	69.6	14.1	2.52	3.05	11.6	3.9	1.03	1.05	.85	.247
	⅞	33.1	9.73	62.5	12.5	2.53	3.00	10.5	3.5	1.04	1.00	.85	.253
	¾	28.7	8.44	54.9	10.9	2.55	2.95	9.4	3.1	1.05	.95	.85	.258
	⅝	24.2	7.11	46.9	9.2	2.57	2.91	8.1	2.6	1.07	.91	.86	.262
	⁹⁄₁₆	21.9	6.43	42.8	8.4	2.58	2.88	7.4	2.4	1.07	.88	.86	.265
	½	19.6	5.75	38.5	7.5	2.59	2.86	6.7	2.2	1.08	.86	.86	.267
	⁷⁄₁₆	17.2	5.06	34.1	6.6	2.60	2.83	6.0	1.9	1.09	.83	.87	.269
7 x 4	⅞	30.2	8.86	42.9	9.7	2.20	2.55	10.2	3.5	1.07	1.05	.86	.318
	¾	26.2	7.69	37.8	8.4	2.22	2.51	9.1	3.0	1.09	1.01	.86	.324
	⅝	22.1	6.48	32.4	7.1	2.24	2.46	7.8	2.6	1.10	.96	.86	.329
	⁹⁄₁₆	20.0	5.87	29.6	6.5	2.24	2.44	7.2	2.4	1.11	.94	.87	.332
	½	17.9	5.25	26.7	5.8	2.25	2.42	6.5	2.1	1.11	.92	.87	.335
	⁷⁄₁₆	15.8	4.62	23.7	5.1	2.26	2.39	5.8	1.9	1.12	.89	.88	.337
	⅜	13.6	3.98	20.6	4.4	2.27	2.37	5.1	1.6	1.13	.87	.88	.339
6 x 4	⅞	27.2	7.98	27.7	7.2	1.86	2.12	9.8	3.4	1.11	1.12	.86	.421
	¾	23.6	6.94	24.5	6.3	1.88	2.08	8.7	3.0	1.12	1.08	.86	.428
	⅝	20.0	5.86	21.1	5.3	1.90	2.03	7.5	2.5	1.13	1.03	.86	.435
	⁹⁄₁₆	18.1	5.31	19.3	4.8	1.90	2.01	6.9	2.3	1.14	1.01	.87	.438
	½	16.2	4.75	17.4	4.3	1.91	1.99	6.3	2.1	1.15	.99	.87	.440
	⁷⁄₁₆	14.3	4.18	15.5	3.8	1.92	1.96	5.6	1.9	1.16	.96	.87	.443
	⅜	12.3	3.61	13.5	3.3	1.93	1.94	4.9	1.6	1.17	.94	.88	.446
	⁵⁄₁₆	10.3	3.03	11.4	2.8	1.94	1.92	4.2	1.4	1.17	.92	.88	.449
6 x 3½	½	15.3	4.50	16.6	4.2	1.92	2.08	4.3	1.6	.97	.83	.76	.344
	⅜	11.7	3.42	12.9	3.2	1.94	2.04	3.3	1.2	.99	.79	.77	.350
	⁵⁄₁₆	9.8	2.87	10.9	2.7	1.95	2.01	2.9	1.0	1.00	.76	.77	.352
	¼	7.9	2.31	8.9	2.2	1.96	1.99	2.3	0.85	1.01	.74	.78	.355
5 x 3½	¾	19.8	5.81	13.9	4.3	1.55	1.75	5.6	2.2	.98	1.00	.75	.464
	⅝	16.8	4.92	12.0	3.7	1.56	1.70	4.8	1.9	.99	.95	.75	.472
	½	13.6	4.00	10.0	3.0	1.58	1.66	4.1	1.6	1.01	.91	.75	.479
	⁷⁄₁₆	12.0	3.53	8.9	2.6	1.59	1.63	3.6	1.4	1.01	.88	.76	.482
	⅜	10.4	3.05	7.8	2.3	1.60	1.61	3.2	1.2	1.02	.86	.76	.486
	⁵⁄₁₆	8.7	2.56	6.6	1.9	1.61	1.59	2.7	1.0	1.03	.84	.76	.489
	¼	7.0	2.06	5.4	1.6	1.61	1.56	2.2	.83	1.04	.81	.76	.492

Table 9. (continued)

Size	Thick-ness	Weight per Foot	Area	AXIS X-X				AXIS Y-Y				AXIS Z-Z	
				I	$\frac{I}{c}$	r	y	I	$\frac{I}{c}$	r	x	r	Tan α
In.	In.	Lb.	In.²	In.⁴	In.³	In.	In.	In.⁴	In.³	In.	In.	In.	
5 x 3	½	12.8	3.75	9.5	2.9	1.59	1.75	2.6	1.1	.83	.75	.65	.357
	7/16	11.3	3.31	8.4	2.6	1.60	1.73	2.3	1.0	.84	.73	.65	.361
	3/8	9.8	2.86	7.4	2.2	1.61	1.70	2.0	.89	.84	.70	.65	.364
	5/16	8.2	2.40	6.3	1.9	1.61	1.68	1.8	.75	.85	.68	.66	.368
	¼	6.6	1.94	5.1	1.5	1.62	1.66	1.4	.61	.86	.66	.66	.371
4 x 3½	5/8	14.7	4.30	6.4	2.4	1.22	1.29	4.5	1.8	1.03	1.04	.72	.745
	½	11.9	3.50	5.3	1.9	1.23	1.25	3.8	1.5	1.04	1.00	.72	.750
	7/16	10.6	3.09	4.8	1.7	1.24	1.23	3.4	1.4	1.05	.98	.72	.753
	3/8	9.1	2.67	4.2	1.5	1.25	1.21	3.0	1.2	1.06	.96	.73	.755
	5/16	7.7	2.25	3.6	1.3	1.26	1.18	2.6	1.0	1.07	.93	.73	.757
	¼	6.2	1.81	2.9	1.0	1.27	1.16	2.1	.81	1.07	.91	.73	.759
4 x 3	5/8	13.6	3.98	6.0	2.3	1.23	1.37	2.9	1.4	.85	.87	.64	.534
	½	11.1	3.25	5.1	1.9	1.25	1.33	2.4	1.1	.86	.83	.64	.543
	7/16	9.8	2.87	4.5	1.7	1.25	1.30	2.2	1.0	.87	.80	.64	.547
	3/8	8.5	2.48	4.0	1.5	1.26	1.28	1.9	.87	.88	.78	.64	.551
	5/16	7.2	2.09	3.4	1.2	1.27	1.26	1.7	.73	.89	.76	.65	.554
	¼	5.8	1.69	2.8	1.0	1.28	1.24	1.4	.60	.90	.74	.65	.558
3½ x 3	½	10.2	3.00	3.5	1.5	1.07	1.13	2.3	1.1	.88	.88	.62	.714
	7/16	9.1	2.65	3.1	1.3	1.08	1.10	2.1	.98	.89	.85	.62	.718
	3/8	7.9	2.30	2.7	1.1	1.09	1.08	1.9	.85	.90	.83	.62	.721
	5/16	6.6	1.93	2.3	.95	1.10	1.06	1.6	.72	.90	.81	.63	.724
	¼	5.4	1.56	1.9	.78	1.11	1.04	1.3	.59	.91	.79	.63	.727
3½x2½	½	9.4	2.75	3.2	1.4	1.09	1.20	1.4	.76	.70	.70	.53	.486
	7/16	8.3	2.43	2.9	1.3	1.09	1.18	1.2	.68	.71	.68	.54	.491
	3/8	7.2	2.11	2.6	1.1	1.10	1.16	1.1	.59	.72	.66	.54	.496
	5/16	6.1	1.78	2.2	.93	1.11	1.14	.94	.50	.73	.64	.54	.501
	¼	4.9	1.44	1.8	.75	1.12	1.11	.78	.41	.74	.61	.54	.506
3 x 2½	½	8.5	2.50	2.1	1.0	.91	1.00	1.3	.74	.72	.75	.52	.667
	7/16	7.6	2.21	1.9	.93	.92	.98	1.2	.66	.73	.73	.52	.672
	3/8	6.6	1.92	1.7	.81	.93	.96	1.0	.58	.74	.71	.52	.676
	5/16	5.6	1.62	1.4	.69	.94	.93	.90	.49	.74	.68	.53	.680
	¼	4.5	1.31	1.2	.56	.95	.91	.74	.40	.75	.66	.53	.684
3 x 2	½	7.7	2.25	1.9	1.0	.92	1.08	.67	.47	.55	.58	.43	.414
	7/16	6.8	2.00	1.7	.89	.93	1.06	.61	.42	.55	.56	.43	.421
	3/8	5.9	1.73	1.5	.78	.94	1.04	.54	.37	.56	.54	.43	.428
	5/16	5.0	1.47	1.3	.66	.95	1.02	.47	.32	.57	.52	.43	.435
	¼	4.1	1.19	1.1	.54	.95	.99	.39	.26	.57	.49	.43	.440
	3/16	3.07	.90	.84	.41	.97	.97	.31	.20	.58	.47	.44	.446
2½ x 2	3/8	5.3	1.55	.91	.55	.77	.83	.51	.36	.58	.58	.42	.614
	5/16	4.5	1.31	.79	.47	.78	.81	.45	.31	.58	.56	.42	.620
	¼	3.62	1.06	.65	.38	.78	.79	.37	.25	.59	.54	.42	.626
	3/16	2.75	.81	.51	.29	.79	.76	.29	.20	.60	.51	.43	.631

Table 10. Properties of Areas

Square		$I_x = \dfrac{b^4}{12}$	$Z = \dfrac{b^3}{6}$
Rectangle		$I_x = \dfrac{bh^3}{12}$	$Z = \dfrac{bh^2}{6}$
Hollow rectangle		$I_x = \dfrac{bh^3 - b_1 h_1{}^3}{12}$	$Z = \dfrac{bh^3 - b_1 h_1{}^3}{6h}$
Equal rectangles		$I_x = \dfrac{b(h^3 - h_1{}^3)}{12}$	$Z = \dfrac{b(h^3 - h_1{}^3)}{6h}$
Circle		$I_x = \dfrac{\pi\,d^4}{64}$	$Z = \dfrac{\pi\,d^3}{32}$
Hollow circle		$I_x = \dfrac{\pi\,(d^4 - d_1{}^4)}{64}$	$Z = \dfrac{\pi\,(d^4 - d_1{}^4)}{32d}$

Table 11. Beam Deflection Equations

1.

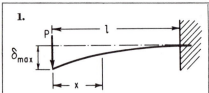

$$\delta_x = \frac{P}{6EI}\,(2l^3 - 3l^2x + x^3)$$

$$\delta_{\max} = \frac{Pl^3}{3EI}$$

2.

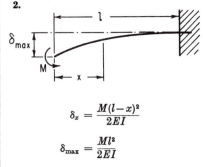

$$\delta_x = \frac{M(l-x)^2}{2EI}$$

$$\delta_{\max} = \frac{Ml^2}{2EI}$$

3.

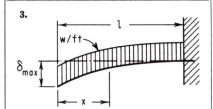

$$\delta_x = \frac{w}{24EI}\,(x^4 - 4l^3x + 3l^4)$$

$$\delta_{\max} = \frac{wl^4}{8EI}$$

4.

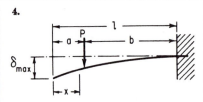

$$\delta_x = \frac{Pb^2}{6EI}\,(3l - 3x - b) \quad \text{for } x < a$$

$$\delta_x = \frac{P(l-x)^2}{6EI}\,(3b - l + x) \quad \text{for } x > a$$

$$\delta_{\max} = \frac{Pb^2}{6EI}\,(3l - b)$$

5.

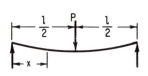

$$\delta_x = \frac{Px}{48EI}\,(3l^2 - 4x^2) \quad \text{for } x < \frac{l}{2}$$

$$\delta_{\max} = \frac{Pl^3}{48EI} \qquad \text{at } x = \frac{l}{2}$$

6.

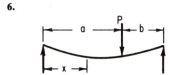

$$\delta_x = \frac{Pbx}{6lEI}\,(l^2 - x^2 - b^2) \quad \text{for } x < a$$

$$\delta_x = \frac{Pb}{6lEI}\left[\frac{l}{b}(x-a)^3 + (l^2 - b^2)x - x^3\right]$$
$$\text{for } x > a$$

$$\delta = \frac{Pb}{48EI}(3l^2 - 4b^2) \quad \text{at center if } a > b$$

$$\delta_{\max} = \frac{Pb(l^2 - b^2)^{3/2}}{9\sqrt{3l}EI} \qquad \text{at } x = \sqrt{\frac{l^2 - b^2}{3}}$$

Table 11. (continued)

7.

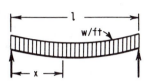

$$\delta_x = \frac{wx}{24EI}\,(l^3 - 2lx^2 + x^3)$$

$$\delta_{max} = \frac{5wl^4}{384EI}\ \text{at center}$$

8.

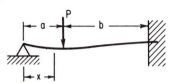

$$\delta_x = \frac{Pb^2x}{12EIl^3}\,(3al^2 - 2lx^2 - ax^2)\ \text{for}\ x<a$$

$$\delta_x = \frac{Pa(l-x)^2}{12EIl^3}\,(3l^2x - a^2\,x - 2a^2l)$$
$$\text{for}\ x>a$$

$$\delta = \frac{Pa^2b^3}{12EIl^3}\,(3l+a)\ \text{at point of load}$$

9.

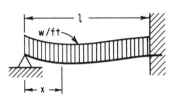

$$\delta_x = \frac{wx}{48EI}\,(l^3 - 3lx^2 + 2x^3)$$

$$\delta_{max} = \frac{wl^4}{185EI}\ \text{at}\ x = 0.422l$$

10.

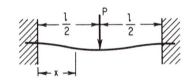

$$\delta_x = \frac{Px^2}{48EI}\,(3l - 4x)$$

$$\delta_{max} = \frac{Pl^3}{192EI}\ \text{at center}$$

11.

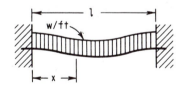

$$\delta_x = \frac{wx^2}{24EI}\,(l-x)^2$$

$$\delta_{max} = \frac{wl^4}{384EI}\ \text{at center}$$

12.

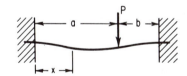

$$\delta_x = \frac{Pb^2x^2}{6EIl^3}\,(3al - 3ax - bx)\ \text{for}\ x<a$$

$$\delta_{max} = \frac{2Pa^3b^2}{3EI(3a+b)^2}\ \text{at}\ x = \frac{2al}{3a+b},\, a>b$$

$$\delta = \frac{Pa^3b^3}{3EIl^3}\ \text{at point of load}$$

INDEX